J. Prawda

AF701070

ITERATIVE METHODS FOR NONLINEAR OPTIMIZATION PROBLEMS

Prentice-Hall
Series in Automatic Computation
George Forsythe, editor

ANDREE, *Computer Programming Techniques, Analysis, and Mathematics*
ARBIB, *Theories of Abstract Automata*
ANSELONE, *Collectively Compact Operator Approximation Theory and Applications to Integral Equations*
BATES AND DOUGLAS, *Programming Language/One*, 2nd ed.
BLUMENTHAL, *Management Information Systems*
BOBROW AND SCHWARTZ, *Computers and the Policy-Making Community*
BOWLES, editor, *Computers in Humanistic Research*
CESCHINO AND KUNTZMAN, *Numerical Solution of Initial Value Problems*
CRESS, et al., *FORTRAN IV with WATFOR and WATFIV*
DANIEL, *The Approximate Minimization of Functionals*
DESMONDE, *A Conversational Graphic Data Processing System*
DESMONDE, *Computers and Their Uses*, 2nd ed.
DESMONDE, *Real-Time Data Processing Systems*
EVANS, et al., *Simulation Using Digital Computers*
FIKE, *Computer Evaluation of Mathematical Functions*
FIKE, *PL/I for Scientific Programmers*
FORSYTHE AND MOLER, *Computer Solution of Linear Algebraic Systems*
GAUTHIER AND PONTO, *Designing Systems Programs*
GEAR, *Numerical Initial Value Problems in Ordinary Differential Equations*
GOLDEN, *FORTRAN IV Programming and Computing*
GOLDEN AND LEICHUS, *IBM/360 Programming and Computing*
GORDON, *System Simulation*
GREENSPAN, *Lectures on the Numerical Solution of Linear, Singular and Nonlinear Differential Equations*
GRUENBERGER, editor, *Computers and Communications*
GRUENBERGER, editor, *Critical Factors in Data Management*
GRUENBERGER, editor, *Expanding Use of Computers in the 70's*
GRUENBERGER, editor, *Fourth Generation Computers*
HARTMANIS AND STEARNS, *Algebraic Structure Theory of Sequential Machines*
HULL, *Introduction to Computing*
JACOBY, et al., *Iterative Methods for Nonlinear Optimization Problems*
JOHNSON, *System Structure in Data, Programs and Computers*
KANTER, *The Computer and the Executive*
KIVIAT, et al., *The SIMSCRIPT II Programming Language*
LORIN, *Parallelism in Hardware and Software: Real and Apparent Concurrency*
LOUDEN AND LEDIN, *Programming the IBM 1130*, 2nd ed.
MARTIN, *Design of Real-Time Computer Systems*
MARTIN, *Future Developments in Telecommunications*
MARTIN, *Man-Computer Dialogue*

MARTIN, *Programming Real-Time Computing Systems*
MARTIN, *Systems Analysis for Data Transmission*
MARTIN, *Telecommunications and the Computer*
MARTIN, *Teleprocessing Network Organization*
MARTIN AND NORMAN, *The Computerized Society*
MATHISON AND WALKER, *Computers and Telecommunications: Issues in Public Policy*
MCKEEMAN, et al., *A Compiler Generator*
MINSKY, *Computation: Finite and Infinite Machines*
MOORE, *Interval Analysis*
PLANE AND MCMILLAN, *Discrete Optimization: Integer Programming and Network Analysis for Management Decisions*
PRITSKER AND KIVIAT, *Simulation with GASP II: a FORTRAN-Based Simulation Language*
PYLYSHYN, editor, *Perspectives on the Computer Revolution*
RICH, *Internal Sorting Methods: Illustrated with PL/I Programs*
RUSTIN, editor, *Computer Networks*
RUSTIN, editor, *Debugging Techniques in Large Systems*
RUSTIN, editor, *Formal Semantics of Programming Languages*
SACKMAN AND CITRENBAUM, editors, *Online Planning: Towards Creative Problem-Solving*
SALTON, editor, *The SMART Retrieval System: Experiments in Automatic Document Processing*
SAMMET, *Programming Languages: History and Fundamentals*
SCHULTZ, *Digital Processing: A System Orientation*
SCHWARZ, et al., *Numerical Analysis of Symmetric Matrices*
SHERMAN, *Techniques in Computer Programming*
SIMON AND SIKLOSSY, editors, *Information Processing Systems: Experiments with Representation and Meaning*
SNYDER, *Chebyshev Methods in Numerical Approximation*
STERLING AND POLLACK, *Introduction to Statistical Data Processing*
STOUTMEYER, *PL/I Programming for Engineering and Science*
STROUD, *Approximate Calculation of Multiple Integrals*
STROUD AND SECREST, *Gaussian Quadrature Formulas*
TAVISS, editor, *The Computer Impact*
TRAUB, *Iterative Methods for the Solution of Polynomial Equations*
VAN TASSEL, *Computer Security Management*
VARGA, *Matrix Iterative Analysis*
VAZSONYI, *Problem Solving by Digital Computers with PL/I Programming*
WAITE, *Implementing Software for Non-Numeric Application*
WILKINSON, *Rounding Errors in Algebraic Processes*
ZIEGLER, *Time-Sharing Data Processing Systems*

ITERATIVE METHODS FOR NONLINEAR OPTIMIZATION PROBLEMS

S.L.S. Jacoby

Boeing Computer Services, Inc.

J. S. Kowalik

Boeing Computer Services, Inc.

J. T. Pizzo

Oceanographic Commission of Washington

Prentice-Hall, Inc., Englewood Cliffs, New Jersey

Library of Congress Cataloging in Publication Data

Jacoby, Samuel L. S.
Iterative methods for nonlinear optimization problems

(Prentice-Hall series in automatic computation)
Bibliography: P.
1. Mathematical optimization. 2. Nonlinear programming. 3. Iterative methods (Mathematics) I. Kowalik, Janusz S., joint author. II. Pizzo, J. T., joint author. III. Title.
QA402.5/J33 519.7'6 72-3485
ISBN 0-13-508119-X

Dedicated to the late
George E. Forsythe

© 1972
by PRENTICE-HALL, INC.
Englewood Cliffs, New Jersey

All rights reserved. No part of this book may be reproduced in any form or by any means without permission in writing from the publisher.

10 9 8 7 6 5 4 3 2 1

Printed in the United States of America

PRENTICE-HALL INTERNATIONAL, INC., *London*
PRENTICE-HALL OF AUSTRALIA, PTY., LTD., *Sydney*
PRENTICE-HALL OF CANADA, LTD., *Toronto*
PRENTICE-HALL OF INDIA PRIVATE LIMITED, *New Delhi*
PRENTICE-HALL OF JAPAN, INC., *Tokyo*

CONTENTS

PREFACE

This book is devoted mainly to motivating and presenting, in algorithmic fashion, currently available techniques of mathematical programming and unconstrained optimization. Because of the fast growth of the theory and applications of optimization in the past decade, there has been a need for critical review in this area, which would serve at least two purposes: to be a manual for practitioners and an orderly description of the state of the art. We hope that the book can partially fulfill this goal also.

The choice of methods, which are presented and evaluated in this book, reflects our experience and familiarity with the selected algorithms. In cases where we cannot offer our own judgment we offer the criticism of other researchers in the field and refer our readers to the relevant literature. G. Zoutendijk's paper entitled, "Nonlinear Programming: A Numerical Survey," published in the *SIAM Journal* in 1966, has been particularly helpful and served as a pattern for the structure of our chapters.

We wish to acknowledge the contribution of Mr. P. F. Nesdore, who produced most of the numerical results and collected the information related to the computer programs. We also wish to acknowledge the cooperation of Dr. A.C.R. Newbery and Dr. T. S. Chow who carefully proofread the manuscript, and Mrs. M. E. Brotherton for her patience and efficiency in preparing the typescript. Our thanks go also to the management of the Computer Services Division of the Boeing Company (now: Boeing Computer Services, Inc.) for supporting our activity in the area of optimization. Special thanks are due to the editor of this series, Professor G. E. Forsythe, and to Mr. R. P. Brent for their review and suggestions.

S.L.S. Jacoby
J. S. Kowalik
J. T. Pizzo

1 INTRODUCTION

1-1 MOTIVATION AND STRUCTURE OF THE BOOK

Engineers, economists, operations researchers, managers, and practicing computer scientists are often interested in achieving optimal solutions to their problems. These problems may be to determine designs, programs, trajectories, allocation of resources, or approximations of functions. In practice one may obtain near optimal solutions to such problems by means of *parametric* or "trade-off" *studies*. In these studies different designs or programs, all satisfying the conditions arising from the actual situation, are compared, and one is chosen that also is best in terms of an optimality criterion. *Optimization techniques*, if properly applied, will automatically examine the cost or profit associated with a large number of different designs or plans and select an optimum. Sometimes this is done without solving the complete design or planning problem at every step. Under certain circumstances all possible designs or plans are taken into consideration and a "global" optimum is selected.

Classical optimization methods of differential calculus and calculus of variations can be applied to certain optimization problems originating with the physical or mathematical sciences. There are, however, many problems to which classical methods are not suitable or may be too cumbersome to apply. To these problems, one can apply *iterative techniques.* Optimization problems requiring iterative techniques were practically insurmountable before the advent of modern computers, because of their complexity and the vast amount of computation required. Recently, however, interest in optimization techniques and their application has been considerable, primarily as a result of the developments in computing technology over the last two decades. The Association of Computing Machinery has recognized mathematical optimization as a part of computing science. Consequently, recommendations for an undergraduate program, recently published by their curriculum committee, include a course in optimization techniques.* It is proposed that this course deal primarily with algorithms and their mathematical foundations, and that the presentation be oriented toward the computer user and the computing scientist. Computability considerations, numerical analysis aspects, and flowcharts should all be emphasized. This book is intended as a supplementary text for such a course at the senior or graduate level.

The development of optimization technology has proceeded sufficiently that it is possible to summarize and present many techniques in a unified manner. We have attempted such a presentation in this book. Further development in optimization techniques is inevitable; we have pointed out some open questions for research. It is very stimulating and rewarding to participate in this development. Of particular interest are applications of optimization technology to practical problems. This book is intended to aid analysts in their model building and optimization. The book should bring to their attention the techniques (including generally available computer programs) and proper ways of applying them. The level of the presentation is geared to engineers, economists, operations researchers, managers, and practicing computer scientists. Thus, it is assumed that the reader has taken undergraduate courses in matrix algebra and numerical analysis, and has some knowledge of computer programming. A glossary of terms is included to remind him of some of these concepts.

The elements normally involved in system optimization include a system model, an optimality criterion, and an optimizer. System models of interest here are mathematical models. To obtain a mathematical model, the parameters of the system or the *independent variables* are identified. Then functions of these variables are determined, which represent the different characteristics of the system. The *mathematical model* is the set of these functions that describes the behavior of the system. The optimality criterion, or *objective function*, is a measure of merit or cost. It is a function of the same independent variables and yields a number corresponding to each setting of these variables. The *optimizer* can be a computer, which is programmed to perform the operations of an *optimization technique*. This book is concerned with different optimization techniques, which are applicable to mathematical models and objective functions of different types. It is usually not practical to use the same, very general, optimization technique for all problems. We have attempted to cover techniques applicable to many of the most frequently

**Comm. ACM*, Vol. 8, No. 9, Sept. 1965, p. 543 and Vol. 11, No. 3, March 1968, p.151.

occurring general types of models and objective functions. In addition, we have mentioned some of the more specialized techniques, which can take advantage of rather special properties of the mathematical model or objective function, and result in quite efficient optimizers that are somewhat more limited in their applicability. The techniques selected for discussion should prove adequate for most applications.

The book starts with a general problem statement and a discussion of optimization problems and associated terminology and theory; examples of applications and descriptions of numerical test problems are included. Subsequent chapters are arranged according to technique applications and complexity. Thus, we start with techniques for optimization along a line (Chap. 3) and continue with search techniques for unconstrained problems requiring function evaluations only (Chap. 4). Next we present descent and higher-order techniques requiring first- or higher-order differentiation and conjugate direction techniques (Chap. 5). Then we discuss special techniques for constrained problems transformed into unconstrained problems (Chap. 6) and techniques that handle constraints explicitly (Chap. 7).

Each chapter starts with preliminary considerations including introductory remarks and the theoretical basis for all techniques discussed in the chapter, followed by a description of the major techniques themselves structured according to a standard format: An introductory section prepares the reader to understand the algorithm; restrictions on the starting point, if any, are described. A typical *iteration step* is described whereby the $(k + 1)$st point is obtained, given the kth, and *convergence criteria* for the algorithm as well as problem properties used in the convergence proof are presented in a mixture of verbal and mathematical expressions believed most suitable as a basis for coding or for the understanding of an existing code.

We have described "pure" techniques; that is, those involving one type of iteration. Many practical computing systems use "combination" or accelerated techniques, however; thus, following the description of the technique is a discussion of *modifications* to avoid difficulties (such as numerical instability), and of special *devices for accelerating convergence*. (Note that some of the techniques described are acceleration devices themselves and should be used only as such.) The discussion also includes advantages and disadvantages of the algorithm in terms of applicability, convergence, computational work, core storage requirements, and coding; and open questions for research. Generally available computing systems (computer programs, codes) are identified in a separate section; in addition to general information as a basis for evaluating each system, certain operational details are given, including system capacity, programming language, and operational experience.

Several of the more important techniques have been applied to sample problems, outlined in this chapter, for illustration and comparison. Each chapter includes test results obtained using the techniques of that chapter. Some of the more significant and comprehensive tests reported in the technical literature are also reviewed.

Optimization techniques in the main body of the book frequently make use of miscellaneous auxiliary techniques for linear programming, numerical differentiation, interpolation and least-squares problems. Brief descriptions of these auxiliary techniques are given in the appendix.

1-2 EXAMPLES OF OPTIMIZATION PROBLEMS

Consider the design of the composition of a vehicle fleet [Serfozo, 1967] comprising, for example, trucks, railroad cars, ships, aircraft, or mixtures of these. The fleet is assembled to perform a set of given tasks in a given period of time. These tasks could be of a transportation or logistics nature. Suppose that the total time period of interest is subdivided into smaller intervals $\nu = 1, \ldots, N$, and that there are vehicles of types $\lambda = 1, \ldots, \Lambda$ performing tasks $\mu = 1, \ldots, M$. In short, any vehicle in the fleet is identifiable by the indices λ, μ, ν.

Let the *problem variables*, the components of a *variable vector* $\mathbf{x}$, be the numbers of vehicles of a certain type assigned to perform a specified task in a given time interval. That is, $x_{\lambda\mu\nu}$ is the number of vehicles of type λ assigned to perform task μ during time interval ν. A solution of the problem —a fleet composition—consists of a set of appropriate values $x_{\lambda\mu\nu}$; $\lambda = 1, \ldots, \Lambda$; $\mu = 1, \ldots, M$; and $\nu = 1, \ldots, N$. And

$$\mathbf{x} = (x_{111}, \ldots, x_{11N}, x_{121}, \ldots x_{\Lambda,M,1}, \ldots, x_{\Lambda MN})^T$$

The *design objective* is to minimize the total cost of the fleet. The cost of each vehicle $\lambda\mu\nu$ is $c_{\lambda\mu\nu}$ and depends on a number of factors, including the part of the procurement cost of the vehicle of type λ assigned to perform task μ, which is amortized during time interval ν and normalized to account for any possible changes in the value of money over the design period; the number of vehicles of this and of any other type purchased; and vehicle operating and maintenance costs during task performance. In general then, $c_{\lambda\mu\nu}$ depends on all the variables $\mathbf{x}$, and we formulate the so-called "objective function" of the minimization problem as

$$\min_{\mathbf{x}} \left[\sum_{\lambda=1}^{\Lambda} \sum_{\mu=1}^{M} \sum_{\nu=1}^{N} c_{\lambda\mu\nu}(\mathbf{x}) x_{\lambda\mu\nu} \right] \tag{1-1}$$

The expression in brackets, to be minimized, is the cost of the program. This is, in general, a nonlinear function of the variables $\mathbf{x}$.

The problem specified by expression (1-1) is an *unconstrained nonlinear optimization problem*. However, the design under consideration is *constrained* by additional relations between the problem variables $\mathbf{x}$, or restrictions representing a model of the physical problem on hand. In particular, consider what the fleet is producing in return for the above cost. Suppose that each vehicle $\lambda\mu\nu$ is capable of producing $w_{\lambda\mu\nu}$ units and that there is a total requirement for $r_{\mu\nu}$ units of task μ during time period ν. In general, $w_{\lambda\mu\nu}$ may depend on all the variables $\mathbf{x}$, because vehicles doing one task may be supported by and depend on vehicles doing other tasks. The first type of (nonlinear inequality) *constraint function* arises from the requirement that a sufficient number of units of each task be produced during each time period. That is, we have MN constraints

$$\sum_{\lambda=1}^{\Lambda} w_{\lambda\mu\nu}(\mathbf{x})x_{\lambda\mu\nu} - r_{\mu\nu} \geq 0 \qquad \mu = 1, \ldots, M; \qquad \nu = 1, \ldots, N \qquad (1\text{-}2)$$

The sum on the left is the total fleet productivity on task $\mu\nu$, generally a nonlinear function of the variables $\mathbf{x}$. The number of vehicles of any type in the problem under consideration must obviously be nonnegative. There may also be a limitation, say $u_{\lambda\mu\nu}$, on the number of vehicles that can be procured. Hence we have *lower/upper bound constraints* on the variables $\mathbf{x}$:

$$\begin{aligned} x_{\lambda\mu\nu} &\geq 0 \qquad \lambda = 1, \ldots, \Lambda; \qquad \mu = 1, \ldots, M; \qquad \nu = 1, \ldots, N \\ & \qquad\qquad\qquad\qquad\qquad\qquad\qquad\qquad\qquad\qquad (1\text{-}3) \\ u_{\lambda\mu\nu} - x_{\lambda\mu\nu} &\geq 0 \qquad \lambda = 1, \ldots, \Lambda; \qquad \mu = 1, \ldots, M; \qquad \nu = 1, \ldots, N \end{aligned}$$

The problem described in expressions (1-1) through (1-3) is a *constrained nonlinear optimization problem*. However, if the amortized procurement cost and the operation and maintenance costs are fixed for each vehicle—that is, the costs $c_{\lambda\mu\nu}$ are constants, independent of the number of vehicles $x_{\lambda\mu\nu}$—the objective function (1-1) is linear in the variables $\mathbf{x}$. Similarly, if the productivities $w_{\lambda\mu\nu}$ of the vehicles are independent of the number of vehicles $x_{\lambda\mu\nu}$, the constraint functions (1-2) also are linear in the variables $\mathbf{x}$, provided the tasks are independent and may be performed simultaneously, and each vehicle type is assigned to no more than one task in any given time period. The constraint functions (1-3) are always linear in the variables $\mathbf{x}$. Thus, given the above conditions, the total problem under consideration is *linear*.

Suppose now that the costs $c_{\lambda\mu\nu}$ and productivities $w_{\lambda\mu\nu}$ depend *only* on $x_{\lambda\mu\nu}$—the number of vehicles of the corresponding type, task, and time period—rather than on several (or all) the variables $\mathbf{x}$. If $c_{\lambda\mu\nu} = c_{\lambda\mu\nu}(x_{\lambda\mu\nu})$ and $w_{\lambda\mu\nu} = w_{\lambda\mu\nu}(x_{\lambda\mu\nu})$, our problem is *separable*. The total problem is *quadratic* if the constraint functions (1-2) are linear in the variables $\mathbf{x}$ and if the costs $c_{\lambda\mu\nu}$ are linear functions of the components of the variable vector $\mathbf{x}$. Note that the problem variables can assume only integer values, because only "whole" vehicles can perform the given tasks; thus, the problem is, strictly speaking, an *integer optimization problem*. If, however, the number of vehicles is likely to be fairly large (several score or more), one may solve the problem as if any values of the variables are admissible and then obtain the integer solution by rounding to the nearest integer.

Another optimization problem is illustrated by the design of a loaded three-bar truss (Fig. 1-1) [Karnes and Tocher, 1969]. The lengths of the

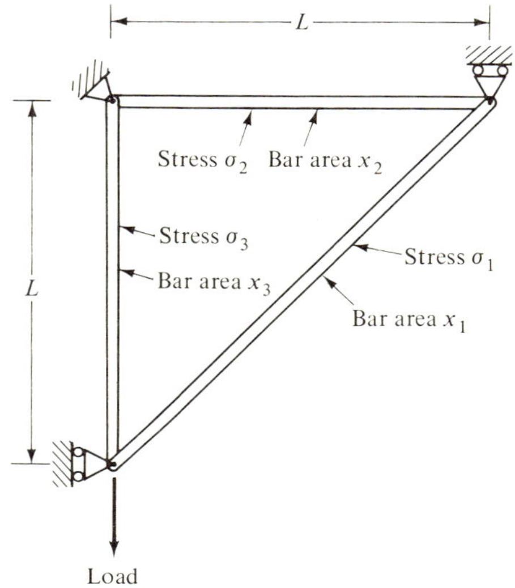

Figure 1-1 Three-bar truss

three bars are fixed to $L\sqrt{2}$, L, and L, but their cross-sectional areas x_1, x_2, and x_3 are variable. These cross-sectional areas must be such that the total weight of the structure is minimized but without exceeding prescribed allowable stresses. The weight M is related to the variable vector $\mathbf{x} = (x_1, x_2, x_3)^T$ by

$$M(\mathbf{x}) = \rho L(x_1\sqrt{2} + x_2 + x_3) \tag{1-4}$$

where ρ is the specific weight of the material. The actual stresses σ_1, σ_2, and σ_3 in the bars are related to their cross-sectional areas, the geometry, and the load. Let these relations be given by

$$\sigma_i = g_i(x_1, x_2, x_3, L, \text{load}) \qquad i = 1, 2, 3 \tag{1-5}$$

where g_i are known functions (possibly nonlinear). But these stresses are bounded by a given minimum/maximum allowable stress σ_a:

$$-\sigma_a \leq \sigma_i \leq \sigma_a \qquad i = 1, 2, 3 \tag{1-6}$$

Combining (1-5) and (1-6) we obtain the six nonlinear *constraint functions* for our minimization problem:

$$\begin{aligned} \sigma_a - g_i(\mathbf{x}, L, \text{load}) &\geq 0 \qquad i = 1, 2, 3 \\ \sigma_a + g_i(\mathbf{x}, L, \text{load}) &\geq 0 \qquad i = 1, 2, 3 \end{aligned} \tag{1-7}$$

Additional (linear) constraints might be stated in the form of lower and upper bounds on the cross sectional areas of the bars. Thus,

$$\begin{aligned} x_i - b_i &\geq 0 \qquad i = 1, 2, 3 \\ -x_i + u_i &\geq 0 \qquad i = 1, 2, 3 \end{aligned} \tag{1-8}$$

In summary, the problem is to minimize $M(\mathbf{x})$ defined by (1-4), a linear function, subject to nonlinear constraints (1-7) and linear constraints (1-8). This is a nonlinear constrained minimization problem with upper and lower bounded variables. The weight savings and consequent cost savings resulting from this design procedure may be considerable if many of these three-bar trusses are to be manufactured. Karnes and Tocher [1969] discuss how much more complex structures such as airframes may be designed for minimum weight.

1-3 PROBLEM STATEMENT

In this section we present a precise statement of the problems for which solution techniques are proposed in this book. A general *optimization problem* (sometimes referred to as a *mathematical* or *nonlinear programming problem*, or in short a *nonlinear program*) can be formulated as follows: Find the values of n variables $x_1, x_2, \ldots, x_n$, denoted for brevity by $\mathbf{x}$, which satisfy m equations and/or inequalities (*constraint functions*) $g_j(\mathbf{x})\{\geq, =\}\, 0, j = 1, \ldots, m_2$, $l_j(\mathbf{x})\{\geq, =\}\, 0, j = m_2 + 1, \ldots, m$ and optimize (minimize or maximize) the *objective function* $M(\mathbf{x})$. Since the problems of minimizing $M(\mathbf{x})$ and maximizing $[-M(\mathbf{x})]$ are equivalent, the general optimization problem can be written as:

$$\min M(\mathbf{x})$$

subject to

$$\begin{aligned} g_j(\mathbf{x}) &\geq 0 \qquad j = 1, 2, \ldots, m_1 \\ g_j(\mathbf{x}) &= 0 \qquad j = m_1 + 1, m_1 + 2, \ldots, m_2 \\ l_j(\mathbf{x}) &\geq 0 \qquad j = m_2 + 1, m_2 + 2, \ldots, m_3 \\ l_j(\mathbf{x}) &= 0 \qquad j = m_3 + 1, m_3 + 2, \ldots, m \end{aligned} \tag{1-9}$$

If there are no constraints ($m = 0$), the problem is said to be *unconstrained.* In general, the m_2 functions $g_j(\mathbf{x}), j = 1, \ldots, m_2$ are nonlinear in $\mathbf{x}$, and the $m - m_2$ functions $l_j(\mathbf{x})$, $j = m_2 + 1, \ldots, m$ are linear in $\mathbf{x}$. The linear constraints may also include as a special case, lower or upper bounds on the variables—that is, $b_i \leq x_i$ or $x_i \leq u_i$, $i = 1, \ldots, n$. In this case we shall use the notation $\mathbf{b} \leq \mathbf{x}$ and $\mathbf{x} \leq \mathbf{u}$. An optimization problem is said to be *linear* if the objective function $M(\mathbf{x})$ and all the constraint functions are linear in the n variables $x_1, \ldots, x_n$. In this case, $m_2 = 0$ in Prob. (1-9). Otherwise, the optimization problem is said to be *nonlinear*. A problem is said to be *quadratic* if the objective function is quadratic

$$M(\mathbf{x}) = a + \mathbf{b}^T\mathbf{x} + \tfrac{1}{2}\mathbf{x}^T A\mathbf{x} \tag{1-10}$$

where a is a constant, $\mathbf{b}$ an n-dimensional column vector, and A an $n \times n$ matrix, and if the constraints are linear (as in the case of a linear problem).

A problem is said to be *separable* if the nonlinear functions $M(\mathbf{x})$ and $g_j(\mathbf{x})$, $j = 1, \ldots, m_2$ can be rewritten as sums of other functions of one variable. That is, if $m_i(x_i)$ and $g_{ij}(x_i)$, $i = 1, \ldots, n$, and $j = 1, \ldots, m_2$ are these functions, then:

$$\begin{aligned} M(\mathbf{x}) &= \sum_{i=1}^{n} m_i(x_i) \\ g_j(\mathbf{x}) &= \sum_{i=1}^{n} g_{ij}(x_i) \qquad j = 1, \ldots, m_2 \end{aligned} \tag{1-11}$$

This form can sometimes be obtained directly or by variable transformations. If some (or all) of the variables are restricted to a set of integer (or discrete) values, the problem is said to be an *integer* (or a *discrete*) optimization problem.

It is possible, and sometimes even desirable or necessary, to transform an optimization problem of one type into a problem of another type. These transformations and what one can possibly gain by each problem reformulation are the subject of Chap. 2.

1-4 THE FEASIBLE REGION AND ITS BOUNDARY

Any set of values of the variables $\mathbf{x} = (x_1, x_2, \ldots, x_n)^T$ may be referred to as a *point* in an n-dimensional space. The subset of all points that satisfy the constraints of the problem is the *feasible region* within which any point is referred to as a *feasible point*. A feasible point must lie at the intersection of all the hypersurfaces corresponding to the equality constraints and simultaneously on the proper sides of all the hypersurfaces corresponding to the inequality constraints. A feasible point is said to be a *boundary point* if it lies on at least one hypersurface, corresponding to a constraint; otherwise, it is an *interior point*. The constraint hypersurfaces on which a boundary point lies are said to be *active* at that point. There are different types of feasible points, corresponding to the type of constraints involved. The feasible region R is defined as

$$R = \{\mathbf{x} \mid [g_j(\mathbf{x}) \geq 0, \quad j = 1, \ldots, m_1; \quad g_j(\mathbf{x}) = 0, \quad j = m_1 + 1, \ldots, m_2; \\ l_j(\mathbf{x}) \geq 0, \quad j = m_2 + 1, \ldots, m_3; \quad l_j(\mathbf{x}) = 0, \quad j = m_3 + 1, \ldots, m]\} \tag{1-12}$$

If there are no equality constraints—that is, $m_2 - m_1 = m - m_3 = 0$ —we refer to the *interior of the feasible region*:

$$R^0 = \{\mathbf{x} \mid [g_j(\mathbf{x}) > 0, \quad j = 1, \ldots, m_1; \\ \text{and} \quad l_j(\mathbf{x}) > 0, \quad j = m_1 + 1, \ldots, m]\} \tag{1-13}$$

In this case, the boundary $\bar{R}$ of the feasible region R is $R - R^0$:

$$\bar{R} = R - R^0 = \{\mathbf{x} \mid [g_j(\mathbf{x}) = 0, \quad \text{for at least one value of} \quad j = 1, \ldots, m_1; \\ \text{or} \quad l_j(\mathbf{x}) = 0, \quad \text{for at least one value of} \quad j = m_3 + 1, \ldots, m]\} \tag{1-14}$$

In general, when equality constraints are included in the problem, the distinction between the boundary and the interior of the feasible region does not apply: all feasible points are boundary points, and we have $\bar{R} = R$.

In some cases it is of interest to distinguish between points that are feasible with respect to the nonlinear (R_{nl}) or the linear (R_l) constraints, but not both. We then define

$$R_{nl} = \{\mathbf{x} \mid [g_j(\mathbf{x}) \geq 0, \quad j = 1, \ldots, m_1; \\ \text{and} \quad g_j(\mathbf{x}) = 0, \quad j = m_1 + 1, \ldots, m_2]\} \tag{1-15}$$

$$R_l = \{\mathbf{x} \mid [l_j(\mathbf{x}) \geq 0, \quad j = m_2 + 1, \ldots, m_3; \\ \text{and} \quad l_j(\mathbf{x}) = 0, \quad j = m_3 + 1, \ldots, m]\} \tag{1-16}$$

Clearly the feasible region R consists of all points common to R_{nl} and R_l: $R = R_{nl} \cap R_l$. Interior and boundary points may be considered R_{nl}^0 and $\bar{R}_{nl}$ or R_l^0 and $\bar{R}_l$ in the sense of (1-13) and (1-14), respectively. Again, this distinction between boundary and interior points applies only when the problem does not include any equality constraints.

We shall be concerned with movement from one point in the n-dimensional space of the variables $x_1, \ldots, x_n$ to another. In this context, directions are important. A direction $\mathbf{s} = (s_1, s_2, \ldots, s_n)^T$ is said to be a *feasible direction* at a feasible point $\mathbf{x} = (x_1, x_2, \ldots, x_n)^T$ if a *steplength* $\hat{\lambda} > 0$ exists such that for all λ, $0 \leq \lambda \leq \hat{\lambda}$, the points $\mathbf{x} + \lambda\mathbf{s}$ are also feasible. Thus, a feasible direction is one in which one can move a finite, positive distance from a feasible point without violating any constraint. It is customary to assume that regions R under consideration satisfy the *Kuhn-Tucker constraint qualification*: For every boundary point $\mathbf{x}^*$ and for every small displacement $\delta\mathbf{x}$ from $\mathbf{x}^*$ in the direction of the (positive) gradient of the active constraints, the vector $\delta\mathbf{x}$ lies in the feasible region R. As pointed out by Kuhn and Tucker [1951], this assumption rules out singularities on the boundary of R, such as outward pointing cusps (See also Sec. 1-6.). The reader is referred to Mangasarian [1969] and Evans [1970] for a detailed discussion of different constraint qualifications.

1-5 CONVEXITY OF SETS, FUNCTIONS, AND PROBLEMS

Sets and Functions

We say that the set of points S is *convex* if $\{\theta\mathbf{x} + (1-\theta)\mathbf{y}\} \in S, 0 \leq \theta \leq 1$, whenever $\mathbf{x}, \mathbf{y} \in S$. The function $f(\mathbf{x})$ defined on the convex set S is convex if

$$f[\theta\mathbf{x} + (1-\theta)\mathbf{y}] \leq \theta f(\mathbf{x}) + (1-\theta)f(\mathbf{y}) \tag{1-17}$$

for all $\mathbf{x}, \mathbf{y} \in S$ and $0 \leq \theta \leq 1$. If the sign in (1-17) is replaced by a strict inequality, the above expression defines a *strictly convex function.* The sum of convex functions is convex. Therefore, any function that can be written as a sum of convex functions is convex. A function is said to be *concave* if its negative is convex.

A simple relationship between the convex set of points and a convex function $f(\mathbf{x})$ is established by the following equivalence: The function $f(\mathbf{x})$ is convex if the set $\{(\mathbf{x}, \alpha) \mid -f(\mathbf{x}) \geq \alpha\}$ is convex for any constant α. Another definition of function convexity equivalent to (1-17) can be stated if $f(\mathbf{x})$ is differentiable with respect to $\mathbf{x}$ on the convex set S; that is, $f(\mathbf{x})$ is convex if its gradient satisfies

$$\nabla f(\mathbf{x})^T(\mathbf{y} - \mathbf{x}) \leq f(\mathbf{y}) - f(\mathbf{x}) \tag{1-18}$$

for all $\mathbf{x}, \mathbf{y} \in S$, and *strictly* convex if strict inequality holds in (1-18). If the function $f(\mathbf{x})$ is twice differentiable, then it is convex if (and only if) its *Hessian matrix*, a matrix whose elements are $h_{ij} = (\partial^2 f(\mathbf{x})/\partial x_i \partial x_j)$ is *positive semidefinite* for all $\mathbf{x} \in S$ (convex set). This follows from the second-order Taylor series approximation to $f(\mathbf{y})$ in the neighborhood of $\mathbf{x}$ given by

$$f(\mathbf{y}) = f(\mathbf{x}) + \nabla f(\mathbf{x})^T(\mathbf{y} - \mathbf{x}) + \tfrac{1}{2}(\mathbf{y} - \mathbf{x})^T \operatorname{Hess}(\mathbf{x})(\mathbf{y} - \mathbf{x})$$

If Hess $(\mathbf{x})$ is positive semidefinite, we have by definition

$$\tfrac{1}{2}(\mathbf{y} - \mathbf{x})^T \operatorname{Hess}(\mathbf{x})(\mathbf{y} - \mathbf{x}) \geq 0 \tag{1-19}$$

and consequently $\quad f(\mathbf{y}) - f(\mathbf{x}) - \nabla f(\mathbf{x})^T(\mathbf{y} - \mathbf{x}) \geq 0$

The last relationship is the same as (1-18). Thus, if the *Hessian matrix* is *positive definite*, the function $f(\mathbf{x})$ is strictly convex.

In many cases it is rather difficult to check the convexity of a function over a particular set of points. The following rules may be of some help:

1 A linear function of $\mathbf{x}$ is both convex and concave (but not strictly convex or concave) for all $\mathbf{x}$.

2 The quadratic function $f(\mathbf{x}) = a + \mathbf{b}^T\mathbf{x} + \frac{1}{2}\mathbf{x}^T A\mathbf{x}$ is convex for all $\mathbf{x}$ if the matrix A is positive definite.

3 The function

$$f(\mathbf{x}) = \sum_{l=1}^{L} f_l(\mathbf{x})$$

is convex on some convex set S if all $f_l(\mathbf{x})$ are convex on S.

4 If $-f(\mathbf{x})$ is convex, then $[f(\mathbf{x})]^{-1}$ is convex on the set $S = \{\mathbf{x} \mid f(\mathbf{x}) > 0\}$.

Since the convexity of $f(\mathbf{x})$ and the positive semidefiniteness of its Hessian, Hess($\mathbf{x}$) imply each other, we may check for convexity by testing the Hessian matrix of $f(\mathbf{x})$. Hess($\mathbf{x}$) is positive semidefinite if any one of the following conditions is satisfied:

1 There exists any matrix A such that Hess $= A^T A$.

2 Hess is a diagonal matrix with nonnegative entries.

3 All the eigenvalues of Hess are nonnegative, for which a sufficient condition is

$$h_{jj} \geq \sum_{\substack{i=1 \\ i \neq j}}^{n} |h_{ij}| \quad \text{for all } j$$

4 The first r determinants D_k, $k = 1, \ldots, r < n$, consisting of the upper-left-hand elements of the Hessian (possibly after appropriate row and column interchanges) are positive, while the remaining $n - r$ such determinants are zero; that is,

$$|D_k| = \begin{vmatrix} h_{11} & \cdots & h_{1k} \\ \cdots & & \cdots \\ h_{k1} & \cdots & h_{kk} \end{vmatrix} = \begin{cases} > 0 & k = 1, \ldots, r < n \\ = 0 & k = r + 1, \ldots, n \end{cases}$$

Similarly, criteria by which to determine whether Hess($\mathbf{x}$) is positive definite are:

1 There exists an $m \times n$ matrix A of rank n, where $m \geq n$ such that Hess $= A^T A$. As a special case, A could be square ($n \times n$) and nonsingular, that is, of rank n.

2 Hess is a diagonal matrix with positive elements.

3 All the eigenvalues of Hess are positive.

4 The n determinants D_k, $k = 1, \ldots, n$, consisting of the upper-left-hand elements of the Hessian, are positive; that is,

$$|D_k| = \begin{vmatrix} h_{11} & \cdots & h_{1k} \\ \cdots & & \cdots \\ h_{k1} & \cdots & h_{kk} \end{vmatrix} > 0 \qquad k = 1, \ldots, n$$

A property related to convexity is quasiconvexity. We say that the function $f(\mathbf{x})$ is quasiconvex in some region S if $f[\theta\mathbf{x} + (1 - \theta)\mathbf{y}] \leq \max\{f(\mathbf{x}), f(\mathbf{y})\}$ for all $\mathbf{x}, \mathbf{y} \in S$, and for any $0 \leq \theta \leq 1$.

Certain quasiconvex functions can be made convex by a nonlinear change of scale in $f(\mathbf{x})$. Wolfe [1967] gives an example of $f(\mathbf{x}) = x^2(1 + x^2)^{-1}$ (quasiconvex) which is transformed to $g[f(\mathbf{x})] = 1 + x^2$ (convex) by means of $g[f(\mathbf{x})] = [1 - f(\mathbf{x})]^{-1}$. A quasiconvex function has the property given in Lemma 1-1.

Lemma 1-1 A function $f(\mathbf{x})$ is *quasiconvex* if and only if the set $S = \{\mathbf{x} \mid -f(\mathbf{x}) \geq \alpha\}$ is convex for any scalar α. If $f(\mathbf{x})$ is quasiconvex, then $-f(\mathbf{x})$ is *quasiconcave*.

Note that every convex function is quasiconvex, but the converse is not necessarily true.

Another related function property is pseudoconvexity. We say that a differentiable function $f(\mathbf{x})$ is *pseudoconvex* in some region if for all points $\mathbf{x}$ and $\mathbf{y}$ belonging to this region

$$\nabla f(\mathbf{x})^T(\mathbf{y} - \mathbf{x}) \geq 0 \tag{1-20}$$

implies that

$$f(\mathbf{y}) \geq f(\mathbf{x})$$

If $f(\mathbf{x})$ is pseudoconvex, then $-f(\mathbf{x})$ is *pseudoconcave*. Note again that any convex function is pseudoconvex, but the converse does not follow. It is also true that a pseudoconvex function is quasiconvex but not conversely. A pseudoconvex function has the property given in Lemma 1-2.

Lemma 1-2 Let $f(\mathbf{x})$ be pseudoconvex on S, then

$$\nabla f(\hat{\mathbf{x}}) = 0 \qquad \hat{\mathbf{x}} \in S$$

is a necessary and sufficient condition that

$$f(\hat{\mathbf{x}}) = \min_{\mathbf{x}} \{f(\mathbf{x}) \mid \mathbf{x} \in S\}$$

Pseudoconvexity and quasiconvexity are used to extend the applicability of the theoretically sufficient conditions for optimality. This is possible because the class of pseudoconvex functions includes convex functions, and the class of quasiconvex functions includes both pseudoconvex and convex functions. In practice, however, it is difficult to test whether or not a function has these properties. They are frequently ignored, therefore, and one tests for convexity instead. Another disadvantage is that linear combinations of functions belonging to these broader classes of pseudo- or quasiconvex functions are not necessarily pseudo- or quasiconvex. Recall that a linear combination of convex functions is convex, and this may be taken advantage of in certain practical situations. The properties of pseudo- and quasiconvexity are included here mainly for completeness of the discussion.

The Convex Optimization Problem

The constrained optimization problem is said to be a *convex optimization problem* if the objective function is convex over the convex feasible region (convex set of feasible points) defined by the constraints. For example, the problem min $M(\mathbf{x})$ subject to

$$\begin{aligned} &g_j(\mathbf{x}) \geq 0 && j = 1, \ldots, m_1 \\ &l_j(\mathbf{x}) = \mathbf{a}_j^T\mathbf{x} - b_j \geq 0 && j = m_1 + 1, \ldots, m \end{aligned}$$

is convex if $M(\mathbf{x})$ and $-g_j(\mathbf{x}), j = 1, \ldots, m_1$ are convex and $l_j(\mathbf{x})$ are linear functions of $\mathbf{x}$. A more general definition of the convex optimization involves pseudo- and quasiconvex functions.

1-6 TYPES OF SOLUTIONS AND OPTIMALITY CONDITIONS

The value of the objective function at the solution point of the mathematical programming problem (1-9) is referred to as the *minimum*. A minimum $M(\mathbf{x}_G)$ is called a *global* (absolute) *minimum* if for every feasible point $\mathbf{x}$, $M(\mathbf{x}_G) \leq M(\mathbf{x})$. A minimum $M(\mathbf{x}_L)$ is a *local* (relative) *minimum* if there is a feasible neighborhood of $\mathbf{x}_L$ (defined by $||\mathbf{x}_L - \mathbf{x}|| < \delta$, $\mathbf{x}$ feasible and $\delta > 0$) such that for all $\mathbf{x}$ in that neighborhood, $M(\mathbf{x}_L) \leq M(\mathbf{x})$. Clearly a global minimum is a local one, but the converse is generally not true.

In this book, we are concerned with methods for determining points at which local or global minima occur *if they exist*. It is of theoretical interest and practical importance to describe *optimality conditions*, which are necessary and sufficient for a point $\hat{\mathbf{x}}$ to be a solution of the optimization problem (1-9). Sometimes it is convenient to reformulate the original problem using these conditions. The reformulated problem is then solved; that is, a point satisfying the optimality conditions is found instead of solving problem (1-9) directly. Thus, the equivalence between the optimality conditions and problem solutions may be used. In general, however, the direct approach to problem (1-9) is more practical. In this case, the optimality conditions are frequently used as termination criteria for the iterative processes that solve the optimization problems directly. Unfortunately, while in most practical cases they would not assure that the minimum has been found (because they are only necessary conditions), they become also sufficient for convex problems and determine, in this case, whether or not a given point is optimal.

The Fritz John Necesary Conditions

The most general optimality condition for a mathematical programming problem is the Fritz John criterion. The original theory has recently been extended by Mangasarian and Fromovitz [1967]. In this criterion, distinction between the equality and inequality constraints is essential and distinction between linearity or nonlinearity is not. For this reason, we introduce the simplified formulation of the nonlinear programming problem (1-9):

$$\min M(\mathbf{x})$$

$$\text{subject to} \quad \begin{aligned} g_j(\mathbf{x}) &\geq 0 \qquad j = 1, 2, \ldots, m \\ h_j(\mathbf{x}) &= 0 \qquad j = 1, 2, \ldots, k \end{aligned} \tag{1-21}$$

It is assumed that the objective and constraint functions are differentiable.

If $\hat{\mathbf{x}}$ is a solution of (1-21), vectors $\mathbf{u}^T = (u_0, u_1, \ldots, u_m)$ and $\mathbf{v}^T = (v_1, v_2, \ldots, v_k)$ exist such that

$$
\begin{aligned}
-u_0\nabla M(\hat{\mathbf{x}}) + \sum_{j=1}^{m} u_j \nabla g_j(\hat{\mathbf{x}}) + \sum_{j=1}^{k} v_j \nabla h_j(\hat{\mathbf{x}}) &= 0 \\
\sum_{j=1}^{m} u_j g_j(\hat{\mathbf{x}}) &= 0 \\
\mathbf{u} &\geq 0 \\
\begin{bmatrix} \mathbf{u} \\ \mathbf{v} \end{bmatrix} &\neq 0
\end{aligned}
\qquad (1\text{-}22)
$$

For unconstrained problems, these conditions reduce to $\nabla M(\hat{\mathbf{x}}) = 0$.

Because the conditions (1-22) are necessary and not sufficient, they may also be satisfied at feasible points other than the minimum. Mangasarian and Fromovitz [1967], for example, specify conditions under which (1-22) are satisfied at a feasible point, which is not necessarily the minimum.

The Kuhn-Tucker Necessary Conditions

The necessary conditions of Kuhn and Tucker are better known, but they require some kind of a constraint qualification (Section 1-4) that guarantees that $u_0 > 0$ in the first of conditions (1-22). It is assumed that the objective and constraint functions are differentiable. If $\hat{\mathbf{x}}$ is a solution of (1-21), vectors $\mathbf{u}^T = (u_1, \ldots, u_m)$ and $\mathbf{v}^T = (v_1, \ldots, v_k)$ exist such that

$$
\begin{aligned}
-\nabla M(\hat{\mathbf{x}}) + \sum_{j=1}^{m} u_j \nabla g_j(\hat{\mathbf{x}}) + \sum_{j=1}^{k} v_j \nabla h_j(\hat{\mathbf{x}}) &= 0 \\
\sum_{j=1}^{m} u_j g_j(\hat{\mathbf{x}}) &= 0 \\
\mathbf{u} &\geq 0
\end{aligned}
\qquad (1\text{-}23)
$$

For unconstrained problems these conditions reduce to $\nabla M(\hat{\mathbf{x}}) = 0$.

The conditions (1-23) may be interpreted geometrically as follows. At the optimal point, the gradient of $M(\mathbf{x})$ is expressible as a linear combination of the gradients of the active constraints, where the nonnegative multipliers u_j correspond to the active inequalities. The multipliers v_j corresponding to equations can be negative, positive, or zero. The second relation (1-23) says that the multipliers corresponding to the inactive (inequality) constraints are zero. This is so because inactive constraints are nonzero (Sec. 1-4). Figure 1-2 illustrates this condition, where we assumed that $m = 3$, $k = 0$, and also indicates the nature of the Kuhn-Tucker constraint qualification noted in Sec. 1-4. The gradient of $M(\mathbf{x})$ has to lie in the cone formed by the gradients of the active constraints. Constraints that produce outward pointing cusps do not satisfy the Kuhn-Tucker qualification. Figure 1-3 shows that the Kuhn-Tucker conditions clearly cannot be satisfied at point $\hat{\mathbf{x}}$ unless $\nabla M(\hat{\mathbf{x}}) = 0$.

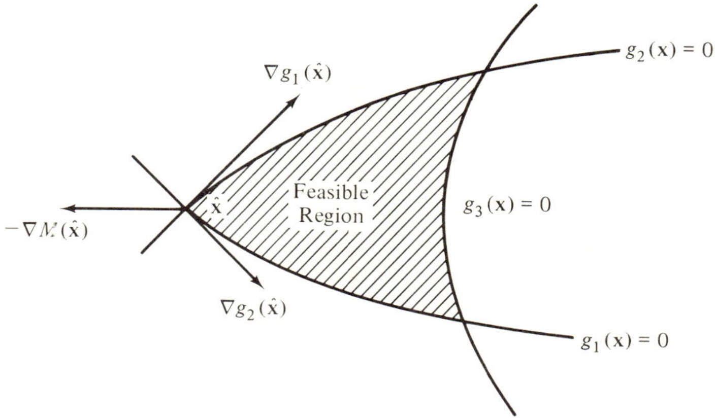

Figure 1-2 Kuhn-Tucker necessary conditions

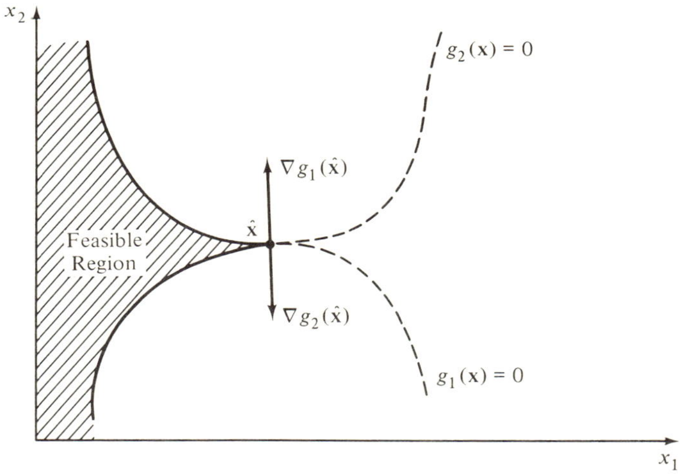

Figure 1-3 Constraints that do not satisfy the constraint qualification condition

Note that this constraint qualification requirement has only theoretical significance: the constraints that do not satisfy this condition rarely occur in practice, and even if they do, such irregularities cannot be easily detected or removed. It was mentioned earlier that the Kuhn-Tucker conditions are also sufficient for a point to be optimal if the objective function $M(\mathbf{x})$ and the feasible region satisfy certain conditions. This equivalence between a point $\hat{\mathbf{x}}$ satisfying the Kuhn-Tucker conditions and also being a solution of the problem (1-21), is stated in Theorem 1-1.

Theorem 1-1 If in a nonlinear programming problem with differentiable objective function and constraints the objective function is pseudoconvex, the feasible region is convex, and the point $\hat{\mathbf{x}}$ satisfies the Kuhn-Tucker optimality conditions, then $\hat{\mathbf{x}}$ is optimal (global minimum) for the problem.

Remark A feasible region is convex if, for example, its definition is $S = \{\mathbf{x} \mid g_j(\mathbf{x}) \geq 0, \text{ for all } j\}$ where all functions $-g_j(\mathbf{x})$ are quasiconvex.

1-7 THE LAGRANGIAN FUNCTION AND DUALITY

The *Lagrangian function* is defined as

$$L(\mathbf{x}, \mathbf{u}) = M(\mathbf{x}) - \sum_{j=1}^{m} u_j g_j(\mathbf{x}) \tag{1-24}$$

corresponding to the constrained minimization problem

$$\min_{\mathbf{x}} \{M(\mathbf{x}) \mid [g_j(\mathbf{x}) \geq 0, \qquad j = 1, \ldots, m]\}$$

The vector $\mathbf{u} = (u_1, \ldots, u_m)^T$ is a vector of *Lagrange multipliers*. This function plays a fundamental role in the so-called *duality theory*. We can look at the Lagrangian as a function of the two sets of variables $\mathbf{x}$ and $\mathbf{u}$, and define the *primal* function

$$L_p(\mathbf{x}) = \max_{\mathbf{u} \geq 0} L(\mathbf{x}, \mathbf{u}) \tag{1-25}$$

and the *dual* function

$$L_d(\mathbf{u}) = \min_{\mathbf{x}} L(\mathbf{x}, \mathbf{u}) \tag{1-26}$$

The primal function $L_p(\mathbf{x})$ reduces to $M(\mathbf{x})$ since in the feasible region $\{g_j(\mathbf{x}) \geq 0\}$ the maximum of $L(\mathbf{x}, \mathbf{u})$ is achieved if $\mathbf{u} = 0$. The infeasible region can lead to an infinite value of $L_p(\mathbf{x})$ for $\mathbf{u} \longrightarrow \infty$ and is disregarded. Since the primal and dual functions are functions of $\mathbf{x}$ and $\mathbf{u}$, respectively, the following two optimization problems can be formulated: the *primal* problem

$$\begin{aligned} &\min_{\mathbf{x}} L_p(\mathbf{x}) \equiv \min M(\mathbf{x}) \\ \text{subject to} \quad &g_j(\mathbf{x}) \geq 0 \qquad j = 1, 2, \ldots, m \end{aligned} \tag{1-27}$$

and the *dual* problem

$$\begin{aligned} &\max_{\mathbf{u}} L_d(\mathbf{u}) \\ \text{subject to} \quad &\mathbf{u} \geq 0 \end{aligned} \tag{1-28}$$

We also define a *saddlepoint* of the Lagrangian as such a pair $(\mathbf{x}^*, \mathbf{u}^*)$ that

$$L(\mathbf{x}^*, \mathbf{u}) \leq L(\mathbf{x}^*, \mathbf{u}^*) \leq L(\mathbf{x}, \mathbf{u}^*) \qquad \text{for all } \mathbf{u} \geq 0 \quad \text{and all } \mathbf{x} \tag{1-29}$$

Finding the saddlepoint is referred to as the *saddlepoint problem.* Assuming the constraint qualification, differentiability, and convexity of $M(\mathbf{x})$ and all $-g_j(\mathbf{x})$, the *equivalence* theorem (1-2) of Kuhn and Tucker [1951] gives a necessary and sufficient condition for a point to be a minimum for the nonlinear programming problem.

Theorem 1-2 The point $\hat{\mathbf{x}}$ is a solution to the primal problem if, and only if, $\hat{\mathbf{x}}$ and some $\hat{\mathbf{u}}$ constitute a saddlepoint for the Lagrangian function.

More explicitly, if $M(\mathbf{x})$ and $g_j(\mathbf{x})$, $j = 1, \ldots, m$ are differentiable functions, then $(\mathbf{x}^*, \mathbf{u}^*)$ is the saddlepoint of $L(\mathbf{x}, \mathbf{u})$ as defined by (1-29) if, and only if,

$$\begin{aligned} \nabla_x L(\mathbf{x}^*, \mathbf{u}^*) &= 0 \\ \nabla_u L(\mathbf{x}^*, \mathbf{u}^*) &\leq 0 \\ (\mathbf{u}^*)^T \nabla_u L(\mathbf{x}^*, \mathbf{u}^*) &= 0 \\ \mathbf{u}^* &\geq 0 \end{aligned} \tag{1-30}$$

where ∇_x and ∇_u denote gradients with respect to $\mathbf{x}$ and $\mathbf{u}$. Note that the relations (1-30) agree with the Kuhn-Tucker necessary conditions (1-23). If the original problem and, consequently, (1-24) requires that $\mathbf{x} \geq 0$, the first condition of (1-30) is replaced by the conditions

$$\begin{aligned} &\nabla_x L(\mathbf{x}^*, \mathbf{u}^*) \geq 0 \\ &(\mathbf{x}^*)^T \nabla_x L(\mathbf{x}^*, \mathbf{u}^*) = 0 \\ &\mathbf{x}^* \geq 0 \end{aligned} \tag{1-31}$$

If the original problem is $\min \{M(\mathbf{x}) \mid g_j(\mathbf{x}) = 0, j = 1, \ldots, m\}$, we obtain the classical saddlepoint conditions

$$\nabla_x L(\mathbf{x}^*, \mathbf{u}^*) = 0 \qquad \text{and} \qquad \nabla_u L(\mathbf{x}^*, \mathbf{u}^*) = 0 \tag{1-32}$$

In addition, Wolfe [1961] proved the *duality* theorem 1-3.

Theorem 1-3 If $\hat{\mathbf{x}}$ is the solution of the convex primal problem, then $\hat{\mathbf{u}} \geq 0$ exists, such that $(\hat{\mathbf{x}}, \hat{\mathbf{u}})$ solves the dual problem and the extremes are equal.

That is, in view of (1-25) through (1-28),

$$L(\hat{\mathbf{x}}, \hat{\mathbf{u}}) = \min_{\mathbf{x}} \max_{\mathbf{u} \geq 0} L(\mathbf{x}, \mathbf{u}) = \max_{\mathbf{u} \geq 0} \min_{\mathbf{x}} L(\mathbf{x}, \mathbf{u})$$

Note that the solution is the saddlepoint defined in (1-29); that is, $(\hat{\mathbf{x}}, \hat{\mathbf{u}}) = (\mathbf{x}^*, \mathbf{u}^*)$.

Since we assumed that $M(\mathbf{x})$ and $-g_j(\mathbf{x})$, $j = 1, 2, \ldots, m$ are differentiable and convex, we can return to the dual problem and give a more explicit formulation:

$$\max_{\mathbf{u}} \{L(\mathbf{x}, \mathbf{u}) = M(\mathbf{x}) - \sum_{j=1}^{m} u_j \, g_j(\mathbf{x})\} \tag{1-33}$$

$$\text{subject to} \qquad \nabla_x L(\mathbf{x}, \mathbf{u}) = 0$$
$$\mathbf{u} \geq 0$$

where ∇_x denotes the gradient with respect to $\mathbf{x}$. In view of the convexity of $L(\mathbf{x}, \mathbf{u})$ with respect to $\mathbf{x}$, the relation $\nabla_x L(\mathbf{x}, \mathbf{u}) = 0$ is equivalent to $\min_{\mathbf{x}} L(\mathbf{x}, \mathbf{u})$.

The duality theory permits us to formulate and solve the dual problem instead of the primal problem. To illustrate this possibility, consider the quadratic programming problem

$$\min \{\tfrac{1}{2}\mathbf{x}^T B\mathbf{x} + \mathbf{b}^T\mathbf{x}\}$$
$$\text{subject to} \qquad A\mathbf{x} \geq \mathbf{c} \tag{1-34}$$

where B is an $n \times n$ positive definite matrix, A is an $m \times n$ matrix, and $\mathbf{b}$, $\mathbf{c}$ are vectors of dimensions n and m, respectively. From (1-24) and (1-33), the dual problem is

$$\max_{\mathbf{u}} \{\tfrac{1}{2}\mathbf{x}^T B\mathbf{x} + \mathbf{b}^T\mathbf{x} + \mathbf{u}^T(\mathbf{c} - A\mathbf{x})\}$$
$$\text{subject to} \qquad B\mathbf{x} + \mathbf{b} - A^T\mathbf{u} = 0 \tag{1-35}$$
$$\mathbf{u} \geq 0$$

or equivalently,

$$\max_{\mathbf{u} \geq 0} \min_{\mathbf{x}} \{\tfrac{1}{2}\mathbf{x}^T B\mathbf{x} + \mathbf{b}^T\mathbf{x} + \mathbf{u}^T(\mathbf{c} - A\mathbf{x})\} \tag{1-36}$$

The positive definite matrix B yields a finite solution $\hat{\mathbf{x}}$ to the unconstrained minimization problem and we have:

$$\hat{\mathbf{x}} = -B^{-1}(\mathbf{b} - A^T\mathbf{u}) \tag{1-37}$$

Substituting (1-37) into (1-36), we have

$$\max_{\mathbf{u} \geq 0} \{\tfrac{1}{2}\mathbf{u}^T AB^{-1}A^T\mathbf{u} + \mathbf{u}^T(\mathbf{c} + AB^{-1}\mathbf{b})\} \tag{1-38}$$

This dual problem is also quadratic, but the number of its variables is m instead of n, which can be advantageous. Notice also that the dual problem has very simple constraints. In the case of a linear programming problem, the matrix $B = 0$ in (1-34) and we obtain the dual formulation from (1-35):

$$\max_{\mathbf{u} \geq 0} \{\mathbf{b}^T\mathbf{x} + \mathbf{u}^T(\mathbf{c} - A\mathbf{x})\}$$
$$\text{subject to} \qquad \mathbf{b} - A^T\mathbf{u} = 0 \tag{1-39}$$

or equivalently,

$$\max \mathbf{c}^T\mathbf{u}$$
$$\text{subject to} \qquad A^T\mathbf{u} = \mathbf{b} \quad \text{and} \quad \mathbf{u} \geq 0. \tag{1-40}$$

Another computational advantage of the duality theory is that for every primal feasible $\bar{\mathbf{x}}$ and dual feasible $(\bar{\bar{\mathbf{x}}}, \bar{\bar{\mathbf{u}}})$—that is, satisfying the primal or dual constraints, respectively—we have the dual objective function

$$L(\bar{\bar{\mathbf{x}}}, \bar{\bar{\mathbf{u}}}) \leq M(\bar{\mathbf{x}}) \tag{1-41}$$

whose value provides us with a lower bound to the value of the primal objective function. This relation is of particular value in cases where Wolfe's duality theorem applies and the primal and dual extrema are equal. In such a case we can use the difference $M(\bar{\mathbf{x}}) - \underline{L}(\bar{\bar{\mathbf{x}}}, \bar{\bar{\mathbf{u}}})$ to estimate what can be gained by continuing the iterative optimization process. Cottle [1964] shows how to form a composite program whose objective is to minimize the difference between the primal and dual objective functions. Solutions, if they exist, are constrained to be points common to the intersection of the constraint sets of the primal and dual problems. This is another way of capitalizing on the duality theory.

The foregoing discussion implies that one can solve the primal problem by solving the dual problem. Furthermore, it was shown in the quadratic case that some computational convenience may be gained by doing this. It should be remarked, however, that intermediate dual solutions may not yield primal feasible points (and vice versa). Converse theorems to the duality theorem for nonlinear programs have been established by Mangasarian [1962] and Huard [1962].

1-8 CHARACTERISTICS OF ITERATIVE TECHNIQUES

The minimization problem (1-9) can be attacked by means of the theoretical conditions outlined in Secs. 1-6 and 1-7. One could, for example, attempt to find a saddlepoint $(\mathbf{x}^*, \mathbf{u}^*)$ of the Lagrangian function

$$L(\mathbf{x}, \mathbf{u}) = M(\mathbf{x}) - \sum_{j=1}^{m} u_j g_j(\mathbf{x}) \tag{1-41}$$

This requires first the solution of $m + n$ equations corresponding to

$$\nabla_x L(\mathbf{x}, \mathbf{u}) = 0 \quad \text{and} \quad \nabla_u L(\mathbf{x}, \mathbf{u}) = 0 \tag{1-42}$$

where ∇_x and ∇_u denote gradient with respect to $\mathbf{x}$ and $\mathbf{u}$. Then one would have to search among all the solutions $(\mathbf{x}, \mathbf{u})$ of (1-42) for that $(\mathbf{x}^*, \mathbf{u}^*)$ at which $M(\mathbf{x})$ assumes its minimum and the inequality constraints and $\mathbf{u} \geq 0$ are satisfied. For this purpose, special combinatorial algorithms have been developed that reduce the problem to one with equality constraints only by searching for active inequalities. This approach is frequently impractical, even when the original problem includes no inequality constraints or no constraints at all, in which case one would have to find $\mathbf{x}^*$ satisfying

$$\nabla M(\mathbf{x}) = 0 \tag{1-43}$$

and

$$M(\mathbf{x}^*) = \min_{\mathbf{x}} M(\mathbf{x}) \tag{1-44}$$

The alternative is to use iterative techniques, the subject of this book. One normally starts the computation with $\mathbf{x}^0$, which is as good an approximation to the solution as is initially available. Then a sequence of points is generated by an iterative relation. If a digital computer is used, this sequence will be discrete, corresponding to a process of motion, in discrete steps, in the n-dimensional space of the problem variables $(x_1, \ldots, x_n)$. This case of a *discrete iterative process* is of interest to us here. All the optimization techniques described in this book can be coded, and many have been, for digital computers, and each code determines a discrete iterative process. The mathematical model of the system being optimized is a set of subroutines that evaluate the objective and constraint functions at any point $\mathbf{x}$. These subroutines are called by the optimizing routine that determines the sequence of points $\mathbf{x}^l$ through which the iteration moves. The motion depends on the functions of the system model and in some cases also on their derivatives. In such an iterative process the operation may be on the points $\mathbf{x}^0, \mathbf{x}^1, \ldots,$ $\mathbf{x}^k$ directly, or on related points $\mathbf{z}^0, \mathbf{z}^1, \ldots, \mathbf{z}^k$, the latter being somewhat more general. Transformation of the problem or variables may lead to situations in which the iteration operates on points $\mathbf{z}^k$ related to the problem variables $\mathbf{x}^k$. At a typical step, say the kth, the process either generates a successor point $\mathbf{z}^{k+1}$ to $\mathbf{z}^k$ or it terminates. There may actually be generated a whole set of points $A(\mathbf{z}^k)$, each of which could be a successor point $\mathbf{z}^{k+1}$.

Each step normally involves *data preparation* and *computation*. Data preparation may consist of evaluating the objective function $M(\mathbf{x}^k)$, and its derivatives $\nabla M(\mathbf{x}^k)$, Hess$(\mathbf{x}^k)$, at the current point. Data from previous steps may also be involved. Zangwill [1969*a*] refers to the procedure as *autonomous* if the data needed for the computation involve only the current point. If the iteration number and data from steps previous to the kth also play a role in the computation, the procedure is said to be *nonautonomous*. In general, the computation makes use of the prepared data, either to terminate the iterative process or to generate a set of successor points. The computation may consist of several stages. Many of the techniques discussed in this book involve, for instance, an exploratory or direction finding stage, and then a steplength determination. In each of these stages an auxiliary problem is solved, and this may involve an iterative process in itself.

A detailed and rigorous discussion of convergence of specific algorithms can be found in the paper by Zangwill [1969*a*], and in his book [Zangwill, 1969*b*]. In the book, Zangwill defines convergence and derives necessary and sufficient conditions for an algorithm, in general, to be convergent. This approach is very useful as a model for the exploration of new algorithms. Convergence to a solution in a finite number of steps can be proven theoretically only for certain algorithms and for special problems. In most cases, convergence can be shown to occur only in an asymptotic sense in which limits may be approached. We indicate what is known about convergence in the discussions of the individual techniques. In addition to the theoretical uncertainty, the computational errors may have an effect on convergence in practice. Consequently, some computer programs are designed, for reasons of economy, to terminate after a predetermined number of steps has been performed or a predetermined amount of computer time has been used.

The iteration will always be terminated if the current point $\mathbf{x}^k$ (or the related point $\mathbf{z}^k$, in general) is identified as a solution, or if there are indications that no solution to the problem exists. The decision to terminate must, of course, be based on the data generated by the algorithm at, or prior to, the kth step. Furthermore, if $\mathbf{z}^k$ is a solution, the algorithm should either be incapable of generating a successor point $\mathbf{z}^{k+1}$ (the set of successor points should be empty), or all such successor points which are generated should also be solutions. A *solution point* is a point with given properties that should be recognizable in the computation. Ideally, the point of termination should be $\mathbf{x}_G$, the point at which the objective function $M(\mathbf{x})$ assumes its global minimum. In reality, the process usually provides a "practical" solution. This is at best approximately $\mathbf{x}_L$, a point at which the objective function $M(\mathbf{x})$ assumes a local minimum. It may also just be a point $\hat{\mathbf{x}}$ at which the objective function assumes a smaller value than at the starting point $\mathbf{x}^0$. Termination could occur at a stationary point $\mathbf{x}_S$, which is neither $\mathbf{x}_L$ nor $\mathbf{x}_G$. The point of termination depends on how the set of solution points is defined in the algorithm. This in turn depends on the algorithm itself, because solution points would have to be identified on the basis of information generated in the iteration step. The discussions of specific termination criteria are therefore included in the sections dealing with the individual techniques.

Computational experience indicates that it is advisable, after termination, to restart the computation from a point obtained by a slight perturbation of the termination point. Termination of the new process at a point that differs considerably from the previous point of termination is a possible indication of multiple local minima. Such a problem should be examined carefully. Furthermore, if the problem is constrained, it may be advisable to repeat the solution with slightly perturbed constraints, for example, replacing $g_j(\mathbf{x}) \geq 0$ by $g_j(\mathbf{x}) \geq \pm\epsilon$ where ϵ is a small positive quantity. Termination of the new process at a point which differs considerably from the previous point of termination is also a possible indication of an unstable (ill-posed) problem, for one normally would not expect small changes in the constraint functions to result in considerably changed solutions. Again, the problem should be examined carefully. It might also be worthwhile to perturb the constant parameters in the objective function and repeat the solution process.

It is of interest to consider some of the factors characterizing iterative processes: We define the *order of convergence* as the power $p \geq 1$ (if it exists) in the expression

$$\lim_{k \to \infty} \frac{||\mathbf{x}^{k+1} - \hat{\mathbf{x}}||}{||\mathbf{x}^k - \hat{\mathbf{x}}||^p} = C \tag{1-45}$$

where $\hat{\mathbf{x}}$ is the solution to the problem and $C > 0$ is an asymptotic error constant. The case of $C < 1$ and $p = 1$, or convergence of order 1, is sometimes also referred to as *linear convergence*. The case of $p > 1$, convergence of order greater than 1, is sometimes also referred to as *superlinear convergence*. Another term is used when the iterative process converges like a geometric series: *geometric convergence*, which occurs if there exists a $C < 1$ such that

$$\frac{||\mathbf{x}^{k+1} - \hat{\mathbf{x}}||}{||\mathbf{x}^k - \hat{\mathbf{x}}||} = C \tag{1-46}$$

The difference between geometric and linear convergence is that in the latter

$$\frac{||\mathbf{x}^{k+1} - \hat{\mathbf{x}}||}{||\mathbf{x}^k - \hat{\mathbf{x}}||} = C + \delta_k \qquad \lim_{k \to \infty} \delta_k = 0$$

Not every iterative process is characterized by an order of convergence; in fact, an order of convergence need not exist. Some techniques, when applied to particular types of problems, require (theoretically) a finite number of iterations. A technique is said to have *quadratic convergence*, for example, if the minimum of a quadratic function can be obtained in a finite number of iterations. Notice that quadratic convergence is not convergence of order 2. Furthermore, due to roundoff errors, all finite techniques require a larger number of iterations than theoretically predicted. It is also important to consider the amount of *computational work* (*cost*) *per iteration*. This may vary from method to method for a given problem. Computational work involved in data preparation and computation of the new point can be measured in terms of evaluations of objective and constraint functions (function evaluations) and/or other arithmetic operations performed in every iteration. One measure that may be used in comparing different techniques, when applied to a given problem, is the *computational efficiency*. This is the rate of reduction in the objective function per unit of computational work (cost). Ease of application of a technique is another measure of efficiency. Order of convergence or rate of reduction in the objective function per iteration alone are obviously not adequate measures of technique efficiency. The reader will find, unfortunately, that more research and numerical experimentation are needed to resolve some of the questions associated with the evaluation of the efficiency of iterative techniques. More on computational efficiency of iterative techniques and related topics can be found in Traub [1964].

1-9 SCALING

The efficiency and rate of convergence of optimization methods depend critically on the scales used for the variables. Even the simplest linear transformations of the variables $\hat{x}_i = a_i x_i$, $a_i > 0$ will not leave many important aspects of the optimization algorithm invariant. Consider the problem

$$\min_{\mathbf{x}} \{x_1^2 + x_2^2 \mid x_1 \geq 0, x_2 \geq 0\} \tag{1-47}$$

The objective function isocontours are exactly concentric circles in this case. Using the steepest descent method (Sec. 5-2) and the starting point $\mathbf{x}^0 = (1, 1)^T$, we get the descent direction $\mathbf{s} = -\nabla M(1, 1) = (-2, -2)^T$, and the solution $\mathbf{x}^* = (0, 0)^T$ can be obtained in one step. It is sufficient, however, to introduce the variable $\hat{x}_1 = 0.1x_1$ instead of x_1, to radically change the efficiency of the steepest descent method when applied to the same problem. The isocontours of the new objective function $M(\hat{x}_1, x_2) = 100\hat{x}_1^2 + x_2^2$ form a deep and narrow valley along the x_2 coordinate axis, and the gradient vector calculated at the same point as before, $\mathbf{s} = -\nabla M(1/10, 1) = (-20, -2)^T$, makes now an angle of nearly 90° with the x_2 axis. Thus, we may expect that the steepest descent method will be very inefficient if used to solve the rescaled problem.

A method that does not have this drawback may be called *invariant with respect to (linear) change of scales of the variables*. Such a method retransforms the gradient vector so as to again point at the solution. Consider, for example, the second-order gradient method (Sec. 5-3), which uses the descent direction

$$\mathbf{s} = -[\text{Hess}(\tfrac{1}{10}, 1)]^{-1} \nabla M(\tfrac{1}{10}, 1)] = -\begin{bmatrix} \frac{1}{200} & 0 \\ 0 & \frac{1}{2} \end{bmatrix} \begin{bmatrix} 20 \\ 2 \end{bmatrix} = -(\tfrac{1}{10}, 1)^T \tag{1-48}$$

This shows that the direction $\mathbf{s}$ is oriented toward the solution in our new (scaled) system of variables. We may state, more generally, that any quasi-Newton method that uses the Hessian is invariant under a linear scale change. The variable metric methods (Sec. 5-6) belong to this category, since they utilize weighting matrices that converge to the inverse of the Hessian in the vicinity of the solution. In theory, conjugate direction methods which are quadratically convergent also should not be sensitive to scale changes, at least if quadratic functions are to be minimized. In computational practice, however, due to the roundoff errors, the efficiency of conjugate direction and quasi-Newton methods is greatly influenced by scaling [Fox and Stanton, 1968].

The problem of scaling functions that enter the nonlinear optimization problem is usually very difficult. In the general nonlinear case we can say only that by "well-scaled" problems we intuitively mean problems in which similar changes in the variables lead to similar changes in the objective function. More insight can be gained by considering the quadratic approximation to $M(\mathbf{x})$ in the neighborhood of the minimum solution $\hat{\mathbf{x}}$,

$$Q(\mathbf{x}) = M(\hat{\mathbf{x}}) + \nabla M(\hat{\mathbf{x}})^T(\mathbf{x} - \hat{\mathbf{x}}) + \tfrac{1}{2}(\mathbf{x} - \hat{\mathbf{x}})^T \operatorname{Hess}(\hat{\mathbf{x}})(\mathbf{x} - \hat{\mathbf{x}}) \qquad (1\text{-}49)$$

We can further simplify $Q(\mathbf{x})$ by shifting the origin of the coordinate system to $\hat{\mathbf{x}}$. Since $\nabla M(\hat{\mathbf{x}}) = 0$ and the constant $M(\hat{\mathbf{x}})$ can be dropped, we may limit our attention to the function

$$Q(\mathbf{x}) = \tfrac{1}{2}\mathbf{x}^T \operatorname{Hess}(\hat{\mathbf{x}})\mathbf{x} \qquad (1\text{-}50)$$

where $\operatorname{Hess}(\hat{\mathbf{x}})$ is a positive definite matrix. The matrix $\operatorname{Hess}(\hat{\mathbf{x}})$ can be decomposed into

$$\operatorname{Hess}(\hat{\mathbf{x}}) = V^T D V \qquad (1\text{-}51)$$

where V is an orthogonal matrix and D is the diagonal matrix, whose elements are the positive eigenvalues of Hess. We see that by introducing the scale change $\mathbf{y} = \sqrt{D}V\mathbf{x}$, we can bring $Q(\mathbf{x})$ into the form

$$Q(\mathbf{y}) = \tfrac{1}{2}\mathbf{y}^T\mathbf{y} \qquad (1\text{-}52)$$

and this is the desirable function with the circular isocontours [see (1-47)].

It is instructive to consider a new variable $\mathbf{z} = V\mathbf{x}$, which transforms the function $Q(\mathbf{x})$ to the form

$$Q(\mathbf{z}) = \tfrac{1}{2}\mathbf{z}^T D\mathbf{z} \qquad (1\text{-}53)$$

Now the isocontours $Q(\mathbf{z}) =$ constant are n-dimensional ellipsoids centered at the origin $\mathbf{x} = 0$, which cut the ith coordinate axis at the points $z_i = \text{constant}(\gamma_i)^{-1/2}$ where γ_i are the components of D. We see that the value z_i (principal axis) associated with the eigenvalue γ_i is inversely proportional to $\sqrt{\gamma_i}$, so that small γ_i corresponds to elongation of the ellipsoid in the direction z_i. Such elongation implies that the variable z_i is poorly determined in the optimization calculation.

Returning to our numerical example with $M(x_1, x_2) = 100x_1^2 + x_2^2$, we see that this function is not sensitive to changes of x_2. It follows that scaling can be characterized by the *condition number* C_N, defined as

$$C_N = \gamma_{\max}/\gamma_{\min} \qquad (1\text{-}54)$$

where $\gamma_{\max} = \max_i \gamma_i$ and $\gamma_{\min} = \min_i \gamma_i$. We may now ask the following question: Given a quadratic form $Q(\mathbf{x}) = 1/2\mathbf{x}^T H\mathbf{x}$, can we find a transformation $\mathbf{x} = D\mathbf{y}$ (where D is diagonal) that minimizes (or decreases) the condition number of DHD and can be obtained using a simple numerical algorithm? The answer to the second part of our question is essential since a transformation to obtain (1-52) is too time consuming to be practical. Unfortunately we have no satisfactory practical solution to the problem of scaling quadratic functions except for a few special types of matrices. The same is also true of the more general nonlinear case.

The problem of scaling in the constrained nonlinear problems may have another interesting aspect. Suppose we have a two-dimensional problem where $10^{-5} \leq x_1 \leq 10^{-2}$ and $10^2 \leq x_2 \leq 10^3$ so that the feasible region has the shape of a very narrow band. It would be difficult to search for a minimum in such a band, because the steplength must be very short to avoid violations of the constraint imposed on x_1. Another difficulty would be introduced by performing arithmetic operations with numbers too different in order of magnitude. The way to overcome such difficulties is to scale the variables and make them of the same order of magnitude. Assume that x_i is in the range of $b_i \leq x_i \leq u_i$. The rescaled variable x_i^R can be introduced by

$$x_i^R = \frac{2C_i x_i - C_i(b_i + u_i)}{u_i - b_i} \tag{1-55}$$

where x_i^R varies in the range $(-C_i, C_i)$. If we decide, for example, that all $C_i = 1$, all the variables are located within a hypercube of length 2 about the origin of the new coordinate system.

1-10 NUMERICAL TEST PROBLEMS

To illustrate some of the more important algorithms, the following numerical test problems have been selected:

1 Test function for unconstrained optimization methods [credited to C. E. Wood of Westinghouse Research Laboratory; see Colville, 1968]:

$$M(\mathbf{x}) = 100(x_2 - x_1^2)^2 + (1 - x_1)^2 + 90(x_4 - x_3^2)^2 + (1 - x_3)^2 + 10.1\{(x_2 - 1)^2 + (x_4 - 1)^2\} + 19.8(x_2 - 1)(x_4 - 1)$$

This test function is an extension of the famous problem of Rosenbrock [1960]. It is designed to cause a possible slow convergence to the solution or premature stop at nonoptimal points. This function has a minimum $M = 0$ at $\mathbf{x} = (1, 1, 1, 1)^T$. The initial point is $\mathbf{x}^0 = (-3, -1, -3, -1)^T$, which gives $M(\mathbf{x}^0) = 19{,}192.000$.

2 To test the methods that can handle constraints, the same function can be used in conjunction with the constraints:

$$-10 \leq x_1 \leq 10$$
$$-10 \leq x_2 \leq 10$$
$$-10 \leq x_3 \leq 10$$
$$-10 \leq x_4 \leq 10$$

The solution is the same as the solution to Prob. 1, and so is the initial point.

3 Another constrained problem has been designed by Rosen and Suzuki [1965]:

$$M(\mathbf{x}) = x_1^2 + x_2^2 + 2x_3^2 + x_4^2 - 5x_1 - 5x_2 - 21x_3 + 7x_4$$

subject to

$$-x_1^2 - x_2^2 - x_3^2 - x_4^2 - x_1 + x_2 - x_3 + x_4 + 8 \geq 0$$
$$-x_1^2 - 2x_2^2 - x_3^2 - 2x_4^2 + x_1 + x_4 + 10 \geq 0$$
$$-2x_1^2 - x_2^2 - x_3^2 - 2x_1 + x_2 + x_4 + 5 \geq 0$$

This function has a minimum $M = -44$ at $\mathbf{x} = (0, 1, 2, -1)^T$. The initial approximation is $\mathbf{x}^0 = (0, 0, 0, 0)^T$, which gives $M(\mathbf{x}^0) = 0$.

4 Colville [1968] cites another constrained problem:

$$\begin{aligned} M(\mathbf{x}) = & -15x_1 - 27x_2 - 36x_3 - 18x_4 - 12x_5 \\ & +30x_1^2 - 40x_1x_2 - 20x_1x_3 + 64x_1x_4 - 20x_1x_5 \\ & +39x_2^2 - 12x_2x_3 - 62x_2x_4 + 64x_2x_5 \\ & +10x_3^2 - 12x_3x_4 - 20x_3x_5 + 39x_4^2 - 40x_4x_5 + 30x_5^2 \\ & +4x_1^3 + 8x_2^3 + 10x_3^3 + 6x_4^3 + 2x_5^3 \end{aligned}$$

subject to ten linear constraints $A\mathbf{x} \geq \mathbf{b}$ and lower bounds $\mathbf{x} \geq 0$, where

$$\mathbf{b}^T = (-40, -2, -.25, -4, -4, -1, -40, -60, 5, 1)$$

and

$$A = \begin{bmatrix} -16 & 2 & 0 & 1 & 0 \\ 0 & -2 & 0 & .4 & 2 \\ -3.5 & 0 & 2 & 0 & 0 \\ 0 & -2 & 0 & -4 & -1 \\ 0 & -9 & -2 & 1 & -2.8 \\ 2 & 0 & -4 & 0 & 0 \\ -1 & -1 & -1 & -1 & -1 \\ -1 & -2 & -3 & -2 & -1 \\ 1 & 2 & 3 & 4 & 5 \\ 1 & 1 & 1 & 1 & 1 \end{bmatrix}$$

This function has a minimum $M = -32.34868$ at $\mathbf{x} = (0.3, 0.33347, 0.4, 0.42831, 0.22396)^T$. The initial approximation is $\mathbf{x}^0 = (0, 0, 0, 0, 1)^T$, which gives $M(\mathbf{x}^0) = 20$.

References

Colville, A. R.: *A Comparative Study on Nonlinear Programming Codes*, Report No. 320–2949, IBM Scientific Center, 1968.

Cottle, R. W.: *Nonlinear Programs with Positively Bounded Jacobians*, ORC 64–12 (RR), Operations Research Center, University of California, Berkeley, 1964.

Evans, J. P.: "On Constraint Qualifications in Nonlinear Programming," *Naval Research Logistics Quarterly*, Vol. 17, 281, 1970.

Fox, R. L. and E. L. Stanton: "Development in Structural Analysis by Direct Energy Minimization," *AIAA Journal*, Vol. 6, 1036, 1968.

Huard, P.: "Dual Programs," *IBM Journal of Research and Development*, Vol. 6, 137, 1962.

Karnes, R. N. and J. L. Tocher: "Trends in Automated Structural Design," *Proceedings of the Aerospace Structures Design Conference*, Pacific Science Center, Seattle Professional Engineering Employees Association, Seattle, Wash., 1969.

Kuhn, H. W. and A. W. Tucker: "Nonlinear Programming," in J. Neyman (ed.), *Proceedings of the Second Berkeley Symposium on Mathematical Statistics and Probability*, University of California Press, Berkeley, 481, 1951.

Mangasarian, O. L.: "Duality in Nonlinear Programming," *Quarterly of Applied Math.*, Vol. 20, 300, 1962.

———: *Nonlinear Programming*, McGraw-Hill Book Company, New York, 1969.

——— and S. Fromovitz: "The Fritz John Necessary Optimality Conditions in the Presence of Equality and Inequality Constraints," *Journal of Math. Analysis and Applications*, Vol. 17, 37, 1967.

Rosen, J. B. and S. Suzuki: "Construction of Nonlinear Programming Test Problems," *Communications of the ACM*, Vol. 8, 113, 1965.

Rosenbrock, H. H.: "An Automatic Method for Finding the Greatest or Least Value of a Function," *Computer Journal*, Vol. 3, 175, 1960.

Serfozo, R. F.: *A Method for Determining the Optimal Composition of a Military Aircraft Fleet*, D6–2651, The Boeing Company, Renton, Wash., 1967.

Traub, J. F.: *Iterative Methods for the Solution of Equations*, Prentice-Hall, Inc., Englewood Cliffs, N. J., 1964.

Wolfe, P.: "A Duality Theorem for Nonlinear Programming," *Quarterly of Applied Math.*, Vol. 19, 239, 1961.

———: "Methods of Nonlinear Programming," in J. Abadie (ed.), *Nonlinear Programming*, North-Holland Publishing Company, Amsterdam, Chap. VI, 97, 1967.

Zangwill, W. I.: "Convergence Conditions for Nonlinear Programming Algorithms," *Management Science*, Vol. 16, 1, 1969*a*.

———: *Nonlinear Programming: A Unified Approach*, Prentice-Hall, Inc., Englewood Cliffs, N. J., 1969*b*.

2 PROBLEM AND VARIABLE TRANSFORMATIONS

It is possible to transform problems of certain types into problems of certain other types and solve them as such. These transformations are occasionally of considerable practical value. For example, a computer program for the solution of the transformed problem may be readily available or the transformed problem may require less computational work. In some cases, computer storage limitations make it impossible to solve the original problem as such. The reformulated version of the problem may also give the analyst much insight and information that the original problem version cannot.

2-1 TRANSFORMATIONS OF THE GENERAL PROBLEM TO SPECIAL PROBLEMS

Here we discuss various transformations of the general problem (1-9) to special problems, including transformations to unconstrained problems for solution by exterior and interior point techniques, a transformation to an unconstrained problem by change of variable, a transformation to partially constrained problems, a transformation to a problem with a linear objective function, and a linearization transformation.

Transformations to an Unconstrained Problem

Solution by Exterior Point Techniques. The iterative process of exterior point techniques computes a sequence of points that are not necessarily in the feasible region of the original problem. The original general problem therefore is transformed into an unconstrained problem by means of a transformation $P(\mathbf{x}, \mathbf{r})$. Let $E[G]$ be the *Heaviside step function* defined as

$$E[G]\begin{cases} = 0 & \text{if } G \geq 0 \\ = 1 & \text{if } G < 0 \end{cases} \tag{2-1}$$

We employ these functions to define new constraint functions:

$$\bar{g}_j(\mathbf{x})\begin{cases} = E[g_j(\mathbf{x})]g_j(\mathbf{x}) \equiv \min\,[g_j(\mathbf{x}), 0] \text{ for inequality constraints} \\ \quad g_j(\mathbf{x}) \geq 0 \text{ or } l_j(\mathbf{x}) \geq 0 \\ = g_j(\mathbf{x}) \text{ for equality constraints } g_j(\mathbf{x}) = 0 \text{ or } l_j(\mathbf{x}) = 0 \end{cases} \tag{2-2}$$

Let $K(\mathbf{r}_k)$ be a diagonal matrix of *penalty coefficients* $k_j \geq 0$ (along the diagonal). These coefficients are in general functions of a parameter $r_j > 0$, such that given $r_{j1} > r_{j2} > \ldots$ we have $0 < k_j(r_{j1}) < k_j(r_{j2}) < \ldots$ for any j. We denote $\mathbf{r}_k = (r_{1k}, \ldots, r_{mk})^T$. The transformation is then

$$P(\mathbf{x}, \mathbf{r}) = M(\mathbf{x}) + \tfrac{1}{2} \sum_{j=1}^{m} k_j(r_j)[\bar{g}_j(\mathbf{x})]^2 \tag{2-3}$$

or equivalently,

$$P(\mathbf{x}, \mathbf{r}) = M(\mathbf{x}) + \tfrac{1}{2}[\bar{\mathbf{g}}(\mathbf{x})]^T K(\mathbf{r})\bar{\mathbf{g}}(\mathbf{x}) \tag{2-4}$$

where $\bar{\mathbf{g}} = (\bar{g}_1, \ldots, \bar{g}_m)^T$, and the second term may be referred to as the *penalty function*; that is, one is "penalized" for the violation of the constraints, although such a violation is permitted throughout the exterior point iteration. This transformation was first suggested by Courant [1943] and later generalized by Zangwill [1967]. Zangwill has shown a penalty transformation

$$P(\mathbf{x}, r) = M(\mathbf{x}) - r^{-1} \sum_{i=1}^{m} \min\,[0, g_i(\mathbf{x})]$$

where $g_i(\mathbf{x}) \geq 0$, $i = 1, \ldots, m$ with a special property. If $\mathbf{x}(r)$ minimizes $P(\mathbf{x}, r)$, a positive R can be found such that $\mathbf{x}(r)$ is a minimum solution to the original problem if $0 < r \leq R$; that is, the problem can be solved by a single unconstrained minimization.

A modification of the definition (2-3) of the penalty function has been proposed recently by Morrison [1968]:

$$P_k(\mathbf{x}, M_k) = [M(\mathbf{x}) - M_k]^2 + \sum_{j=1}^{m} [\bar{g}_j(\mathbf{x})]^2 \tag{2-5}$$

where

$$M_k \leq M(\hat{\mathbf{x}}), \qquad M_k \leq M_{k+1} \longrightarrow M(\hat{\mathbf{x}}) \quad \text{as} \quad k \longrightarrow \infty \tag{2-6}$$

and $\hat{\mathbf{x}}$ is the solution to the original minimization problem (1-9). Notice that the problem of minimizing $P_k(\mathbf{x}, M_k)$ as defined in (2-5) is a nonlinear least-squares problem and can be solved by use of special methods (see appendix). If the functions of the problem (1-9) satisfy certain regularity conditions, the sequence of solutions $\mathbf{x}^k$ of Prob. (2-5), corresponding to the sequence M_k (2-6), will converge to the solution $\hat{\mathbf{x}}$ of (1-9).

Another modification of the definition (2-3) of the penalty function has been proposed by Pietrzykowski [1969]:

$$P(\mathbf{x}, r) = rM(\mathbf{x}) - \sum_j \min [0, g_j(\mathbf{x})] + \sum_\mu |g_\mu(\mathbf{x})|$$

where $g_j(\mathbf{x}) \geq 0$ are inequality and $g_\mu(\mathbf{x}) = 0$ are equality constraints. He shows that under certain conditions $P(\mathbf{x}, r)$ for sufficiently small $r \geq 0$ assumes its minimum at a point $\hat{\mathbf{x}}$, which is a solution to the original constrained problem (1-9). He also proposes a method for solving the transformed problem.

Huang and Naqvi [1970], Miele and Levy [1970] and Miele, et al. [1971] address themselves to the minimization problem with nonlinear equality constraints

$$\min M(\mathbf{x}) \text{ subject to } g_j(\mathbf{x}) = 0, \qquad j = 1, \ldots, m$$

In this case use can be made of the Lagrangian defined in (1-24) as

$$L(\mathbf{x}, \mathbf{u}) = M(\mathbf{x}) - \mathbf{u}^T\mathbf{g}(\mathbf{x})$$

where $\mathbf{u} = (u_1, \ldots, u_m)^T$ and $\mathbf{g} = (g_1, \ldots, g_m)^T$. A necessary condition for $(\mathbf{x}^*, \mathbf{u}^*)$ to be a solution is (1-32)

$$\nabla_x L(\mathbf{x}^*, \mathbf{u}^*) = 0 \quad \text{and} \quad \nabla_u L(\mathbf{x}^*, \mathbf{u}^*) = \mathbf{g}(\mathbf{x}^*) = 0.$$

Therefore, the original constrained problem can be transformed into the unconstrained problem of minimizing $P(\mathbf{x}, \mathbf{u})$ given by

$$P(\mathbf{x}, \mathbf{u}) = [\nabla_x L(\mathbf{x}, \mathbf{u})]^T \nabla_x L(\mathbf{x}, \mathbf{u}) + [\mathbf{g}(\mathbf{x})]^T\mathbf{g}(\mathbf{x}).$$

Huang and Naqvi [1970] and Miele, et al. [1971] propose and test the application of some algorithms for unconstrained problems to the transformed problem, treating it as a problem in $(\mathbf{x}, \mathbf{u})$.

Solution by Interior Point Techniques. The iterative process of interior point techniques computes a sequence of feasible solutions to the original problem. In this case, the transformation prevents the solutions from approaching the inequality constraints. This restraint is gradually relaxed, however, and by the end of the iterative process the inequality constraints can theoretically be approached with any desired accuracy. The constrained problem $\min [M(\mathbf{x}) \mid g_j(\mathbf{x}) \geq 0, j = 1, \ldots, m]$ is transformed into an unconstrained problem by means of

$$P(\mathbf{x}, r_k) = M(\mathbf{x}) + r_k \sum_{j=1}^{m} [g_j(\mathbf{x})]^{-1} \tag{2-7}$$

where $r_k > 0$ is a parameter and the second term may be referred to as the *barrier function.* The sequence of parameters r_k is computed by means of $r_{k+1} = r_k/c$, $c > 1$ so that $r_1 > r_2 > \ldots > r_k > r_{k+1} > \ldots$. This type of transformation has been developed and studied extensively by Fiacco and McCormick [1968] and also by Pomentale [1965]. Alternative formulations of the barrier function include

$$P(\mathbf{x}, r_k) = M(\mathbf{x}) - r_k \sum_{j=1}^{m} \log [g_j(\mathbf{x})] \tag{2-8}$$

where r_k is as before. In general, barrier functions in the form of $\sum_{j=1}^{m} \phi[g_j(\mathbf{x})]$, are added to the objective function:

$$P(\mathbf{x}, r_k) = M(\mathbf{x}) + r_k \sum_{j=1}^{m} \phi[g_j(\mathbf{x})] \tag{2-9}$$

The barrier function ϕ is convex and nonincreasing in the interval $(0, \infty)$ and with the property that

$$\lim \phi[g_j] = +\infty \quad \text{if } g_j(\mathbf{x}) \longrightarrow 0 \tag{2-10}$$

which ensures that for every $r_k > 0$, P is convex for the convex problem. One is normally restricted to starting the process with a feasible point $\mathbf{x}^0 \in R$ [Eq. (1-12), Sec. 1-4].

Problems with both inequality and equality constraints have to be handled by functions of a mixed penalty-barrier type. The resulting transformations are mixed interior-exterior:

$$P(\mathbf{x}, r) = M(\mathbf{x}) + r_k \sum_{j} [g_j(\mathbf{x})]^{-1} + r_k^{-1/2} \sum_{\mu} [g_\mu(\mathbf{x})]^2$$

where $g_j(\mathbf{x}) \geq 0$ and $g_\mu(\mathbf{x}) = 0$, and where $r_k > 0$ is as before.

Transformation by Change of Variable. Problems with linear constraints of the type

$$\begin{aligned}
&1 \quad x_i \geq 0 \\
&2 \quad 0 \leq x_i \leq 1 \\
&3 \quad b_i \leq x_i \leq u_i \\
&4 \quad 0 \leq x_i \leq x_j \leq x_k \\
&5 \quad -1 \leq x_i \leq 1
\end{aligned} \tag{2-11}$$

can be transformed to unconstrained problems by means of the following corresponding change of variable:

$$\begin{aligned}
&1 \quad x_i = y_i^2, \qquad x_i = e^{y_i}, \qquad x_i = |y_i| \\
&2 \quad x_i = \sin^2 y_i, \qquad x_i = \frac{e^{y_i}}{e^{y_i} + e^{-y_i}} \\
&3 \quad x_i = b_i + (u_i - b_i) \sin^2 y_i \\
&4 \quad x_i = y_i^2, \qquad x_j = y_i^2 + y_j^2, \qquad x_k = y_i^2 + y_j^2 + y_k^2 \\
&5 \quad x_i = \sin y_i
\end{aligned} \tag{2-12}$$

These transformations do not introduce additional local optima, but may "lose" local optima by sending them to infinity; for example, $x_i = e^{y_i}$ with optimum at $x_i = 0$. The periodicity introduced in cases (**2**), (**3**) and (**5**) will not cause any computational problems if counteracted by appropriate measures such as step size adjustment.

Transformations to a Partially Constrained Problem

Elimination of the $\bar{\bar{m}}$ equality constraints (nonlinear as well as linear) can frequently be achieved by selecting $\bar{\bar{m}} < n$ of the n variables and fixing them so that these constraints are always satisfied. This would involve the solution of a set of $\bar{\bar{m}}$, in general nonlinear, equations in $\bar{\bar{m}}$ of the problem variables. The remaining optimization problem involves $n - \bar{\bar{m}}$ variables and $m - \bar{\bar{m}}$ inequality constraints. Both problems will be resolved at every step.

One other way of eliminating equality constraints results in a larger problem. It involves replacing each equality $g_j(\mathbf{x}) = 0$ by two inequalities $g_j(\mathbf{x}) \geq 0$ and $-g_j(\mathbf{x}) \geq 0$, and thereby doubling the number $\bar{\bar{m}}$ of constraints.

Elimination of $\overset{\leq}{m}$ inequality constraints can be achieved by adding $\overset{\leq}{m}$ slack variables x_{n+i}, $i = 1, \ldots, \overset{\leq}{m}$ to the set of n problem variables. These variables are restricted to be nonnegative; they are used to reformulate the inequality constraints as follows:

$$g_j(\mathbf{x}) - x_{n+j} = 0 \tag{2-13}$$

One of the transformations discussed in the preceding section may be used to eliminate the inequalities introduced by the nonnegativity requirement on the slack variables.

Elimination of any subset of constraints can also be achieved, of course, by use of any of the transformations discussed previously. The eliminated constraints would be included in the transformation function or the transformed variable and others not. In most such cases, one would then apply a technique that handles constraints explicitly to minimize the transformed function subject to the constraints that were not eliminated. Suppose, for example, that we have a problem with linear and nonlinear constraints. One can then combine the original function and the nonlinear constraints into a transformed function and minimize this function subject to the linear constraints (Sec. 6-3).

The linear constraints of the minimization problem $\{\min [1/2\mathbf{x}^T A\mathbf{x} - \mathbf{b}^T\mathbf{x}] \mid B\mathbf{x} = \mathbf{c}\}$ can be eliminated using Lagrange multipliers. The equivalent problem is a set of linear equations

$$\begin{bmatrix} A & B^T \\ B & 0 \end{bmatrix} \begin{bmatrix} \mathbf{x} \\ \mathbf{y} \end{bmatrix} = \begin{bmatrix} \mathbf{b} \\ \mathbf{c} \end{bmatrix}$$

This set can be solved directly or transformed into an unconstrained minimization problem (Secs. 2-3 and 5-5).

Transformation to a Problem with a Linear Objective Function

Some solution techniques require the objective function $M(\mathbf{x})$ to be linear in $\mathbf{x}$. The general problem (1-9) can be transformed to an equivalent problem with a linear objective function for the application of such a technique. This is done by adding one variable x_{n+1} to the set of n problem variables and setting the objective function of the transformed problem equal to this variable, and introducing an additional constraint requiring this variable to be greater than the original objective function:

$$\min\{x_{n+1}\} \quad \text{and} \quad x_{n+1} - M(\mathbf{x}) \geq 0 \tag{2-14}$$

Linearization by Change of Variables

Nonlinear problems can be linearized by replacing the variables $\mathbf{x}$ by

$$\mathbf{x} = \sum_{l=0}^{L} a_l \mathbf{x}_l \tag{2-15}$$

where $$a_l \geq 0 \qquad l = 0, \ldots, L \qquad \sum_{l=0}^{L} a_l = 1$$

and $\{\mathbf{x}_l, l = 0, \ldots, L\}$ is a set of preselected grid points. Many nonlinear functions $F(\mathbf{x})$ can be conveniently linearized on the grid $\{\mathbf{x}_l\}$ by means of

$$F(\mathbf{x}) = \sum_{l=0}^{L} a_l F(\mathbf{x}_l) \tag{2-16}$$

and the nonlinear problem

$$\min_{\mathbf{x}} \{M(\mathbf{x}) \mid g_j(\mathbf{x}) \geq 0, \qquad j = 1, \ldots, m\}$$

can be transformed to

$$\min_{\mathbf{a}} \sum_{l=0}^{L} a_l M(\mathbf{x}_l)$$

subject to

$$a_l \geq 0, \qquad l = 0, \ldots, L \tag{2-17}$$

$$\sum_{l=0}^{L} a_l = 1$$

$$\sum_{l=0}^{L} a_l g_j(\mathbf{x}_l) \geq 0, \qquad j = 1, \ldots, m$$

If $\mathbf{a}^* = (a_0^*, a_1^*, \ldots, a_L^*)^T$ is the solution of (2-17), then under certain conditions

$$\mathbf{x}^* = \sum_{l=0}^{L} a_l^* \mathbf{x}_l \tag{2-18}$$

approximates the solution of the original problem. For further information concerning this technique, see Miller [1963], Beale [1967, 1968], and Wolfe [1967].

2-2 TRANSFORMATIONS OF SOME SPECIAL PROBLEMS

In this section we discuss the transformations of certain special problems. Details include the transformation of a problem and an arbitrary initial solution into an equivalent problem and a feasible initial solution, transformation of separable problems into special linear problems, geometric program transformation, transformation of quadratic programs into linear programs, and decomposition of specially structured nonlinear programs.

Transformation to a Problem with a Feasible Solution

Some of the iterative techniques discussed in this book do not pose any restrictions on the starting point $\mathbf{x}^0$. Certain techniques, however, require that the starting point be interior feasible $\mathbf{x}^0 \in R^0$ or feasible $\mathbf{x}^0 \in R$ as defined in (1-13) and (1-12), respectively. In certain cases, it may be difficult to generate a feasible $\mathbf{x}^0$. A feasible point can then be obtained from an infeasible one by solving a sequence of auxiliary problems reducing the infeasibilities one at a time by repeated application of the technique on hand. An alternative method is to transform the original problem (1-9), assuming it has inequalities only, to an equivalent problem involving one additional variable x_{n+1}, restricted to $x_{n+1} \geq 0$:

$$\min\,[M(\mathbf{x}) + \rho_0 x_{n+1}], \qquad g_j(\mathbf{x}) + \rho_j x_{n+1} \geq 0, \qquad j = 1, \ldots, m_1 \tag{2-19}$$

where ρ_0 is a large number and

$$\rho_j \begin{cases} = 1 & \text{if } g_j(\mathbf{x}^0) \leq 0 \\ = 0 & \text{otherwise} \end{cases}$$

and where $\mathbf{x}^0$ is the infeasible starting point. It is easy to see that x_{n+1}^0 can be found such that $[(\mathbf{x}^0)^T, x_{n+1}^0]^T$ is feasible in the modified problem. It can be proven that for sufficiently large ρ_0 the modified problem (2-19) has the same solution as the original problem (1-9). The value of x_{n+1} at this solution will be zero or very near zero. If equality constraints are included in the original problem, they could be eliminated first, through transformation to a partially constrained problem as described in the preceding section.

The same technique is also useful if equality constraints have to be replaced by inequalities when using an optimization method which cannot handle equality constraints directly. If $g_j(\mathbf{x}) = 0$ has to be satisfied and we have to start with a feasible point $[(\mathbf{x}^0)^T, x_{n+1}]^T$, and $g_j(\mathbf{x}^0) < 0$, the equation is replaced by the two inequalities $-g_j(\mathbf{x}) \geq 0$ and $g_j(\mathbf{x}) + x_{n+1} \geq 0$.

Transformation of the Separable Problem (1-11) into a Special Linear Problem

The constraint and objective functions of the separable problem (1-11) can be expressed as sums of other functions, each involving only one variable. We transform only that part of the problem which is nonlinear in x_i, $i = 1, \ldots, n$, which is assumed to be all of (1-11). The linear part of the problem remains as is. In the transformation under consideration the summand functions are approximated by polygons. For this purpose the interval of variation $[0, u_i]$ of each of the problem variables x_i is divided into L_i (depending on the desired accuracy) subintervals by means of $L_i + 1$ meshpoints $x_{i0}, \ldots, x_{il}, \ldots, x_{iL_i}$. The functions $\hat{m}_i(x_i)$ and $\hat{g}_{ij}(x_i)$ assume the same values as the functions $m_i(x_i)$ and $g_{ij}(x_i)$ in (1-11) at these meshpoints, and approximate them by straight lines on the subintervals between the meshpoints. Thus, linearizing the problem by a change of variables [Miller, 1963; Hadley, 1964],

$$\begin{aligned} m_i(x_i) &\cong \hat{m}_i(x_i) = \sum_{l=0}^{L_i} \alpha_{il} m_i(x_{il}) \\ g_{ij}(x_i) &\cong \hat{g}_{ij}(x_i) = \sum_{l=0}^{L_i} \alpha_{il} g_{ij}(x_{il}) \\ \sum_{l=0}^{L_i} \alpha_{il} &= 1 \qquad \alpha_{il} \geq 0 \qquad \text{for all } i, l \end{aligned} \tag{2-20}$$

where for a given i at any time two α_{il}, at most, may be positive and these must be adjacent. To demonstrate that (2-20) is the alleged approximation, consider $\hat{m}_i$ and suppose that $x_{i\mu} \leq x_i \leq x_{i(\mu+1)}$. Then $\alpha_{i\mu}$ and $\alpha_{i(\mu+1)}$ are nonzero, and $\alpha_{i(\mu+1)} = 1 - \alpha_{i\mu}$. The approximation is $\hat{m}_i(x_i) = \alpha_{i\mu} m_i(x_{i\mu}) + (1 - \alpha_{i\mu}) m_i(x_{i(\mu+1)})$, and this is indeed the straight line between $m_i(x_{i\mu})$ and $m_i(x_{i(\mu+1)})$.

We use (2-20) in (1-11) to obtain the special linear problem corresponding to (1-11)

$$\min \sum_{i=1}^{n} \sum_{l=0}^{L_i} \alpha_{il} m_i(x_{il})$$

subject to

$$\begin{aligned} &\sum_{i=1}^{n} \sum_{l=0}^{L_i} \alpha_{il} g_{ij}(x_{il}) \begin{cases} \geq 0, & j = 1, \ldots, m_1 \\ = 0, & j = m_1 + 1, \ldots, m_2 \end{cases} \\ &\sum_{l=0}^{L_i} \alpha_{il} = 1, \qquad i = 1, \ldots, n, \qquad \alpha_{il} \geq 0, \qquad \text{for all } i, l \end{aligned} \tag{2-21}$$

where for a given i at any time at most two α_{il} may be positive and these must be adjacent. Problem (2-21) is a linear problem in the variables α_{il}, $i = 1, \ldots, n$, and $l = 0, \ldots, L_i$. The last restriction on the variables makes it a special linear problem. The values of the variables x_i, $i = 1, \ldots, n$ at the solution may be obtained by means of

$$x_i = \sum_{l=0}^{L_i} \alpha_{il} x_{il} \tag{2-22}$$

With these restrictions imposed on the set of variables α_{il} we may introduce an extraneous local minimum into (1-11). In this case, a solution to (2-21) may not approximate a local solution of (1-11). Notice that Prob. (1-11) is a nonlinear problem in n variables and m_2 constraints. Problem (2-21) is linear in $n + \sum_{i=1}^{n} L_i$ variables and $m_2 + n$ constraints. The nonnegativity requirements on the variables are satisfied automatically by the linear programming procedure.

The approximation (2-20) can be transformed to a somewhat smaller linear problem with upper bounded variables, for which a computer program may be available. For this purpose, we denote by Δ_l an interval between the lth and the $(l - 1)$st abcissa or ordinate. We also introduce a set of new variables δ_{il} restricted to $0 \leq \delta_{il} \leq 1$ for all i and l, where

$$\delta_{il} = \frac{x_i - x_{i(l-1)}}{x_{il} - x_{i(l-1)}} = \frac{x_i - x_{i(l-1)}}{\Delta x_{il}}$$

for $x_{i(l-1)} \leq x_i \leq x_{il}$. Instead of (2-20) one can then write [Hadley, 1964]:

$$\begin{aligned} m_i(x_i) = \hat{m}_i(x_i) &= \sum_{l=1}^{L_i} \delta_{il}[m_i(x_{il}) - m_i(x_{i(l-1)})] + m_i(x_{i0}) \\ &= \sum_{l=1}^{L_i} \delta_{il} \Delta m_i(x_{il}) + m_i(x_{i0}) \\ g_{ij}(x_i) = \hat{g}_{ij}(x_i) &= \sum_{l=1}^{L_i} \delta_{il}[g_{ij}(x_{il}) - g_{ij}(x_{i(l-1)})] + g_{ij}(x_{i0}) \\ &= \sum_{l=1}^{L_i} \delta_{il} \Delta g_{ij}(x_{il}) + g_{ij}(x_{i0}) \end{aligned} \tag{2-23}$$

where if for a given i and l, $\delta_{il} > 0$, then $\delta_{i\mu} = 1$ for $\mu = 1, \ldots, l - 1$ and the given i. To demonstrate that (2-23) is the alleged approximation, consider $\hat{m}_i$ supposing that $x_{i(\mu-1)} \leq x_i \leq x_{i\mu}$. Then $\delta_{i1} = 1, \delta_{i2} = 1, \ldots, \delta_{i(\mu-1)} = 1$, $0 < \delta_{i\mu} < 1$, $\delta_{i(m+1)} = 0, \ldots, \delta_{iL_i} = 0$; and $\hat{m}_i(x_i)$ is determined by linear interpolation between $m_i(x_{i(\mu-1)})$ and $m_i(x_{i\mu})$. We use (2-23) in (1-11) to obtain the special linear problem corresponding to (1-11)

$$\min \sum_{i=1}^{n} \sum_{l=1}^{L_i} \delta_{il} \Delta m_i(x_{il}) + m_i(x_{i0}) \tag{2-24}$$

subject to

$$\sum_{i=1}^{n} \sum_{l=1}^{L_i} \delta_{il}\Delta g_{ij}(x_{il}) + g_{ij}(x_{i0})\begin{cases} \geq 0 & j = 1, \ldots, m_1 \\ = 0 & j = m_1 + 1, \ldots, m_2 \end{cases}$$

$$0 \leq \delta_{il} \leq 1 \quad \text{for all } i, l$$

where if at any time and for any given i, $\delta_{il} > 0$ then $\delta_{i\mu} = 1$ for $\mu = 1, \ldots, l - 1$ and the given i. Otherwise, δ_{il} is not allowed to become positive. Problem (2-24) is a linear problem in the variables δ_{il}, $i = 1, \ldots, n$ and $l = 1, \ldots, L_i$. The last restriction on the variables makes it a special linear problem, to which not every linear programming algorithm can be applied. The values of the variables x_i, $i = 1, \ldots, n$ at the solution may be obtained by means of

$$x_i = \sum_{l=1}^{L_i} \delta_{il}\Delta x_{il} \tag{2-25}$$

Note that (1-11) is a nonlinear problem in n variables and m_2 constraints; (2-24) is linear in $\sum_{i=1}^{n} L_i$ upper bounded variables and m_2 constraints. The nonnegativity requirements on the variables are satisfied automatically by the linear programming procedure.

Transformation of the Geometric Programming Problem

In a geometric programming problem the objective function and the constraint functions (inequalities only) are generalized polynomials with real coefficients of the following form:

$$f_j(\mathbf{x}) = \sum_{l=1}^{L_j} c_{jl} \prod_{i=1}^{n} x_i^{a_{jli}} \qquad j = 0, \ldots, m \tag{2-26}$$

where $\prod_{i=1}^{n} x_i^{a_{jli}}$ denotes the product $x_1^{a_{jl1}} x_2^{a_{jl2}} \ldots x_n^{a_{jln}}$

and where the exponents a_{jli} are arbitrary real numbers. We shall restrict our attention to cases where the polynomial coefficients $c_{jl} > 0$ for all j and l, although situations with negative c_{jl} can also be handled [Wilde and Beightler, 1967]. In these cases, however, the introduction of additional variables to the problem is required, and the transformed problem is more complicated. The geometric programming (primal) problem is

$$\min_{\mathbf{x}} \{M(\mathbf{x}) \mid [\sigma_j[1 - g_j(\mathbf{x})] \geq 0$$
$$\text{and} \quad \sigma_j = \pm 1, \quad j = 1, \ldots, m; \quad x_i > 0, \quad i = 1, \ldots, n]\} \tag{2-27}$$

where $M(\mathbf{x}) = f_0(\mathbf{x})$ and $g_j(\mathbf{x}) = f_j(\mathbf{x})$, $j = 1, \ldots, m$, and $f_j(\mathbf{x})$, $j = 0, \ldots, m$ are the generalized polynomials defined by (2-26). The geometric programming problem which is defined by (2-26) and (2-27) can be transformed to an equivalent problem with a nonlinear objective function and linear constraints. The transformed problem is usually easier to solve than the original problem. Details of the transformation can be found in Duffin, et al. [1967], Wilde and Beightler [1967], and Zangwill [1969]. To understand in principle how this transformation can be made, first introduce the new variable

$$x_0 = f_0(\mathbf{x}) = M(\mathbf{x})$$

In addition, let

$$e^{w_i} = x_i \qquad i = 0, \ldots, n \tag{2-28}$$

where e^{w_i} denotes the exponential function of w_i. Furthermore, define

$$\begin{aligned} u_{0l} &= \frac{c_{0l}}{x_0} \prod_{i=1}^{n} x_i^{a_{0li}} > 0 \qquad l = 1, \ldots, L_0 \\ u_{jl} &= c_{jl} \prod_{i=1}^{n} x_i^{a_{jli}} \qquad j = 1, \ldots, m; \qquad l = 1, \ldots, L_j \end{aligned} \tag{2-29}$$

If these variable transformations are used, the geometric programming problem (2-27) can be rewritten as

$$\min_{\mathbf{w},\,\mathbf{u}} w_0$$

subject to

$$\begin{aligned} &-w_0 + \sum_{i=1}^{n} a_{0li} w_i = \ln\left(\frac{u_{0l}}{c_{0l}}\right) \qquad l = 1, \ldots, L_0 \\ &\sum_{i=1}^{n} a_{jli} w_i = \ln\left(\frac{u_{jl}}{c_{jl}}\right) \qquad j = 1, \ldots, m; \qquad l = 1, \ldots, L_j \\ &\sum_{l=1}^{L_0} u_{0l} = 1 \\ &\sigma_j\left(1 - \sum_{l=1}^{L_j} u_{jl}\right) \geq 0 \qquad j = 1, \ldots, m \\ &u_{jl} > 0 \qquad j = 1, \ldots, m; \qquad l = 1, \ldots, L_j \end{aligned} \tag{2-30}$$

where ln denotes natural logarithm. The new objective function of (2-30) follows from the definition of x_0 and (2-28). The first two constraints arise from (2-29). The third constraint results from the definitions of x_0 and u_{0l}, and the fourth constraints are a reformulation of the constraints of the original problem. The variables of Prob. (2-30) are w_i, $i = 0, \ldots, n$, and u_{jl}, $j = 0, \ldots, m$; $l = 1, \ldots, L_j$. For brevity, these are denoted by $\mathbf{w}$ and $\mathbf{u}$, respectively. Note that it is necessary to adopt the convention $b^b = b^{-b} = 1$ for $b = 0$.

The dual of the geometric programming problem may be obtained from the Lagrangian function (Sec. 1-7) of the transformed problem (2-30). Derivation details can be found in Wilde and Beightler [1967] or Duffin, et al. [1967].

The dual problem is

$$W(\mathbf{v}) = \prod_{j=0}^{m} \prod_{l=1}^{L_j} \left[\frac{c_{jl}}{v_{jl}} \sum_{\mu=1}^{L_j} v_{j\mu} \right]^{\sigma_j v_{jl}} \tag{2-31}$$

subject to the linear constraints

$$\sum_{j=0}^{m} \sum_{l=1}^{L_j} \sigma_j a_{jli} v_{jl} = 0 \qquad \sigma_0 = 1, \qquad i = 1, \ldots, n$$

$$\sum_{l=1}^{L_j} v_{jl} \geq 0 \qquad j = 0, \ldots, m$$

$$\sum_{l=1}^{L_0} v_{0l} = 1$$

where the variables of the dual problem $\mathbf{v} = (v_{01}, \ldots, v_{0L_0}, v_{11}, \ldots, v_{1L_1}, \ldots, v_{mL_m})^T$ are the Lagrange multipliers. One and only one variable v_{jl} is associated with each term in the primal problem; the coefficients c_{jl} and the a_{jli} appearing in the dual problem also appear in the primal problem.

In the case that $\sigma_j = 1$, $j = 1, \ldots, m$, the constraints of the problem become $g_j(\mathbf{x}) \leq 1$, $j = 1, \ldots, m$, and the functions f_j, $j = 0, \ldots, m$, defined by (2-26) are strictly convex functions of $\mathbf{w} = (w_0, \ldots, w_n)^T$. The latter is true because each e^{w_i} is a convex function of w_i and the $f_j(\mathbf{w})$ are given by

$$f_j(\mathbf{w}) = \sum_{l=1}^{L_j} c_{jl} \exp\left(\sum_{i=1}^{n} a_{jli} w_i\right) \qquad j = 0, \ldots, m$$

where $\exp(y)$ denotes e^y and with $c_{jl} > 0$ for all j and l. So, by definition (Sec. 1-5), the transformed geometric programming (primal) problem (2-30) is convex. It can also be shown that the natural logarithm of the objective function (2-26) of the dual is strictly concave with respect to the variables $\mathbf{u}$. Therefore, it is possible to find the global minimum of the primal by maximizing the dual function $W(\mathbf{v})$ (or its natural logarithm) subject to the linear constraints of problem (2-31), and for all primal feasible $\mathbf{x}$ and dual feasible $\mathbf{v}$

$$M(\mathbf{x}) \geq M(\mathbf{x}_G) = W(\mathbf{v}_G) \geq W(\mathbf{v})$$

where $\mathbf{x}_G$ is the point of global minimum of the primal problem and $\mathbf{v}_G$ is the point of global maximum of the dual problem. When the solution $\hat{\mathbf{v}}$ to the dual problem (2-31) with $\sigma_j = 1$ for all j has been obtained, the corresponding values $\hat{\mathbf{x}}$ of the primal variables can be computed by means of

$$c_{0l} \prod_{i=1}^{n} \hat{x}_i^{a_{0li}} = \hat{x}_0 \hat{v}_{0l} \qquad l = 1, \ldots, L_0$$

$$c_{jl} \prod_{i=1}^{n} \hat{x}_i^{a_{jli}} = v_{jl} \left(\sum_{l=1}^{L_j} \hat{v}_{jl}\right)^{-1} \qquad j = 1, \ldots, m; \qquad l = 1, \ldots, L_j$$

recalling that $\hat{x}_0 = M(\hat{\mathbf{x}}) = W(\hat{\mathbf{v}})$. One should probably use the more convenient equivalent form:

$$\sum_{i=1}^{n} a_{0li} \ln \hat{x}_i = \ln\left(\frac{\hat{x}_0 \hat{v}_{0l}}{c_{0l}}\right) \qquad l = 1, \ldots, L_0 \tag{2-32}$$

$$\sum_{i=1}^{n} a_{jli} \ln \hat{x}_i = \ln \frac{\hat{v}_{jl}}{\left(c_{jl} \sum_{l=1}^{L_j} \hat{v}_{jl}\right)} \qquad j = 1, \ldots, m; \qquad l = 1, \ldots, L_j \tag{2-33}$$

Note that the number of dual variables v_{jl}, $j = 0, \ldots, m$; $l = 1, \ldots, L_j$ is

$$N = \sum_{j=0}^{m} L_j \tag{2-34}$$

This is the total number of terms in the generalized polynomials f_j, $j = 0, \ldots, m$ defined by (2-26). Usually, $N > n$, the number of primal variables, and it should be possible to select n equations from (2-32) and (2-33) and to solve these for $\hat{x}_1, \ldots, \hat{x}_n$.

In the case that some of the $\sigma_j = -1$ (except $\sigma_0 = 1$ by definition) we have some constraints of the form $g_j \leq 1$ for j such that $\sigma_j = 1$ and constraints of the form $g_j \geq 1$ for j such that $\sigma_j = -1$. Therefore, one cannot make any statement concerning the convexity of the geometric programming problem. Wilde and Beightler [1967] show how to use the dual of the problem in this case. The computation of several local optimum points and a search are involved, however, and it seems that the added complication and computational labor make this approach somewhat less practical.

Transformation of Quadratic Programs into Linear Programs

In Sec. 1-3 we stated the quadratic program (1-10) as a nonlinear problem with a quadratic objective function and linear constraints. Let us consider this problem in n variables and m constraints

$$\min_{\mathbf{x}} \{\mathbf{b}^T\mathbf{x} + \tfrac{1}{2}\mathbf{x}^T B \mathbf{x} \mid A\mathbf{x} \leq \mathbf{c}\} \tag{2-35}$$

where B is an $n \times n$ positive semidefinite (or definite), symmetric matrix, A is an $m \times n$ matrix, $\mathbf{b}$ and $\mathbf{x}$ are n-dimensional vectors, and $\mathbf{c}$ is an m-dimensional vector. Nonnegativity requirements on the variables may be included in the constraints. The procedure of Wolfe [1959] starts with the application of the Kuhn-Tucker conditions (1-23) to this problem. This yields the problem

$$\begin{aligned} &B\mathbf{x} + A^T\mathbf{u} - \mathbf{q} = -\mathbf{b} \\ &A\mathbf{x} + \mathbf{y} = \mathbf{c} \\ &\mathbf{y} \geq 0, \quad \mathbf{u} \geq 0, \quad q_j = b_j v_j, \quad \text{and} \quad \mathbf{v} = 0 \end{aligned} \tag{2-36}$$

where $\mathbf{u} = (u_1, \ldots, u_m)^T$ are the Lagrange multipliers associated with the constraints that have been modified by introduction of the slack variables

$\mathbf{y} = -A\mathbf{x} + \mathbf{c}$ (equality constraints could be handled in their original form without introducing slack variables), and where $\mathbf{v} = (v_1, \ldots, v_n)^T$ are artificial variables introduced into the problem. The latter will have to be zero at the solution. The problem constraints are linear and hence satisfy the Kuhn-Tucker constraint qualification (Sec. 1-6). Also, because of the positive definiteness of the matrix A entering the objective function, the conditions of problem (2-36) are necessary and sufficient. We can, therefore, replace the original problem (2-35) in $\mathbf{x}$ by problem (2-36) in $\mathbf{x}$, $\mathbf{y}$, $\mathbf{u}$, and $\mathbf{v}$. The set of values $\mathbf{x} = \mathbf{u} = 0$, $\mathbf{y} = \mathbf{c}$, and $\mathbf{v} = 1$ satisfies that problem with the exception of the condition $\mathbf{v} = 0$. Now Wolfe simply proposes to apply a linear programming procedure (see appendix) to solve

$$\min\left\{\sum_{i=1}^{n} v_i \,\middle|\, [B\mathbf{x} + A^T\mathbf{u} - \mathbf{q} = -\mathbf{b}, \qquad A\mathbf{x} + \mathbf{y} = \mathbf{c}, \qquad \mathbf{y} \geq 0, \qquad \mathbf{u} \geq 0]\right\} \tag{2-37}$$

with the modified basis entry rule that if $u_i \neq 0$, then y_i (same index i) must not enter the basis (remains $y_i = 0$) and vice versa. This rule follows from the fact that no more than $m + n$ components of the solution vectors $\mathbf{x}$, $\mathbf{v}$, $\mathbf{u}$, $\mathbf{y}$ must be nonzero. This process will converge to a solution with $\mathbf{v} = 0$ in a finite number of steps, as linear programming processes will normally do.

Another algorithm that reduces quadratic to linear programs was given by Beale [1959, 1967]. Unlike Wolfe's algorithm, it reduces automatically to the ordinary simplex method when the objective function is linear—that is, $B = 0$.

Decomposition of Large Nonlinear Problems

The size of an optimization problem is determined roughly by the sum of the number of variables and constraints. Many practical optimization problems are of considerable size in relation to the available computer capability. What exactly is a large-scale problem depends also on the complexity of the constraint and objective functions and on the available methods of solution. The following table shows the approximate maximum size of different optimization problems that can be solved currently by available computers:

Type of Problem	*Maximum Size*
Implicity defined nonlinear functions involved	50
Nonlinear differentiable functions involved	200
Quadratic	below 10^3
Linear	10^4
Structured linear	over 10^6

It is fortunate that most large problems that arise in practical situations usually possess a special structure that can be exploited in obtaining a solution. The two fundamental approaches to solving large-scale optimization are *problem manipulation* and *solution strategy* [Geoffrion, 1971]. Problem manipulation based on the problem structure is a technique of obtaining an alternative and equivalent problem which is relatively easy to solve. Examples of problem manipulation techniques given by Geoffrion include partitioning, linearization, and dualization. The use of these techniques typically leads to decomposition of the large problem into a number of smaller subproblems with their own constraints and objective functions. The subproblems are solved one at a time in such a way that the optimum of the large problem is obtained. A solution strategy reduces a large optimization problem to a sequence of simpler problems with the same solution. Two examples of solution strategies are the penalty methods and feasible direction methods, which are discussed in Chapters 6 and 7, respectively.

To illustrate how to take advantage of the problem structure, we consider an optimization problem with a block diagonal *structure matrix*. A structure matrix S is one whose (i, j)th element is one if variable j appears in constraint i, and zero otherwise. Assume that the problem to be solved is:

$$\min \{M(\mathbf{x}) \,|\, g_j(\mathbf{x}) \leq 0, \qquad j = 1, \ldots, m\} \tag{2-38}$$

where $M(\mathbf{x})$ and $g_j(\mathbf{x})$ are convex. We partition the set of constraints into two subsets, say one with indices $j = 1, 2$ and the other with $j = 3, \ldots, m$, and form the subproblem

$$\min \{\bar{M}(\mathbf{x}) = M(\mathbf{x}) + \lambda_1 g_1(\mathbf{x}) + \lambda_2 g_2(\mathbf{x}) \,|\, g_j(\mathbf{x}) \leq 0, \qquad j = 3, \ldots, m\}$$

where λ_1, λ_2 are Lagrange multipliers. It can be demonstrated that if, given λ_1, λ_2, a vector $\mathbf{x}$ solves the subproblem and is feasible with respect to all inequalities, and in addition satisfies $\lambda_i g_i(\hat{\mathbf{x}}) = 0$, $i = 1, 2$, then $\hat{\mathbf{x}}$ solves (2-38).

Suppose now that Prob. (2-38) has the following structure:

$$\begin{aligned}
M(\mathbf{x}) &= M_1(x_1, x_2) + M_2(x_3, x_4) + M_3(x_5) \\
g_1(\mathbf{x}) &= g_a(x_1, x_2) + g_b(x_3, x_4) + g_c(x_5) \leq 0 \\
g_2(\mathbf{x}) &= g_2(x_1, x_2) \leq 0, \\
g_3(\mathbf{x}) &= g_3(x_3, x_4) \leq 0, \\
g_4(\mathbf{x}) &= g_4(x_5) \leq 0.
\end{aligned}$$

Introducing a single Lagrange multiplier λ, we form the subproblem

$$\min \{M(\mathbf{x}) + \lambda g_1(\mathbf{x}) \,|\, g_j(\mathbf{x}) \leq 0, \qquad j = 2, \ldots 4\}.$$

This subproblem can be decomposed into three separate optimization problems that are solved independently for a given value of λ. One of these problems is

$$\min \{M_1(x_1, x_2) + \lambda g_a(x_1, x_2)\}$$

subject to $$g_2(x_1, x_2) \leq 0$$

and the other two are similar. What we have just achieved is the decomposition of the larger problem into smaller subproblems that can be solved iteratively until $\hat{\mathbf{x}} = (\hat{x}_1, \hat{x}_2, \hat{x}_3, \hat{x}_4, \hat{x}_5)^T$ and λ satisfy the optimality condition. Such a diagonal structure occurs very frequently in practice, since variables in large-scale problems are usually related within certain subsets, although there may be linking relationships present, such as those represented by $g_1(\mathbf{x})$ in our problem. Problems of this type are called *weakly coupled.* The method we have used to decouple the problem can be classified as Lagrange decomposition or dualization (see Sec. 1.7). The original weakly coupled problem has been partially dualized (only with respect to the linking constraints) and converted into a two-level optimization problem.

Rosen [1963] considered the following block-diagonal structure:

$$\min_{\mathbf{x},\mathbf{y}} \sum_{l=1}^{N} \mathbf{c}_l^T \mathbf{x}_l$$

subject to

$$\begin{array}{llllr} A_1\mathbf{x}_1 & & & \geq \mathbf{b}_1(\mathbf{y}) & \\ & A_2\mathbf{x}_2 & & \geq \mathbf{b}_2(\mathbf{y}) & \\ & & \cdot & \cdot & (2\text{-}39) \\ & & \cdot & \cdot & \\ & & A_N\mathbf{x}_N & \geq \mathbf{b}_N(\mathbf{y}) & \end{array}$$

in which there are N constant submatrices A_l, $l = 1, \ldots, N$ of dimension $m_l \times n_l$. The n problem variables are grouped in $N + 1$ distinct vectors $\mathbf{x}_1, \ldots, \mathbf{x}_n, \mathbf{y}$ of dimension n_l, $l = 1, \ldots, N$ and q, respectively, with

$$\sum_{l=1}^{N} n_l + q = n$$

The constraints are coupled through $\mathbf{y}$, of which the right side $\mathbf{b}_l$ is a given vector function. The vectors $\mathbf{c}_l$, $l = 1, \ldots, N$ appearing in the objective function are of dimension n_l, $l = 1, \ldots, N$, respectively. For any fixed value of $\mathbf{y}$, say $\mathbf{y}^k$, problem (2-39) reduces to a set of N smaller linear subproblems that can be solved independently by minimizing over $\mathbf{x}$ for fixed $\mathbf{y}^k$, then minimizing the result over $\mathbf{y}$. This type of procedure is called *partitioning.* In the first step we solve the N linear subproblems

$$\min_{\mathbf{x}_l}\{\mathbf{c}_l^T\mathbf{x}_l \,|\, [A_l\mathbf{x}_l \geq \mathbf{b}_l(\mathbf{y}^k)]\} \qquad l = 1, 2, \ldots, N$$

with $k = 0$ for an initial $\mathbf{y}^0$. Using the resulting solution, we form a so-called coordinating problem, which has a convex objective function and linear constraints. The next step consists of solving this convex problem to obtain a new vector $\mathbf{y}^{k+1}$. The process is then terminated or continued, depending on the relationship between $\mathbf{y}^k$ and $\mathbf{y}^{k+1}$ and the result of a certain optimality test. If the computation is continued, the modified problem (2-39) is formed and solved again. Rosen's method provides in each iteration a feasible solution for the original problem with a nonincreasing objective function value. Under suitable conditions the algorithm converges in a finite number of iterations.

Another partitioning method, proposed by Benders [1962], solves programming problems of a quite general structure:

$$\min \{\mathbf{c}^T\mathbf{x} + f(\mathbf{y})\}$$

subject to

$$\begin{gathered} A\mathbf{x} + F(\mathbf{y}) \geq \mathbf{b} \\ \mathbf{x} \geq 0 \quad \text{and} \quad \mathbf{y} \in S. \end{gathered}$$

The matrix A is $m \times n$, $\mathbf{x}$ and $\mathbf{c}$ are n vectors, $\mathbf{y}$ is a p vector, f a scalar function of $\mathbf{y}$, F an m vector of functions of $\mathbf{y}$, $\mathbf{b}$ an m constant vector, and S an arbitrary set in the Euclidean space E^p. If S is a set of discrete values, then the problem is continuous-discrete. For the algorithmic details and examples of applications, the reader is referred to Lasdon [1970].

Ritter [1971] has developed a partitioning procedure for problems with a block-diagonal structure and linking constraints and variables:

$$\min_{\mathbf{x},\mathbf{y}} \sum_{i=1}^{N} \mathbf{c}_i^T\mathbf{x}_i + \mathbf{c}_o^T\mathbf{y}$$

subject to

$$\begin{array}{llllll} A_1\mathbf{x}_1 + & A_2\mathbf{x}_2 + & \cdots + & A_N\mathbf{x}_N + & D_0\mathbf{y} = & \mathbf{b}_0 \\ B_1\mathbf{x}_1 & & & + & D_1\mathbf{y} = & \mathbf{b}_1 \\ & B_2\mathbf{x}_2 & & + & D_2\mathbf{y} = & \mathbf{b}_2 \\ & & \ddots & & & \\ & & & B_N\mathbf{x}_N + & D_N\mathbf{y} = & \mathbf{b}_N \end{array}$$

$$\mathbf{x}_i \geq 0, \qquad i = 1, 2, \ldots, N \qquad \text{and} \qquad \mathbf{y} \geq 0,$$

where A_i is $m_0 \times n_i$, B_i is $m_i \times n_i$, D_i is $m_i \times n_0$ and $\mathbf{c}_0$, $\mathbf{c}_i$, $\mathbf{b}_i$, $\mathbf{x}_i$, $\mathbf{y}$ are vectors of appropriate size. In the absence of $\mathbf{y}$, the problem has a block-diagonal structure with coupling constraints only, and can be solved by the Dantzig-Wolfe [1961] decomposition method.

2-3 TRANSFORMATION OF OTHER MATHEMATICAL PROBLEMS TO NONLINEAR PROGRAMS

It is frequently desirable to cast other mathematical problems into the format of the general problem (1-9). In this section we indicate a few examples of such problems and transformations. Many iterative techniques proceed to solve mathematical problems by evaluating trial solutions. A new trial solution is computed based on the error associated with the previous one. In most such cases, one can form an objective function, the value of which corresponds to the errors associated with each of these trial solutions. The problem is then one of minimizing this objective function subject to the associated relations by which the errors are tied to the trial solution. This process will result in a solution with virtually no error. In other cases, the transformation is achieved by a discretization of the original problem. A more complicated transformation may arise from the solution of boundary value problems involving differential equations. In the absence of techniques that are directly applicable to these problems it has been found useful to transform them to the format of (1-9). This is achieved by forming a subsidiary boundary value problem for which a solution scheme is available. This may involve a set of boundary values of a different type than that specified with the original problem. The variables of the minimization problems are the boundary values of the subsidiary problem at selected meshpoints, or parameters from which these values can be calculated, such as curve-fitting parameters. A set of boundary values of the original type corresponding to each trial solution is calculated and compared with the originally specified boundary conditions. The difference between the two sets of boundary conditions is considered as an objective for minimization. This minimization is achieved by examining different solutions of the subsidiary problem which correspond to perturbed sets of boundary conditions. The perturbation is obtained by changing the values of the minimization variables. Another approach [Osborne and Watson, 1969] is based on fitting a suitable approximation to the solution of the boundary value problem and minimizing the maximum residual defined by substituting the approximation into the differential equation. The approximation contains free parameters that become variables in the optimization process.

Rabinowitz [1968] has made a survey of applications of linear programming to numerical analysis. Some of the great variety of problems that can be solved by linear programming include approximations, curve fitting, linear equations and matrix inversion, ordinary and partial differential equations.

Transformation of Systems of Equations into an Equivalent Optimization Problem

Linear and nonlinear equations can be easily transformed to optimization problems whose solutions are equivalent. Let us first consider a system of linear equations

$$A\mathbf{x} - \mathbf{b} = 0 \tag{2-40}$$

where A is positive definite. The solution of this system will be denoted by $\mathbf{h} = A^{-1}\mathbf{b}$. We define a nonnegative convex function

$$M(\mathbf{x}) = \tfrac{1}{2}(\mathbf{x} - \mathbf{h})^T A(\mathbf{x} - \mathbf{h}) \tag{2-41}$$

which possesses as a minimum

$$M(\mathbf{x}) = 0 \quad \text{for} \quad \mathbf{x} = \mathbf{h} = A^{-1}\mathbf{b} \tag{2-42}$$

Substituting $\mathbf{h}$ into (2-41), we see that the problem of solving (2-40) is equivalent to the unconstrained minimization problem $\min_{\mathbf{x}} \{\tfrac{1}{2}\mathbf{x}^T A\mathbf{x} - \mathbf{b}^T\mathbf{x}\}$, which can be solved by the conjugate gradient methods described in Chap. 5.

Similarly, a system of nonlinear algebraic or transcendental equations

$$f_1(x_1, \ldots, x_n) = 0$$
$$\cdots$$
$$f_m(x_1, \ldots, x_n) = 0$$

where $m \geq n$ can be reformulated as an optimization problem $\min_{\mathbf{x}} M(\mathbf{x})$, and $M(\mathbf{x})$ is formulated to measure the length of the residual vector $\mathbf{r} = (r_1, r_2, \ldots, r_m)^T$ defined by $r_j(\mathbf{x}) = f_j(\mathbf{x})$. One of the following definitions of $M(\mathbf{x})$ can be used: $M(\mathbf{x}) = [\mathbf{r}(\mathbf{x})]^T\mathbf{r}(\mathbf{x})$, $M(\mathbf{x}) = \sum_{j=1}^{m} |r_j(\mathbf{x})|$, or $M(\mathbf{x}) = \sum_{j=1}^{m} \ln |r_j(\mathbf{x})|$. We should stress, however, that the direct minimization of residuals by using generally applicable optimization techniques is not recommended. Methods using the Jacobian matrix to minimize the sum of squares are presented in the appendix.

Transformation of Optimal Control Problems into Nonlinear Programs

In optimal control or optimal programming problems one considers an n vector of state variable histories

$$\mathbf{x}(t) = [x_1(t), \ldots, x_n(t)]^T \tag{2-43}$$

and a (scalar)* control variable $u(t)$, bounded such that $0 \leq u(t) \leq U$. There is, in general, the requirement to satisfy a given system of differential

*In many problems we have a vector function $\mathbf{u}(t)$. For this example, we consider the scalar case to simplify the presentation.

equations (of state)

$$\frac{d\mathbf{x}}{dt} = \mathbf{f}[t, \mathbf{x}(t), u(t)] \tag{2-44}$$

where $\mathbf{f}$ is an m vector of known functions of t, $\mathbf{x}(t)$ and $u(t)$, with $m \leq n$, which is assumed everywhere differentiable with respect to $\mathbf{x}$ and u. One is normally interested in $t_0 \leq t \leq T$ and the satisfaction of the following end conditions may be required

$$\begin{aligned} \mathbf{g}_0[t_0, \mathbf{x}(t_0), u(t_0)] &= 0 \\ \mathbf{g}_T[T, \mathbf{x}(T), u(T)] &= 0 \end{aligned} \tag{2-45}$$

where $\mathbf{g}_0$ and $\mathbf{g}_T$ are p vectors ($p \leq n$) of known initial and terminal constraint functions of their arguments. The problem is to find $\mathbf{x}(t)$ and $u(t)$ satisfying equations (2-44) and (2-45); that is, one must find an "admissible" control function, as well as minimize the expression

$$J = [V(t, \mathbf{x})]_{t_0}^{T} + \int_{t_0}^{T} W\left(t, \mathbf{x}, \frac{d\mathbf{x}}{dt}, u\right) dt \tag{2-46}$$

where V and W are known functions of their arguments. The problem just described is the so-called problem of Bolza. Meyer or Lagrange reformulations where $W \equiv 0$ or $V \equiv 0$, respectively, can easily be obtained. In some applications, inequality constraints of the type

$$\mathbf{g}[t, \mathbf{x}(t), u(t)] \leq 0 \tag{2-47}$$

may also arise, where $\mathbf{g}$ is a μ vector of known constraint functions g_j of t, $\mathbf{x}$, and u. For discussion of conditions necessary for the existence of an extremal solution to the extended problem, see Bryson, et al. [1963] or Speyer and Bryson [1968].

The transformation of optimal control problems into optimization problems is obtained by discretization of the original problem with respect to time [Fanning and Jacoby, 1968]. One considers the problem of a set of H abcissas t_η, $\eta = 0, \ldots, H - 1$, with $\Delta t = T/H$, $t_\eta = \eta \Delta t$, $t_H = T$. The state variable vector $\mathbf{x}_\eta = \mathbf{x}(t_\eta)$ satisfies instead of (2-44) the iterative relation

$$\mathbf{x}_{\eta+1} - \mathbf{x}_\eta = \Delta t \mathbf{f}[t_\eta, \mathbf{x}_\eta, u_\eta] \tag{2-48}$$

The initial and terminal constraints (2-45) need not be discretized further with respect to time and therefore remain unchanged. The objective function is redefined as

$$J = V_{H-1} - V_0 + \sum_{\eta=0}^{H-1} W_\eta \Delta t \tag{2-49}$$

where V_η and W_η denote these functions as defined above but evaluated at t_η. The set of constraints (2-47) is similarly discretized and replaced by

$$\mathbf{g}_\eta \leq 0 \qquad \eta = 0, \ldots, H - 1 \tag{2-50}$$

where again $\mathbf{g}_\eta$ denotes the functions g_j as defined above, but evaluated at t_η. It is easy to see that the problem defined by the objective function (2-49), the equality constraints (2-45) and (2-48), the inequality constraints (2-50), and the bounds on u constitute a problem of the type (1-9). The $H(n+1)$ variables of the transformed problem are $x_{i\eta}$ and u_η. This problem has been investigated extensively by Rosen [1967]; computational experience is also available [Fanning and Jacoby, 1968].

A slightly modified transformation explored by Rosen [1967] is based on the assumption that a scheme is available for the solution of the differential equations of state (2-44), provided that a trial solution is available for $u(t)$, $t_0 \leq t \leq T$. The remainder of the problem may then be approached as if $\mathbf{x}(t)$ satisfying the equations of state (2-44) is known for each trial solution $u(t)$. This subproblem is discretized in the preceding described manner, and transformed into an optimization problem of the type (1-9).

Balakrishnan [1968] investigated the problem of combining (2-44) and (2-46) into a penalty function. The problem under consideration is transformed into an unconstrained problem in a fashion similar to the transformation described in Sec. 2-1. The penalty may be defined as

$$P = J + \frac{1}{2\epsilon} \int_{t_0}^{T} \left\| \frac{dx}{dt} - \mathbf{f} \right\| dt \tag{2-51}$$

where the penalty coefficient ϵ is reduced to zero as the second term in (2-51) approaches zero. This problem can then be discretized in the previously described manner.

A more recent treatment of the relation between optimal control and mathematical programming has been given by Beltrami [1970], Cannon, et al. [1970] and by Tabak and Kuo [1971].

References

Balakrishnan, A. V.: "On a New Computing Technique in Optimal Control," *SIAM Journal of Control Theory and Applications*, Vol. 9, 149, 1968.

Beale, E.M.L.: "On Quadratic Programming," *Naval Research Logistics Quarterly*, Vol. 6, 227, 1959.

———: "Numerical Methods," in J. Abadie (ed.), *Nonlinear Programming*, North-Holland Publishing Company, Amsterdam, Ch. VII, 133, 1967.

———: "Separable Programming," *Mathematical Programming in Practice*, John Wiley & Sons, Inc., New York, Section 14, 124, 1968.

Beltrami, E. I.: "An Algorithmic Approach to Nonlinear Analysis and Optimization," *Mathematics in Science and Engineering*, Vol. 63, Academic Press Inc., New York, 1970.

Benders, J. F.: "Partitioning Procedures for Solving Mixed Variables Programming Problems," *Numerische Mathematik*, Vol. 4, 238, 1962.

Bryson, A. E., Jr., W. F. Denham and S. E. Dreyfus: "Optimal Programming Problems with Inequality Constraints I: Necessary Conditions for Extremal Solutions," *AIAA Journal*, Vol. 1, 2544, 1963.

Cannon, M. D., C. D. Cullum, Jr. and E. Polak: *Theory of Optimal Control and Mathematical Programming*, McGraw-Hill Book Company, New York, 1970.

Courant, R.: "Variational Method for the Solution of Problems of Equilibrium and Vibrations," *American Mathematical Society Bulletin*, Vol. 49, 1, 1943.

Dantzig, G. B. and P. Wolfe: "The Decomposition Algorithm for Linear Programming," *Econometrica*, Vol. 29, 767, 1961.

Duffin, R. J., E. L. Peterson and C. Zener: *Geometric Programming-Theory and Application*, John Wiley & Sons, Inc., New York, 1967.

Fanning, M. L. and S.L.S. Jacoby: "A Solution Technique for Optimal Programming Problems," *SIAM Symposium on Optimization*, Toronto, 1968.

Fiacco, A. V. and C. P. McCormick: *Nonlinear Programming*, John Wiley & Sons, Inc., New York, 1968.

Geoffrion, A. M.: "Large Scale Linear and Nonlinear Programming," in D. A. Wismer (ed.), *Optimization Methods for Large-Scale Systems*, McGraw-Hill Book Company, New York, 47, 1971.

Hadley, G.: "Approximate Methods for Solving Problems Involving Separable Functions," *Nonlinear and Dynamic Programming*, Addison-Wesley Publishing Co. Inc., Reading, Mass., Ch. 4, 104, 1964.

Huang, H. Y. and S. Naqvi: "*Unconstrained Approach to the Extremization of Constrained Functions*," Aero-Astronautics Report No. 82, Rice University, Houston, Texas, 1970.

Lasdon, L. S.: *Optimization Theory for Large Systems*, The Macmillan Company, New York, 1970.

Miele, A. and A. V. Levy: "Modified Quasilinearization and Optimal Initial Choice of the Multipliers, Part 1–Mathematical Programming Problems," *Journal of Optimization Theory and Applications*, Vol. 6, 364, 1970.

Miele, A., A. V. Levy and E. E. Cragg: "Modifications and Extentions of the Conjugate Gradient-Restoration Algorithm for Mathematical Programming Problems," *Journal of Optimization Theory and Applications*, Vol. 7, 450, 1971.

Miller, C. E.: "The Simplex Method for Local Separable Programming," in R. L. Graves and P. Wolfe, (eds.) *Recent Advances in Mathematical Programming*, McGraw-Hill Book Company, New York, Ch. 12, 89, 1963.

Morrison, D. D.: "Optimization by Least Squares," *SIAM Journal Numer. Anal.*, Vol. 5, 83, 1968.

Osborne, M. R. and G. A. Watson: "On Estimating Best Approximations of Functions Defined by Differential Equations," *Numerische Mathematik*, Vol. 13, 199, 1969.

Pietrzykowski, T.: "An Exact Potential Method for Solving Constrained Maxima," *SIAM Journal Numer. Anal.*, Vol. 6, 299, 1969.

POMENTALE, T.: "A New Method for Solving Conditional Maxima Problems," *Journal of Math. Analysis and Applications*, Vol. 10, 216, 1965.

RABINOWITZ, P.: "Applications of Linear Programming to Numerical Analysis," *SIAM Review*, Vol. 10, 121, 1968.

RITTER, K.: "A Decomposition Method for Linear Programming Problems with Coupling Constraints and Variables," Mathematics Research Center, University of Wisconsin, Rept. 739, 1967.

ROSEN, J. B.: "Convex Partition Programming," in R. L. Graves and P. Wolfe (eds.), *Recent Advances in Mathematical Programming*, McGraw-Hill Book Company, Ch. 22, 159, 1963.

———: "Optimal Control and Convex Programming," in J. Abadie (ed.), *Nonlinear Programming*, North-Holland Publishing Company, Amsterdam, Ch. XIII, 287, 1967.

SPEYER, J. L. and A. E. BRYSON, Jr.: "Optimal Programming Problems with a Bounded State Space," *AIAA Journal*, Vol. 6, 1488, 1968.

TABAK, D. and B. C. KUO: "Optimal Control by Mathematical Programming," Prentice-Hall Instrumentation and Control Series, Prentice-Hall, Inc., Englewood Cliffs, N.J., 1971.

WILDE, D. J. and C. S. BEIGHTLER: "Polynomial Inequalities: Geometric Programming," *Foundations of Optimization*, Prentice-Hall, Inc., Englewood Cliffs, N.J., Ch. 4, 99, 1967.

WOLFE, P.: "The Simplex Method for Quadratic Programming," *Econometrica*, Vol. 27, 382, 1959.

———: "Methods of Nonlinear Programming," in J. Abadie (ed.), *Nonlinear Programming*, North-Holland Publishing Company, Amsterdam, Ch. VI, 97, 1967.

ZANGWILL, W. I.: "Nonlinear Programming via Penalty Functions," *Management Science*, Vol. 13, 344, 1967.

———: "The Dual Geometric Programming Problem," *Nonlinear Programming: A Unified Approach*, Prentice-Hall, Inc., Englewood Cliffs, N. J., Section 3.3, 68, 1969.

3 METHODS FOR OPTIMIZATION ALONG A LINE

This chapter deals with methods for optimization along a line. These are methods for solving optimization problems in one variable. The one-variable minimization problem in the n-variable space is: given a point $\mathbf{x}^k$ and a search direction $\mathbf{s}^k$, find λ^k that gives

$$M(\mathbf{x}^k + \lambda^k \mathbf{s}^k) = \min_{\lambda} M(\mathbf{x}^k + \lambda \mathbf{s}^k) \tag{3-1}$$

Sequences of one-variable subproblems like (3-1) are generated by a variety of minimization methods for multivariable functions. This discussion, of course, also applies to one-variable minimization problems (in one-variable space), which may arise directly in a mathematical model.

Methods for minimization along a line include sequential and simultaneous (nonsequential) procedures. Further classification may include the specific properties of the objective function—for example, whether the function is continuous or defined by a table of values, and whether one is searching local neighborhoods or globally.

In general, we are looking for an efficient procedure of searching for the solution in the interval $[b, u]$. The search strategy will involve: selecting a number of values of λ distributed on the interval (in some manner), and determining (by a process involving function evaluations) a subinterval wherein the minimum of the objective function will lie. The theoretically most efficient (optimal) algorithm for finding the subinterval containing the minimum of a unimodal function is called the Fibonacci search, and its simplified version is known as the golden section method. It is optimal in the sense that it gives the smallest ratio of final to initial interval for a fixed number of function evaluations. Although "optimal" as defined here, these two search procedures are inefficient if the solution is to be found with high precision.

Nonoptimal sequential search and simultaneous techniques for finding a minimum of a one-dimensional function that is unimodal over a given interval are less efficient than the Fibonacci search (they give a larger ratio of the final to initial interval for a fixed number of function evaluations). Combinations of simultaneous and sequential searches are frequently used in computations.

Another approach to the problem of finding a minimum of one-dimensional function is curve fitting. It is based on the assumption that close enough to the minimum the function $M(\lambda)$ may be approximated by quadratic or cubic polynomials that have well-defined minima. One proceeds to make successive approximations, until the solution is found to the required accuracy.

A brief discussion is given of the application of multivariable methods to univariate problems, and comments are noted on nonoptimal step methods, which solve the problem of finding λ^k satisfying $M(\mathbf{x}^k + \lambda^k\mathbf{s}^k) \leq M(\mathbf{x}^k)$.

3-1 ASSUMPTIONS AND PROCEDURES

The normal procedure for minimization along the line is to look for the first local minimum of $M(\lambda)$. To make the problem more tractable, we assume that an interval $[b, u]$ is known on a line in the direction $\mathbf{s}^k$, from $\mathbf{x}^k$, that contains the minimum of $M(\lambda)$. This interval may change from iteration to iteration so that $b^k \leq \lambda^k \leq u^k$. This region in which a search for the minimum value of the function is made is generally called the *interval of uncertainty*, for while the optimum value lies within the interval, exactly where, is uncertain.

No restrictions are placed on the function M in terms of continuity, smoothness, or whether it possesses derivatives at any specific point. However, the function is assumed to be error free and unimodal in this interval. By *error free* we mean that the function is deterministic and not stochastically derived. Let the minimum of $M(\lambda)$ be at λ^*, and define $\bar{\lambda} < \bar{\bar{\lambda}}$. Then the function $M(\lambda)$ is *unimodal* if $\bar{\bar{\lambda}} < \lambda^*$ implies $M(\bar{\bar{\lambda}}) < M(\bar{\lambda})$, and $\bar{\lambda} > \lambda^*$ implies that $M(\bar{\lambda}) < M(\bar{\bar{\lambda}})$. Unimodality is a reasonable assumption if the function is single valued and the interval of search is small enough. It represents a larger class of functions than the class of convex functions. Unimodality is restrictive insofar as the function can have only one minimum point in the prescribed interval; this uniqueness of the minimum of $M(\lambda)$ in $[b, u]$ makes it possible to locate the minimum by examining values of the function within $[b, u]$ as described by the following example of a *sequential* process.

Let us consider the unimodal function $M(\lambda)$ in $[b, u]$ and any two values of λ, say λ^1 and λ^2, where $b < \lambda^1 < \lambda^2 < u$. Evaluation of M at these two interior points will make it possible to reduce the size of the interval $[b, u]$. If $M(\lambda^1) > M(\lambda^2)$, two situations are possible for $\lambda^1 < \lambda^2$ (Fig. 3-1). That is,

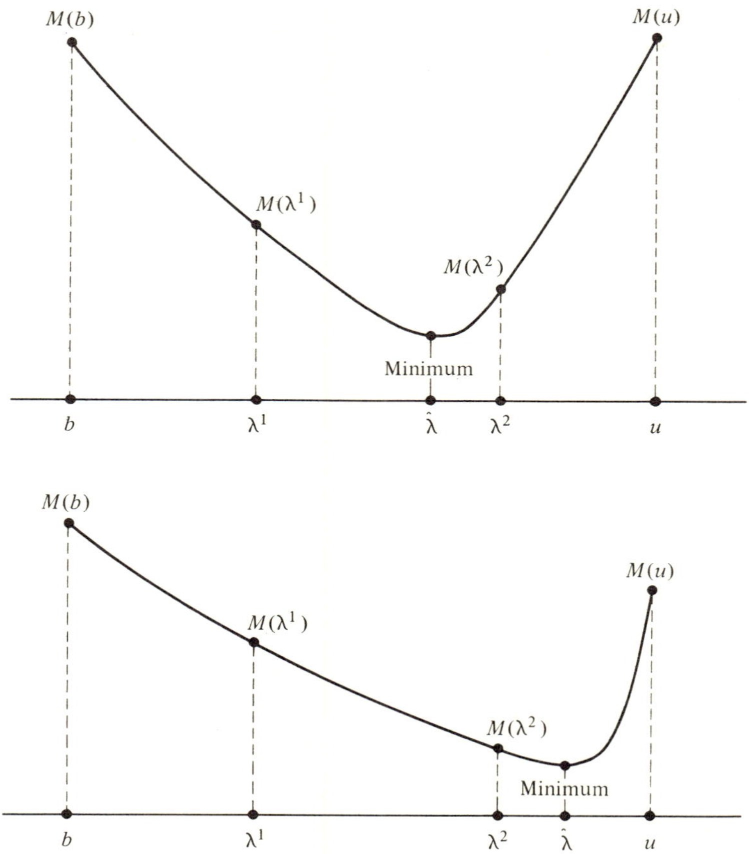

Figure 3-1 Two possible locations $\hat{\lambda}$ of a minimum if $M(\lambda)$ is unimodal and $M(\lambda_1) > M(\lambda_2)$

either the minimum point $\hat{\lambda}$ is within or is outside of $[\lambda^1, \lambda^2]$. If no further information about the function $M(\lambda)$ is given, the interval $[\lambda^1, u]$ contains the minimum. Conversely, if $M(\lambda^1) < M(\lambda^2)$, the interval is $[b, \lambda^2]$. The case where $M(\lambda^1) = M(\lambda^2)$ is less likely in practice and can be included in one of the previous cases. It can be observed that the initial interval $[b, u]$ is reduced to $[b, \lambda^2]$ or $[\lambda^1, u]$, and we have one test point in the interior of the reduced interval that may be used in the next step. Therefore, two function evaluations have to be made in the first step but only one in all subsequent steps. With repetition of this procedure, the minimum can be found with any precision desired.

In *sequential* techniques the function evaluation points are selected during the iterative process, whereas *simultaneous* (nonsequential) techniques require that the entire sequence of function evaluation points be preselected. This is desirable when insufficient time or information is available at the moment the computation is performed. Thus, no information from past data is used to determine new data points. This can happen, for example, if the function evaluation or some part of it cannot be done within the optimization computation, or cannot be included within the actual running of the computer program, since the function evaluation may be too time consuming to be practical. Large-scale engineering problems may give rise to this difficulty, which can be mitigated by approximation assumptions.

Brooks [1959] showed in numerical experiments that sequential methods were superior to the nonsequential methods. In fact, Wilde [1964] points out that as the number of function evaluations grows, sequential search improves the size of the interval containing a minimum exponentially while simultaneous uniform pairs search improves the interval only in direct proportion.

Combinations of simultaneous and sequential searches are also quite feasible, depending on the structure of the solved problem and how it is to be investigated. For a detailed description of these sequential and nonsequential methods, see the excellent treatment by Wilde [1964, 1967], and Pierre [1969].

3-2 SIMULTANEOUS AND NONOPTIMAL SEQUENTIAL SEARCH METHODS

There are a number of nonoptimal procedures of searching an interval $[b, u]$ for the minimum value of an objective function. Some of these are simultaneous searches based on preselecting the set of points, while others take advantage of information generated in the iteration and select the points sequentially. It is assumed in the discussion of these methods that an interval of the one-dimensional function can be selected over which the function is unimodal.

Simultaneous Search Methods

It is sometimes necessary to preselect all the values of λ that one will or can explore. During operation of the optimization program, for example, it may not be possible to select new data points based on present or previously gained information. There are several methods that can be used.

Uniform, or Factorial, Search. In this technique, K interior points are selected so that $[b, u]$ is subdivided into $K + 1$ intervals of equal length. The function $M(\lambda)$ is evaluated at the K equally spaced points, and the relative minimum value $M(\lambda^k)$ is found $[M(\lambda^k) = \min_i \{M(\lambda^i), i = 1, \ldots, K\}]$. The search ends when we find the smallest value of the objective function corresponding to one of the preselected points. If we intend to repeat the procedure, we are aware at this time that the true minimum $M(\lambda^k)$ of the function lies in subinterval $[\lambda^{k-1}, \lambda^k]$ or $[\lambda^k, \lambda^{k+1}]$; the procedure can be repeated in this interval. Pierre [1969] and Brooks [1959] refer to this method as *factorial search.*

Search by Uniform Pairs. This technique is a variation of the uniform search in which the objective function is evaluated at equally spaced pairs of very close points.

Random Search. We can also select the values of λ in an arbitrary manner, thus introducing complete nonbias. This may be achieved by relating random numbers to distance on the interval $[b, u]$. This does not imply an improvement over selecting the points equally spaced, but it does give an alternative choice. Brooks [1959] showed by numerical experiments that random search was not as efficient as uniform (factorial) search. [See Wilde, 1964, for a further discussion on random search].

Nonoptimal Sequential Search

Sequential search has the obvious advantage over simultaneous search that one can learn about the function by the first selection of points and select subsequent sets of points to take advantage of the experience gained in previous function evaluations.

Sequential Halving. The simplest form of sequential search, sequential halving, is similar to a procedure for finding the real roots of an algebraic or transcendental function. The interval to be searched is halved, and the portion wherein the objective functions are smallest is retained; the other portion is rejected, and the process is repeated as follows:

1. The function $M(\lambda)$ is evaluated at a number K of points, $\lambda^1, \lambda^2, \ldots, \lambda^K$ and the value λ^k corresponding to the smallest computed $M(\lambda^k)$ is selected. If $k = 1$ or $k = K$, the interval containing the minimum is $[\lambda^1, \lambda^2]$ or $[\lambda^{K-1}, \lambda^K]$, respectively. In this case, proceed to step 5; otherwise, proceed to step 2.
2. Two points from the original selection of points that bound λ^k are selected—that is, $\lambda^{k-1} \leq \lambda^k \leq \lambda^{k+1}$, $k = 2, \ldots, K - 1$.

3 One new point is calculated:

$$\lambda^{k-1/2} = \frac{\lambda^{k-1} + \lambda^k}{2}$$

4 If $M(\lambda^{k-1/2}) < M(\lambda^k)$, the minimum is in the interval $[\lambda^{k-1}, \lambda^k]$; otherwise, the minimum is in the interval $[\lambda^k, \lambda^{k+1}]$ since the function is unimodal.
5 The interval containing the minimum is renamed and steps 1 through 5 are repeated until max $(|\lambda^{k-1} - \lambda^k|, |\lambda^k - \lambda^{k+1}|) \leq \tau$, where $\tau > 0$ is a predetermined tolerance.

Sequential Search by Uniform Pairs. One selects two points $(\lambda^{k+1}, \lambda^{k+2})$ close to each other ($|\lambda^{k+1} - \lambda^{k+2}| = \epsilon$, where ϵ is a small positive number) in approximately the center of the interval $[b, u]$. Then the function is evaluated at these points. If $M(\lambda^{k+1}) < M(\lambda^{k+2})$, the new subinterval is $[b, \lambda^{k+2}]$; and if $M(\lambda^{k+1}) \geq M(\lambda^{k+2})$, the new subinterval is $[\lambda^{k+1}, u]$. This process is repeated until the subinterval obtained satisfies a predetermined tolerance.

3-3 FIBONACCI SEARCH

The optimal strategy for finding the optimum value of a unimodal function of one variable is the Fibonacci search method. Kiefer [1953] proved by induction that given a number of function evaluations, the sequential search procedure, using Fibonacci numbers, finds the smallest interval containing the optimum. That is, the Fibonacci search method is the best procedure among the class of all sequential nonrandom procedures requiring K observations. Thus, we should expect that the ratio of final to initial interval is minimized for a given number of iterations.

Johnson [1955] independently demonstrated the superiority of the Fibonacci search method. He specified the optimal one-variable sequential search strategy in the following theorem.

Theorem 3-1 Let $M(\lambda)$ be any unimodal function defined on an interval $[0, L_p]$. Let F_p be the largest L_p with the property that we can always locate the minimum (or maximum) of $M(\lambda)$ on a subinterval of length one, by calculating only p values of the function. Then F_p is the pth Fibonacci number, given by the recurrence relation

$$F_p = F_{p-1} + F_{p-2}, \quad p \geq 2, \qquad F_0 = F_1 = 1 \tag{3-2}$$

The solution of (3-2) is

$$F_p = \frac{1}{\sqrt{5}}\left[\left(\frac{1+\sqrt{5}}{2}\right)^{p+1} - \left(\frac{1-\sqrt{5}}{2}\right)^{p+1}\right] \qquad p = 0, 1, 2, \ldots \tag{3-3}$$

Theorem 3-1 can be briefly explained in the following way. Referring to Fig. 3-2, suppose that the values of $M(\lambda)$ have been computed at $\lambda = \lambda_1^p$ and $\lambda = \lambda_2^p$, which are interior points in the interval $[0, F_p]$; F_p is defined as

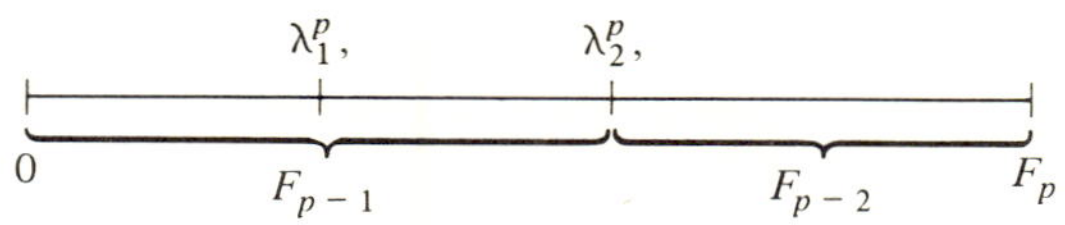

Figure 3-2 Interval partition by the Fibonacci numbers

in the theorem. Due to the unimodality of $M(\lambda)$ we take the minimum to be on $[0, \lambda_2^p]$ if $M(\lambda_1^p) < M(\lambda_2^p)$ and on $[\lambda_1^p, F_p]$ if $M(\lambda_1^p) > M(\lambda_2^p)$. In either case, we are left with a smaller interval with one interior point and the function value corresponding to it. Clearly, these two intervals are equal to F_{p-1}, since we must be able to find the minimum after the remaining $p - 1$ function evaluations.

Suppose now that the minimum occurs in the interval $[0, \lambda_1^p]$ or $[\lambda_2^p, F_p]$. In this case, we have only $p - 2$ function evaluations to reduce them to the unit interval containing the minimum. The lengths of these intervals are therefore F_{p-2}. The following notation is used:

1 b^k, u^k are the lower and upper values of the independent variable λ. The minimum is between these bounds. Superscript k denotes the number of the iteration.

2 λ_1^k and λ_2^k (interior points in the interval $[b^k, u^k]$) are chosen according to the Fibonacci scheme.

3 K is a predetermined number of iterations necessary to locate the minimum with the required accuracy ϵ, the length of the final interval containing the minimum. This length, or *resolution*, must be greater than the minimum separation to be maintained between adjacent function evaluations. The concept of this resolution is discussed in detail by Wilde [1964].

Iteration Steps

1 Two values b^1 and u^1, which bracket the independent variable, are given. Find the following interior points:

$$\lambda_1^1 = \frac{F_{K-2}}{F_K}(u^1 - b^1) + b^1 \tag{3-4}$$

$$\lambda_2^1 = \frac{F_{K-1}}{F_K}(u^1 - b^1) + b^1 \tag{3-5}$$

where $K > 2$ is the total number of iterations. Compute $M(\lambda_1^1)$ and $M(\lambda_2^1)$, set $k = 1$ and proceed to step 3.

2 Find the following interior points:

$$\lambda_1^k = \frac{F_{K-1-k}}{F_{K+1-k}}(u^k - b^k) + b^k \tag{3-6}$$

or

$$\lambda_2^k = \frac{F_{K-k}}{F_{K+1-k}}(u^k - b^k) + b^k \tag{3-7}$$

and compute $M(\lambda_1^k)$ or $M(\lambda_2^k)$.

3 If $M(\lambda_1^k) > M(\lambda_2^k)$, set

$$b^{k+1} = \lambda_1^k, \quad \lambda_1^{k+1} = \lambda_2^k, \quad u^{k+1} = u^k$$

If $k < K - 1$, set $k := k + 1$, and start from step 2, computing λ_2^k only; otherwise, terminate the iteration. If $M(\lambda_1^k) \leq M(\lambda_2^k)$, set

$$u^{k+1} = \lambda_2^k, \quad \lambda_2^{k+1} = \lambda_1^k, \quad b^{k+1} = b^k$$

If $k < K - 1$, set $k := k + 1$, and start from step 2, computing λ_1^k only; otherwise, terminate the iteration.

The use of rules (3-6) and (3-7) makes the last two test points coincident, since for $k = K - 1$ we get

$$\lambda_1^k = \frac{F_0}{F_2}(u^k - b^k) + b^k = \frac{u^k + b^k}{2} \tag{3-8}$$

and

$$\lambda_2^k = \frac{F_1}{F_2}(u^k - b^k) + b^k = \frac{u^k + b^k}{2} \tag{3-9}$$

To determine in which half of the interval $[b^k, u^k]$ the minimum lies, we shift one of the two final test points by a small amount $\hat{\epsilon}$. At the kth iteration, the interval containing the minimum is reduced from $(u^k - b^k)$ to $(\lambda_2^k - b^k)$ or $(u^k - \lambda_1^k)$, both of which are equal to $(F_{K-k}/F_{K+1-k})(u^k - b^k)$ so that the reduction ratio is F_{K-k}/F_{K+1-k}. It follows that after $K - 1$ iterations, the final interval is not greater than

$$\frac{1}{2}\frac{F_2}{F_3}\frac{F_3}{F_4}\cdots\frac{F_{K-1}}{F_K}(u^1 - b^1) + \hat{\epsilon} = \frac{1}{F_k}(u^1 - b^1) + \hat{\epsilon} \tag{3-10}$$

If ϵ is the desired accuracy (the maximum length of the final interval containing the minimum), then

$$\frac{1}{F_K}(u^1 - b^1) + \hat{\epsilon} \leq \epsilon \tag{3-11}$$

This gives us the relationship between the desired accuracy ϵ and the number of iterations K, which has to be determined before the process starts. If the accuracy ϵ, initial interval $[b^1, u^1]$, and $\hat{\epsilon} < \epsilon$ are given, then we take the smallest Fibonacci number F_K satisfying (3-11). Note that by taking the midpoint of the final interval, one can locate the minimum with the accuracy $\epsilon/2$.

Witzgall [1969] considered what, if anything, would be done when the first evaluation in the Fibonacci search is taken at an arbitrary point. He shows how the Fibonacci search could be modified to still give an optimal solution.

Generalized Fibonacci Search

A simple extension of Fibonacci search to multivariate optimization problems has been suggested by Krolak and Cooper [1963] and by Sugie [1964]. Consider the two-dimensional case where a given convex function $M(x_1, x_2)$ is defined over an area such that $b_1 \leq x_1 \leq u_1$ and $b_2 \leq x_2 \leq u_2$. We let λ_1^k and λ_2^k be determined as in (3-6) or (3-7) and search for minima along the lines $\bar{x}_1 = \lambda_1^k$ and $\bar{\bar{x}}_1 = \lambda_2^k$; we use the usual Fibonacci search, because $M(\bar{x}_1, x_2)$ and $M(\bar{\bar{x}}_1, x_2)$ have only one degree of freedom. If x_2^* and x_2^{**} yield $\min_{x_2} M(\bar{x}_1, x_2)$ and $\min_{x_2} M(\bar{\bar{x}}_1, x_2)$, respectively, then the values $M(\bar{x}_1, x_2^*)$ and $M(\bar{\bar{x}}_1, x_2^{**})$ are compared, and the computation proceeds in the manner described in step 3 of the Fibonacci algorithm. Optimization problems of higher dimensionality may be reduced to such nested univariate Fibonacci searches by extending this concept. If we consider the n dimensional problem with the required K_i, $i = 1, 2, \ldots, n$ function evaluations along the ith coordinate x_i, then the total number of evaluations is $\prod_{i=1}^n K_i$, which can be approximated by

$$T = \prod_{i=1}^{n} \left(\frac{\log F_{K_i} + \log \sqrt{5}}{\log \tau} - 1 \right) \tag{3-12}$$

since $F_{K_i} \cong \tau^{K_i+1}/\sqrt{5}$ where $\tau = (1 + \sqrt{5})/2$.

This extension of the efficient one-dimensional Fibonacci search technique does not imply that the technique itself is a good one to use in n space. In fact, Krolak and Cooper [1963] did not prove optimality, but did demonstrate in numerical experiments how well the method performed. Kaupe [1964] shows that there is no guarantee that the method will converge to a preassigned accuracy in a finite number of steps. However, the extended Fibonacci search is more efficient than any other one-dimensional search technique that could be extended to n space, such as Wilde's extension of sequential search by uniform pairs to two-dimensional space [Krolak and Cooper, 1963]. The one-dimensional searches are simply repeated n times, and it is assumed that no variations are introduced to alter the scheme. Kiefer [1957] indicates some of the difficulties in extending Fibonacci-like search to more than one dimension.

Discussion

The Fibonacci search is the most efficient sequential univariate search procedure that *guarantees a prescribed accuracy*. The only condition required is that the function be unimodal in the interval of search. Except for this property, which is normally satisfied in some neighborhood of a minimum, the function may be quite irregular—for example, discontinuous or piecewise differentiable. Another advantage is that the method is simple to code. It is very costly, however, if the location of the minimum is required with great precision. If, for example, $\epsilon/(u^1 - b^1) = 10^{-4}$, we would have to choose $K = 20$. In other words, 20 function evaluations are necessary to assure the relative precision 10^{-4}. Methods based on curve fitting (Sec. 3-5) have been found less costly and tend to give even greater accuracy. In the Fibonacci search, the number of iterations K must be selected in advance. One would prefer a process that can be continued or discontinued at any iteration depending on a flexible criterion. Such a modification (golden section) is discussed in the next section.

The Fibonacci search is recommended when the objective function is likely to be irregular and approximation by a low-order polynomial is inadequate. A typical application would be in solving the problem of minimizing along the lines of descent in an optimization problem transformed by the SUMT technique (Chap. 6); with this transformation the objective function has a very steep slope close to its boundaries.

We do not recommend the extension of the Fibonacci approach to multivariable minimization problems. Considerably more efficient methods for these problems are described in Chaps. 4 and 5.

3-4 GOLDEN SECTION SEARCH

The golden section method is a numerical approximation to Fibonacci search but is not optimal for any fixed K [Kiefer, 1953]; to use it, the number of observations K need not be specified in advance. In a computer this fact can be used to advantage if $M(\lambda)$ is, say, sharply peaked near its minimum and we suddenly want to change or enlarge the original number of evaluations. When the number of function evaluations is not known in advance, we could utilize any of the nonoptimal methods discussed in Sec. 3-2, but we shall see that golden sectioning is preferable.

When the number of function evaluations K becomes relatively large, the second term in (3-3) decreases rapidly and can be neglected. We have, therefore,

$$F_k = \frac{1}{\sqrt{5}}\tau^{k+1} = 0.4472(1.6180)^{k+1}, \qquad \tau = 1.6180 \tag{3-13}$$

which gives the constant ratio of the approximate Fibonacci numbers

$$\frac{F_{K-1-k}}{F_{K+1-k}} = \frac{1}{\tau^2} \cong 0.382 \tag{3-14}$$

and

$$\frac{F_{K-k}}{F_{K+1-k}} = \frac{1}{\tau} \cong 0.618 \tag{3-15}$$

Iteration Steps

1 Two values b^1, u^1 are given that bracket the minimum. Find the following interior points:

$$\lambda_1^1 = 0.382(u^1 - b^1) + b^1 \tag{3-16}$$

$$\lambda_2^1 = 0.618(u^1 - b^1) + b^1 \tag{3-17}$$

Set $k = 1$ and move to step 3.

2 Find the following interior points:

$$\lambda_1^k = 0.382(u^k - b^k) + b^k \tag{3-18}$$

or

$$\lambda_2^k = 0.618(u^k - b^k) + b^k \tag{3-19}$$

and compute $M(\lambda_1^k)$ or $M(\lambda_2^k)$.

3 If $M(\lambda_1^k) > M(\lambda_2^k)$, set $b^{k+1} = \lambda_1^k$, $\lambda_1^{k+1} = \lambda_2^k$, $u^{k+1} = u^k$, $k := k + 1$, and start from step 2, computing λ_2^k only. Otherwise, set $u^{k+1} = \lambda_2^k$, $\lambda_2^{k+1} = \lambda_1^k$, $b^{k+1} = b^k$, $k := k + 1$, and start from step 2, computing λ_1^k only.

4 Repeat above steps until $\lambda_2^k - \lambda_1^k < \epsilon$, where ϵ is a predetermined accuracy. Other termination criteria can also be applied.

Discussion

The golden section method does not use the exact Fibonacci numbers. Hence, these numbers need not be computed or stored in the program, and the total number of iterations K does not have to be fixed in advance, before the search is started. This method reduces the initial interval $[u^1, b^1]$ by a factor τ^{1-K} after K function evaluations. This factor is somewhat smaller than the optimal $1/F_K$ given by the Fibonacci search. We have

$$F_k\tau^{1-K} \cong 1.17, \qquad \tau = 1.6180 \tag{3-20}$$

In the first iteration we compute two function values; only one value need be determined in subsequent iterations. This new point is positioned symmetrically with respect to the existing interval point used in the previous iteration, suggesting that we can replace relationships (3-18) and (3-19) by more economical subtraction operations. Unfortunately, due to the finite precision arithmetic, such a process is unstable (may diverge) and it is not recommended without special precautions [Overholt, 1965].

3-5 SEARCH BY LOW-ORDER POLYNOMIAL APPROXIMATION

The approximate minimum of a function can also be found by curve fitting, which consists of representing the function along a line by a simpler function—a polynomial—and using the minimum of the latter to approximate that of the former. Gauss first considered the general problem of representing a given function by a polynomial. In the approximation theorem (3-2), Weierstrass proved that every continuous function can be represented by a polynomial.

Theorem 3-2 Let $M(\lambda)$ be a continuous function defined on the closed interval $[b, u]$. For every $\epsilon > 0$, there exists a polynomial $P(\lambda)$ such that $|M(\lambda) - P(\lambda)| < \epsilon$ for all $\lambda \in [b, u]$.

The obvious advantage of using polynomial curve fitting is that we can replace a possibly complex function by a very simple function, the minimum of which is known analytically. This will mean that finding the minimum will be a straightforward procedure. We are not concerned here with very precise fits, but would like to approximate the function by a low-order (second- or third-order) polynomial. The use of quadratic or cubic polynomials means that we assume that the approximated function is relatively smooth and possibly unimodal in the interval being investigated. In many practical applications, a small enough interval assures validity of this assumption. Various procedures have been developed for approximating functions and their derivatives (see appendix). Let us examine several iterative procedures formulated for finding the minimum of a function along a line by polynomial approximation.

Powell's Univariate Search Method

Powell [1964], in describing parametric optimization, suggested fitting a quadratic equation $M(\lambda) = \alpha\lambda^2 + \beta\lambda + \gamma$ to the function when searching for the minimum along a line. In this method, the function $M(\lambda)$ is evaluated at the points $\lambda^1 = \lambda^0 + \Delta\lambda^1$ and $\lambda^2 = \lambda^0 + \Delta\lambda^2$, where λ^0 is a starting point. Initially, $\Delta\lambda^1$ is set at zero and $\Delta\lambda^2$ at some preselected step length. The point $\lambda^3 = \lambda^0 + \Delta\lambda$ is chosen to be $\Delta\lambda^3 = 2\Delta\lambda^2$ if $M(\lambda^1) > M(\lambda^2)$, and $\Delta\lambda^3 = -\Delta\lambda^2$ otherwise. The optimum point of the parabola passing through these three points is given by $\lambda^m = \lambda^0 + \Delta\lambda^m$, where

$$\Delta\lambda^m = \frac{M(\lambda^1)[(\Delta\lambda^2)^2 - (\Delta\lambda^3)^2] + M(\lambda^2)[(\Delta\lambda^3)^2 - (\Delta\lambda^1)^2] + M(\lambda^3)[(\Delta\lambda^1)^2 - (\Delta\lambda^2)^2]}{2\{M(\lambda^1)[\Delta\lambda^2 - \Delta\lambda^3] + M(\lambda^2)[\Delta\lambda^3 - \Delta\lambda^1] + M(\lambda^3)[\Delta\lambda^1 - \Delta\lambda^2]\}} \tag{3-21}$$

If λ^m is close enough to the λ^j ($j = 1, 2$, or 3) which yields the lowest value of the objective function, this point is accepted as a good approximation of the minimum. Otherwise, the value of $M(\lambda^m)$ is computed and λ^m replaces λ^j, which yields the largest value of $M(\lambda)$. However, we do not reject it if by discarding some other point we can achieve a bracket on the minimum that would not be obtained otherwise. Then, the quadratic interpolation is repeated. Unfortunately, this procedure has two undersirable features: first, the predicted turning point may be a maximum if

$$\frac{M(\lambda^1)[\Delta\lambda^2 - \Delta\lambda^3] + M(\lambda^2)[\Delta\lambda^3 - \Delta\lambda^1] + M(\lambda^3)[\Delta\lambda^1 - \Delta\lambda^2]}{[\Delta\lambda^1 - \Delta\lambda^2][\Delta\lambda^2 - \Delta\lambda^3][\Delta\lambda^3 - \Delta\lambda^1]} \geq 0 \qquad (3\text{-}22)$$

Second, the predicted step $\Delta\lambda^m$ to the turning point may be too long.

To overcome these difficulties a more careful approach has been suggested by Davies, Swann, and Campey, and described by Box, et al. [1969]. Their method has been found more reliable especially when the initial estimate of the location of the minimum is poor. The iteration steps are as follows.

Given is an initial estimate λ^0 of the location of the minimum and a steplength $\Delta\lambda^0$. Set $k = 0$.

1 Compute $\lambda^{k+1} = \lambda^k + \Delta\lambda^k$ and $M(\lambda^k + \Delta\lambda^k)$.

2 If $M(\lambda^{k+1}) \leq M(\lambda^k)$, set $\Delta\lambda^k := 2\Delta\lambda^k$, and repeat from step 1 with $k := k + 1$.

3 If $M(\lambda^{k+1}) > M(\lambda^k)$ and $k \geq 1$, compute $M[\lambda^{k+1} - (\Delta\lambda^k/2)]$. This yields four equally spaced points, and we use three that bracket the minimum to predict the turning point in step 4. If, however, $M(\lambda^{k+1}) \geq M(\lambda^k)$ and $k = 0$, then we reverse the direction of search by changing the sign of $\Delta\lambda^0$ and compute $M(\lambda^0 - \Delta\lambda^0)$. If this trial fails again, the minimum is bracketed by $\lambda^0 + \Delta\lambda^0$ and $\lambda^0 - \Delta\lambda^0$. Otherwise, repeat step 1 with $\Delta\lambda^1 := -2\Delta\lambda^0$ and $k = 1$.

4 Having the three equidistant points, the quadratic is fitted through these points and the turning point λ^m is predicted. Calling these points $\lambda^1 < \lambda^2 < \lambda^3$, we have from (3-21)

$$\lambda^m = \lambda^2 - \frac{\Delta\lambda}{2}\frac{M(\lambda^3) - M(\lambda^1)}{M(\lambda^1) - 2M(\lambda^2) + M(\lambda^3)} = \lambda^2 + \Delta\lambda^m \qquad (3\text{-}23)$$

where $\Delta\lambda$ is the distance between adjacent points.

5 If $|\Delta\lambda^m| < \tau$, the process is terminated and the current λ^m is accepted as the location of the minimum point. Otherwise, the process is repeated with reduced steplength from whichever of λ^2 and λ^m yields the better value of $M(\lambda)$.

Remark If $M(\lambda)$ can be approximated adequately by a parabola in a neighborhood of the minimum, the described procedure performs much better than the Fibonacci search.

Davidon's Univariate Search Method

Davidon [1959], in discussing his variable metric method, introduced a cubic interpolation procedure for locating the minimum along a (descent) line $\mathbf{s}^k$. His algorithm requires that the function value and its gradient are available at two points which bracket the minimum. That is, λ^1, λ^2 are two values of λ for which $M(\lambda^i)$ and the gradients $\nabla M(\lambda^i)$, $i = 1, 2$, are given.

The turning point of the cubic, which is uniquely determined by $M(\lambda^i)$, $\nabla M(\lambda^i)$, $i = 1, 2$, is found from $\lambda^m = \lambda^1 + \Delta\lambda^m$, where

$$\Delta\lambda^m = \Delta\lambda\left\{1 - \frac{[\nabla M(\lambda^2)]^T\mathbf{s}^k + w - z}{[\nabla M(\lambda^2)]^T\mathbf{s}^k - [\nabla M(\lambda^1)]^T\mathbf{s}^k + 2w}\right\} \tag{3-24}$$

and

$$w = \{z^2 - [\nabla M(\lambda^1)]^T\mathbf{s}^k[\nabla M(\lambda^2)]^T\mathbf{s}^k\}^{1/2} \tag{3-25}$$

$$z = 3\frac{M(\lambda^1) - M(\lambda^2)}{\Delta\lambda} + [\nabla M(\lambda^1)]^T\mathbf{s}^k + [\nabla M(\lambda^2)]^T\mathbf{s}^k \tag{3-26}$$

$$\Delta\lambda = \lambda^2 - \lambda^1$$

An initial choice of $\Delta\lambda$ is given by

$$\Delta\lambda = \min\left\{1, \frac{-2[M(\lambda^1) - M(\lambda^{LB})]}{[\nabla M(\lambda^1)]^T\mathbf{s}^k}\right\}, \tag{3-27}$$

where $M(\lambda^{LB})$ is the lower bound of $M(\mathbf{x})$ provided by the user.

After interpolating, λ^m is accepted as a good approximation of the minimum if both $M(\lambda^1)$ and $M(\lambda^2)$ are larger than $M(\lambda^m)$. If, however, this is not the case, then by testing the sign of $[\nabla M(\lambda^m)]^T\mathbf{s}^k$ the interpolation is repeated over a smaller interval. To select the subinterval where the minimum is located, we test the sign of

$$\left.\frac{dM(\lambda)}{d\lambda}\right|_{\lambda^m} = [\nabla M(\lambda^m)]^T\mathbf{s}^k \tag{3-28}$$

If the sign is negative, the interval $[\lambda^m, \lambda^2]$ is chosen; otherwise, the minimum is located in $[\lambda^1, \lambda^m]$.

Murtagh and Sargent [1970] tested different procedures involving the Davidon technique plus added computation to locate the minimum along the line more accurately. Their tests show that the added computation is not worthwhile. A variation of the Davidon technique is described in Pierre [1969].

Discussion

The repeated interpolation technique by the parabolic or cubic fit is generally more efficient than the Fibonacci search; that is, fewer function evaluations are necessary for the minimum to be located to a specified accuracy. This efficiency is particularly good if the objective function is smooth enough and three ordinates or two ordinates and slopes of the function are available in the iterative process.

Davidon's method may be recommended whenever the objective function gradients are available. If the gradients are not available, we recommend the Davies algorithm for the first interpolation and Powell's method afterward.

These methods can fail or be inefficient if the objective function cannot be well approximated by a low-order polynomial. Thus, there is still an open question of how accurately we must approximate the minimum of the objective function. The computational cost of repeated polynomial approximations and minimum calculation will lessen the appeal of this approach as the requirement for accuracy increases.

3-6 UNIVARIATE SEARCH WITH MULTIVARIATE METHODS

If the objective function $M(\lambda)$ is differentiable, the necessary condition for $\hat{\lambda}$ to be the solution point at which $M(\lambda)$ assumes its minimum is

$$\nabla M(\hat{\lambda}) = \left.\frac{dM(\lambda)}{d\lambda}\right|_{\hat{\lambda}} = 0$$

This suggests that the problem of minimizing a differentiable function of one variable can be replaced by that of finding the zeros of $dM(\lambda)/d\lambda$, which can be attacked by any of the zero- or root-finding methods described in many elementary numerical analysis texts.

One of the descent methods for unconstrained multivariable minimization problems that is particularly suitable is the second-order gradient or Newton method. The multivariable version, its theoretical motivation, and several of its variations are described in Sec. 5-3. This method has second order convergence. When applied to the problem of univariate minimization—finding the zero of $dM(\lambda)/d\lambda$—the iteration proceeds as follows:

1 Initialize, set $k = 0$, evaluate $M(\lambda^0)$; λ^0 is given.

2 Compute

$$\left.\frac{dM(\lambda)}{d\lambda}\right|_{\lambda^k} = [\nabla M(\mathbf{x}^k + \lambda^k \mathbf{s}^k)]^T \mathbf{s}^k$$

3 If $|dM(\lambda^k)/d\lambda| < \tau$, terminate the iteration and accept λ^k as the solution ($\tau > 0$ is a predetermined tolerance). Otherwise, proceed to step 4.

4 Compute

$$\left.\frac{d^2M(\lambda)}{d\lambda^2}\right|_{\lambda^k} = [\mathbf{s}^k]^T \operatorname{Hess}(\mathbf{x}^k + \lambda^k \mathbf{s}^k)\mathbf{s}^k$$

5 Compute

$$\lambda^{k+1} = \lambda^k - \frac{dM(\lambda^k)}{d\lambda} \Big/ \frac{d^2M(\lambda^k)}{d\lambda^2} \tag{3-29}$$

and evaluate $M(\lambda^{k+1})$.

6 If $|M(\lambda^k) - M(\lambda^{k+1})| < \epsilon_1$ and $|\lambda^{k+1} - \lambda^k| < \epsilon_2$, terminate the iteration and accept λ^k as the solution ($\epsilon_1 > 0$ is a predetermined minimal func-

tion improvement and $\epsilon_2 > 0$ a predetermined minimal stepsize). Otherwise, set $k := k + 1$ and proceed from step 2.

3-7 NONOPTIMAL STEP METHODS

Practical experience has shown that the efficiency of certain descent methods for multivariable minimization can be increased if nonoptimal steps are used along the descent lines (see Sec. 5-1). Sometimes the nonoptimal descent step is an appropriate alternative to the optimal step strategy; this alternative is particularly desirable in conjunction with the second-order gradient (Newton) descent. Optimal steplengths are obtained as the solution of the problem of minimization along the line for which methods are described in the preceding sections. The strategy of nonoptimal steps involves steps λ^k, satisfying

$$M(\mathbf{x}^k + \lambda^k \mathbf{s}^k) \leq M(\mathbf{x}^k) \tag{3-30}$$

Different approaches to the solution of this problem have been reported. Most recently, Bard [1970] outlined a number of such strategies. We have found that it is best to adjust the steplength depending on the progress of the iteration. Failure to reduce the objective function should result in a decreased step length and conversely. Specifically, the following algorithm has worked well in conjunction with the second-order gradient iteration (where one can justify theoretically a steplength of $\lambda^k = 1$):

1 Given an initial value for λ^k (e.g., $\lambda^0 = 1$).

2 If $\|\lambda^k \mathbf{s}^k\| \leq \gamma^k \|\mathbf{x}^0\|$, use λ^k; otherwise, replace λ^k by $\lambda^k(\gamma^k \|\mathbf{x}^0\|/\|\lambda^k \mathbf{s}^k\|)$, where $0.1 \leq \gamma^k \leq 2.0$, $\gamma^0 = 1$ and

$$\gamma^{k+1} \begin{cases} = \gamma^k + 0.10 \text{ after two consecutive reductions in } M \text{ without halving} \\ \quad \text{of } \lambda^k \text{[see Step 3]} \\ = \gamma^k - 0.05 \text{ after each failure to reduce } M. \\ = \gamma^k \text{ otherwise} \end{cases}$$

3 If $M(\mathbf{x}^k + \lambda^k \mathbf{s}^k) \geq M(\mathbf{x}^k)$, replace λ^k by $\frac{1}{2}\lambda^k$ and repeat until $M(\mathbf{x}^k + \lambda^k \mathbf{s}^k) < M(\mathbf{x}^k)$ or until λ^k has been halved three times.

3-8 COMPUTING SYSTEMS

Remarks Several ALGOL versions of the Fibonacci and golden section methods have been coded and published [Pike and Pixner, 1965; Pike, et al., 1967; Boothroyd, 1965*a*, 1965*b*; and Overholt, 1967]. A method that combines the Fibonacci search with a cubic approximation technique can be found in Johnson and Myers [1967]. Ferguson [1960] provides a flowchart from which a Fibonacci search can be coded, and Trushel

[1968] offers an extensive review of the Fibonacci search algorithm and a listing of a FORTRAN IV program.

Note also that many of the computer programs listed in Chap. 5 include univariable search subroutines.

References

BARD, Y.: "Comparison of Gradient Methods for the Solution of Nonlinear Parameter Estimation Problems," *SIAM Journal Numer. Anal.*, Vol. 7, 157, 1970.

BOX, M. J., D. Davies and W. H. Swann: *Nonlinear Optimization Techniques*, ICI Ltd., Monograph No. 5, Oliver and Boyd, Ltd., Edinburgh, 1969.

BOOTHROYD, J.: "Algorithm 7 MINIX," *The Computer Bulletin*, Vol. 9, 104, 1965*a*.

———: "Certification of Algorithm 2, FIBONACCI SEARCH," *The Computer Bulletin*, Vol. 9, 105, 1965*b*.

BROOKS, S. H.: "A Comparison of Maximum Seeking Methods," *Operations Research*, Vol. 7, 430, 1959.

DAVIDON, W. C.: "Variable Metric Method for Minimization," *AEC Research and Development*, Report ANL-5990, 1959.

FERGUSON, D.: "Fibonacci Searching," *Communications of the ACM*, Vol. 3, 648, 1960.

JOHNSON, I. L., JR. and G. E. MYERS: One-Dimensional Minimization Using Search by Golden Section and Cubic Fit Methods, N68–18823 (NASA), Manned Spacecraft Center, Houston, Tex., 1967.

JOHNSON, S. M.: *Best Exploration for Maximum Is Fibonacci*, Report No. RM-1590. The RAND Corporation, 1955.

KAUPE, A. F.: "On Optimal Search Techniques," *Communications of the ACM*, Vol. 7, 38, 1964.

KIEFER, J.: "Sequential Minimax Search for a Maximum," *Proceedings of the American Mathematical Society*, Vol. 4, 502, 1953.

———: "Optimum Sequential Search and Approximation Methods Under Minimum Regularity Assumptions," *SIAM Journal*, Vol. 5, 105, 1957.

KROLAK, P. and L. COOPER: "An Extension of Fibonaccian Search to Several Variables, *Communications of the ACM*, Vol. 6, 639, 1963.

MURTAGH, B. A. and R.W.H. SARGENT: "Computational Experience with Quadratically Convergent Minimization Methods," *Computer Journal*, Vol. 13, 185, 1970.

OVERHOLT, K. J.: "An Instability in the Fibonacci and the Golden Section Search Methods," *BIT*, Vol. 5, 284, 1965.

———: "Note on Algorithm 2, Algorithm 16 and Algorithm 17," *Computer Journal*, Vol. 9, 414, 1967.

PIERRE, D. A.: *Optimization Theory with Applications*, John Wiley & Sons, Inc., New York, Ch. 5, 264, 1969.

PIKE, M. C. and J. PIXNER: "Algorithm 2, FIBONACCI SEARCH," *The Computer Bulletin*, Vol. 8, 147, 1965.

———, I. D. Hill and F. D. James: "Note on Algorithm 2, FIBONACCI SEARCH and on Algorithm 7 MINX," *Computer Journal*, Vol. 9, 416, 1967.

POWELL, M.J.D.: "An Efficient Method for Finding the Minimum of a Function of Several Variables Without Calculating Derivatives," *Computer Journal*, Vol. 7, 155, 1964.

SUGIE, N.: "An Extension of Fibonaccian Searching to Multidimensional Cases," *IEEE Transactions On Automatic Control*, AC-9, 105, 1964.

TRUSHEL, P. J.: *A Fibonacci Search Algorithm for the Approximate Minimization of a Function of a Single Variable*, NME, MK–24, National Research Council of Canada, 1968.

WILDE, D. J.: *Optimum Seeking Methods*, Prentice-Hall, Inc., Englewood Cliffs, N. J., 1964.

——— and C. S. BEIGHTLER: *Foundations of Optimization*, Prentice-Hall, Inc., Englewood Cliffs, N. J., 1967.

WITZGALL, C.: *Fibonacci Search with Arbitrary First Evaluation*, Document D1–82–0916, Boeing Scientific Research Laboratories, Seattle, Wash., 1969.

4 DIRECT SEARCH METHODS

This chapter deals with search techniques for the solution of unconstrained minimization problems. The unconstrained minimization problem is one where a value of the variable vector $\mathbf{x} = (x_1, \ldots, x_n)^T$ is sought that minimizes the objective function. The problem min $M(\mathbf{x})$ is a special case of the general nonlinear programming or optimization problem (1-9). Its special feature is that no other conditions or constraints need be satisfied by the solution. Search methods are most suitable for simple unconstrained problems involving a relatively small number of variables. Not much experience is available concerning the performance of search techniques in solving problems resulting from transformations of constrained problems.

Search methods are in general less efficient than descent techniques, but this characteristic may be compatible with the smaller accuracy in the objective function evaluation which the user of the technique may provide. Many search techniques lack a convergence proof, even for objective functions of a special (e.g., quadratic) form. The search techniques require only objective function evaluations and do not use partial derivatives. In addition, information accumulated in earlier iterations may be used by some search techniques.

We have mentioned that the Fibonacci search is the most efficient way of specifying an interval in which the minimum is located (Chap. 3). We may expect, therefore, that an extension of this search to the multivariable problems should provide us with an efficient algorithm. Sugie [1964] has suggested a nested Fibonacci search to minimize convex multivariable functions. It turns out, however, that this method reduces an interval containing the minimum in a more expensive way (in terms of the number of function evaluations) than other direct search methods which do not employ the Fibonacci search concept.

4-1 TYPES OF DIRECT SEARCH METHODS

There are two main types of direct search methods: (1) linear methods, and (2) evolutionary operation methods. In the first category, a set of directions is generated and used to move from the current approximation $\mathbf{x}^k$ toward the minimum. The simplest concept of this type dates back to Gauss, who proposed to adjust one variable x_i at a time. We can either move in the n-dimensional space parallel to each axis in turn or base our choice of axes on some rules. We can, for instance, choose the coordinate x_i such that at the current point $\mathbf{x}^k$,

$$\left|\frac{\partial M}{\partial x_i}\right| \geq \left|\frac{\partial M}{\partial x_l}\right| \qquad l = 1, \ldots, n \tag{4-1}$$

Relaxation methods are used frequently to minimize quadratic forms related to the problem of solving sets of linear algebraic equations. It can be demonstrated that if the quadratic $M(\mathbf{x}) = a + b^T\mathbf{x} + \frac{1}{2}\mathbf{x}^T A\mathbf{x}$ has a positive definite matrix A, convergence of the relaxation method is guaranteed [Kowalik and Osborne, 1968]. More about convergence of the relaxation process for general functions may be found in Schechter [1968]. Experience indicates that the relaxation method works well if $M(\mathbf{x})$ is quadratic and has almost hyperspherical contours. It may be extremely inefficient, however, in more difficult cases such as (1) ill-conditioned quadratic form with principal axes not parallel to the coordinate system; and (2) curved and narrow valleys formed by the isocontours of nonlinear functions.

A more efficient strategy combines the relaxation method with an attempt to find a direction that approximates the local gradient vector. This concept leads to the *pattern search* method of Hooke and Jeeves [1961]. Another possibility is to use a set of n mutually orthogonal search directions and rotate them in the space of variables to avoid the slow convergence of the relaxation method (where the set of search directions, coordinate system, is fixed). In the quadratic case, the *method of rotating coordinates* attempts to align the search directions with the principal axes of the objective function isocontours (hyperellipsoids). Such a method was originated by Rosenbrock [1960] and substantially improved by Swann [Box, et al., 1969].

The *evolutionary operation* methods originated by Box [1957] evaluate the objective function at the vertices of some geometric configuration and move it towards the minimum. One of the simplest forms of this concept is *factorial design* [Box, et al., 1969], in which the objective function is evaluated at the vertices and center of a hypercube generated in the n-dimensional space. This hypercube is either contracted, if the smallest value of $M(\mathbf{x})$ is at the center, or a new hypercube is centered at the vertex that corresponds to the smallest value of $M(\mathbf{x})$. These rules define a single step of the algorithm. The factorial design method is more economical than the well known grid test method and may be useful if we can afford very few function evaluations and are interested in a rough estimation of the minimum solution to the problem with few variables.

A more flexible approach based on the evolutionary operation concept is that of Spendley, et al. [1962]; an improved version has been suggested by Nelder and Mead [1965]. We also describe the random search methods and a method in which the objective function values are used to fit a sequence of quadratics that approximate locally the nonlinear function $M(\mathbf{x})$.

4-2 PATTERN SEARCH METHOD

Each step of the pattern search method comprises two kinds of moves: *exploratory* and *pattern*. In the relaxation method, the exploratory moves are performed by changing only one variable at a time, a procedure known to be often very inefficient when applied to optimization problems. For this reason, Hooke and Jeeves [1961] suggested that the exploratory moves be followed by pattern moves, steps along the approximate gradient direction, which is determined from results of the exploratory move. The search begins at an arbitrarily chosen starting point and with prescribed step lengths $\lambda^i > 0$ in each of the coordinate directions x_i, $i = 1, 2, \ldots, n$.

Iteration Steps

Exploratory Move. Initialize, set $k = 0$, $\mathbf{x}^0$ and λ^i, $i = 1, \ldots, n$ are given. Given a starting base $\mathbf{x}^k$. Find a temporary point $\hat{\mathbf{x}}^{k+1}$, which is given by the following process:

$$\min \{M(x_1^k, \ldots, x_n^k), M(x_1^k \pm \lambda^1, x_2^k, \ldots, x_n^k)\}$$

$$\ldots$$

$$\min \{\min \text{ of step } (r - 1), M(\hat{x}_1^{k+1}, \ldots, \hat{x}_{r-1}^{k+1}, x_r^k \pm \lambda^r, \ldots, x_n^k\}$$

$$\ldots$$

$$\min \{\min \text{ of step } (n - 1), M(\hat{x}_1^{k+1}, \ldots, \hat{x}_{n-1}^{k+1}, x_n^k \pm \lambda^n)\}$$

The point $\hat{\mathbf{x}}^{k+1}$ arrived at by the exploratory move is called a new base-point.

Pattern Move. The pattern move is a simple step of length $\lambda^k = 1$ from the current base $\hat{\mathbf{x}}^{k+1}$ along the approximate gradient direction $\mathbf{s}^k$. The step is

$$\hat{\hat{\mathbf{x}}}^{k+1} = \hat{\mathbf{x}}^{k+1} + \lambda^k \mathbf{s}^k = \hat{\mathbf{x}}^{k+1} + \mathbf{s}^k \tag{4-2}$$

and the direction is given by

$$\mathbf{s}^k = (\hat{x}_1^{k+1} - x_1^k, \hat{x}_2^{k+1} - x_2^k, \ldots, \hat{x}_n^{k+1} - x_n^k) \tag{4-3}$$

An alternative is to solve a one-dimensional minimization problem in the direction $\mathbf{s}^k$ (Chap. 3) and to use the resulting steplength λ^k instead of 1.

Next we perform an exploratory move from $\hat{\hat{\mathbf{x}}}^{k+1}$ to produce the next basepoint, $\hat{\mathbf{y}}^{k+1}$. Two cases are now possible.

1 If $M(\hat{\mathbf{y}}^{k+1}) < M(\hat{\mathbf{x}}^{k+1})$, then $\mathbf{x}^{k+1} = \hat{\mathbf{y}}^{k+1}$ (Fig. 4-1), and the process is repeated with an exploratory move with $k := k + 1$.

2 Otherwise, $\mathbf{x}^{k+1} = \hat{\mathbf{x}}^{k+1}$ (Fig. 4-2), and we start the procedure with an exploratory move with reduced (e.g. halved) steplengths and $k := k + 1$.

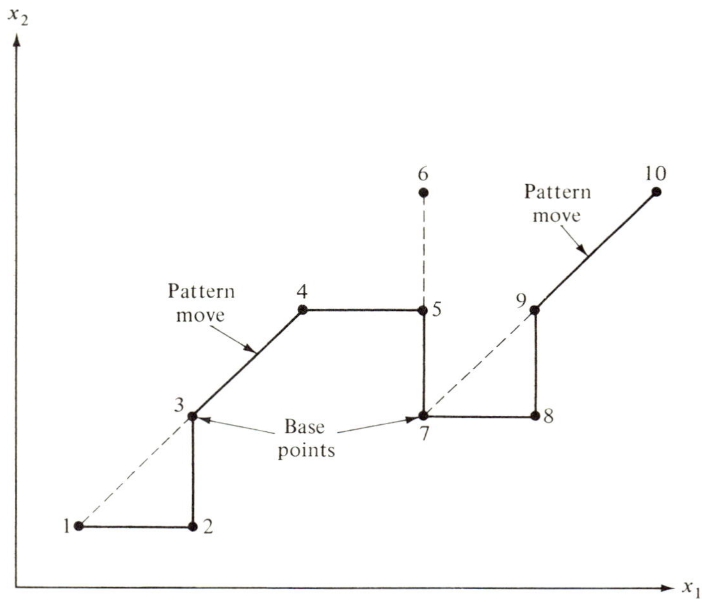

Figure 4-1 Pattern search method, $M(\mathbf{x}^7) < M(\mathbf{x}^3)$

Remarks The steps have to be reduced (e.g. halved) also if the exploratory move fails to produce a new basepoint. Convergence is assumed and the process is terminated when the step sizes fall below a preselected minimum τ; that is,

$$\max_i \lambda^i < \tau \tag{4-4}$$

Discussion

Computational experience has shown that the algorithm can detect and follow a steep valley of a multivariable function. The method is an approxi-

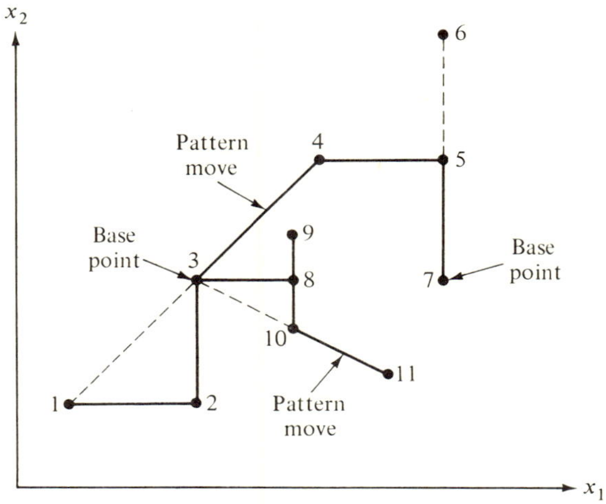

Figure 4-2 Pattern search method, $M(\mathbf{x}^7) > M(\mathbf{x}^3)$

mate steepest descent method (*cf.* Sec. 5-2) and should have many of its disadvantages.

The method does not take advantage of a quadratic or nearly quadratic form of the objective function, nor does it supply any information about the function behavior in the vicinity of the solution. We may expect slow convergence even close to the optimum. The method may be useful to solve problems having a small number of variables or as a component of a mixed algorithm; in the latter case, it should be used at the initial stage of the optimization process.

4-3 THE METHOD OF ROTATING COORDINATES

The method of rotating coordinates is more effective than the pattern search method of Hooke and Jeeves [1961]. Rosenbrock [1960] suggested that the coordinate system be rotated in such a way that one axis is oriented toward the locally estimated direction of the valley (or the ridge if we are maximizing), and the other axes are normal to it and mutually orthogonal (Fig. 4-3). To simplify our notation the index k denoting the number of iteration is deleted. Otherwise, the following notation is used:

1 $S = (\mathbf{s}^0, \mathbf{s}^1, \ldots, \mathbf{s}^{n-1})$ is a set of n (mutually) orthogonal directions.

2 $\boldsymbol{\lambda}^T = (\lambda^0, \lambda^1, \ldots, \lambda^{n-1})$ is a set of steplengths along the directions $\mathbf{s}^j, j = 0. \ldots, n-1$.

3 $\alpha > 1$, $0 < \beta < 1$ are two coefficients used to control the steplengths.

4 A trial step from $\mathbf{x}$ is called *successful* if $M(\mathbf{x} + \lambda^j \mathbf{s}^j) \leq M(\mathbf{x})$.

5 $\Lambda^j \neq 0$ are the algebraic sums of all successful steplengths in the cor-

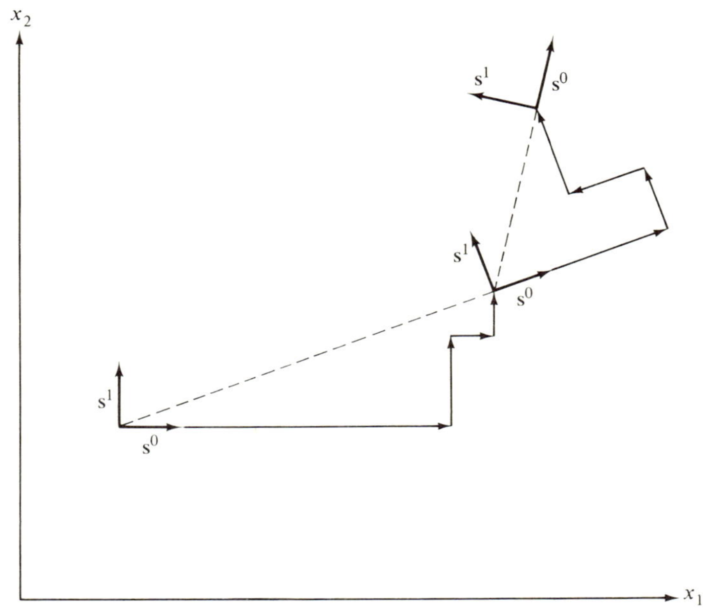

Figure 4-3 The Rosenbrock method of rotating coordinates

responding directions $\mathbf{s}^j$. Note that if $\Lambda^j = 0$ for all j, the process will terminate.

Iteration steps

1 Given is $\mathbf{x}^0$ and the set $S = (\mathbf{s}^0, \mathbf{s}^1, \ldots, \mathbf{s}^{n-1})$ of directions (initially these are the coordinate axes).

2 Find $\mathbf{x}^k$ at which the minimum of the following process occurs:

$$\min\{M(\mathbf{x}^0), M(\mathbf{x}^0 + \lambda^0\mathbf{s}^0)\} \text{ yields } \mathbf{x}^1$$
$$\ldots$$
$$\min\{\text{min of step } (r), M(\mathbf{x}^r + \lambda^r\mathbf{s}^r)\} \text{ yields } \mathbf{x}^{r+1}$$
$$\ldots$$
$$\min\{\text{min of step } (n-1), M(\mathbf{x}^{n-1} + \lambda^{n-1}\mathbf{s}^{n-1})\} \text{ yields } \mathbf{x}^n$$

Each time the trial step is executed we alter the steplength in the following manner:

$$\lambda^j := \alpha\lambda^j \quad \text{if step is successful} \tag{4-5}$$

or otherwise,

$$\lambda^j := -\beta\lambda^j \tag{4-6}$$

This step is repeated until at least one trial has been successful and one has failed in every direction $\mathbf{s}^j$.

3 Compute the following set of independent directions:

$$Q = [\mathbf{q}^0, \mathbf{q}^1, \ldots, \mathbf{q}^{n-1}] = S \begin{bmatrix} \Lambda^0 \\ \Lambda^1 \Lambda^1 \\ \Lambda^2 \Lambda^2 \Lambda^2 \\ \cdots \\ \Lambda^{n-1} \cdots \Lambda^{n-1} \end{bmatrix} \tag{4-7}$$

4 These directions are used to generate a new set of orthogonal directions by means of the Gram-Schmidt orthogonalization procedure:

$$\mathbf{v}^0 = \mathbf{q}^0$$

$$\mathbf{v}^i = \mathbf{q}^i - \sum_{j=0}^{i-1} [(\mathbf{q}^i)^T \hat{\mathbf{s}}^j] \hat{\mathbf{s}}^j \qquad i = 1, 2, \ldots, n-1 \tag{4-8}$$

where
$$\hat{\mathbf{s}}^i = \frac{\mathbf{v}^i}{\|\mathbf{v}^i\|} \qquad i = 0, 1, \ldots, n-1 \tag{4-9}$$

are new search directions.

5 Repeat the procedure from step 1 starting with the best point obtained so far and new search directions.

Remarks Convergence of the above process can be assumed and the computation is terminated when the step sizes fall below a prescribed tolerance, $\max_j |\lambda^j| < \tau$, or if the magnitude $\|\mathbf{q}^0\|$ of the progress at each of several consecutive stages is less than a prescribed minimum.

A derivation of the Gram-Schmidt iterative formula (4-8) is as follows: Suppose that i directions $\hat{\mathbf{s}}^0, \ldots, \hat{\mathbf{s}}^{i-1}$ have been chosen for the next stage of the algorithm and that $\hat{S}_{i-1} = \{\hat{\mathbf{s}}^0, \ldots, \hat{\mathbf{s}}^{i-1}\}$. We generate the direction $\hat{\mathbf{s}}^i$ using the relationship

$$\hat{\mathbf{s}}^i = \mathbf{q}^i - \hat{S}_{i-1}\mathbf{v}^i \tag{4-10}$$

where $\mathbf{v}^i$ is such that $\hat{\mathbf{s}}^i$ is orthogonal to all columns of S_{i-1} (all previous directions). This gives

$$\mathbf{v}^i = (\hat{S}_{i-1}^T \hat{S}_{i-1})^{-1} \hat{S}_{i-1}^T \mathbf{q}^i \tag{4-11}$$

and
$$\hat{\mathbf{s}}^i = P_{i-1}\mathbf{q}^i \tag{4-12}$$

where
$$P_{i-1} = I - \hat{S}_{i-1}(\hat{S}_{i-1}^T \hat{S}_{i-1})^{-1} \hat{S}_{i-1}^T \tag{4-13}$$

is a *projection matrix*. In the unconstrained problem under consideration, the matrix P_{i-1} is simple since

$\hat{S}_{i-1}^T \hat{S}_{i-1} = I$. We have, therefore,

$$\hat{\mathbf{s}}^i = \mathbf{q}^i - \hat{S}_{i-1}\hat{S}_{i-1}^T \mathbf{q}^i = \mathbf{q}^i - \sum_{j=0}^{i-1} [(\hat{\mathbf{s}}^j)^T \mathbf{q}^i]\hat{\mathbf{s}}^j \qquad (4\text{-}14)$$

which is the same as (4-8) except for the length normalization (4-9).

Discussion

Rosenbrock found the values $\alpha = 2$ and $\beta = 0.5$ suitable for this algorithm. The method is quite flexible in following curved and steep valleys, because the coordinate system can rotate freely in the multivariable space. It has proved to be successful in locating a good approximation to the minimum. The method may be useful when combined with other techniques that are more efficient in the vicinity of the solution.

The basic Rosenbrock method has been modified by Davies, Swann, and Campey [Box, et al., 1969]. They suggested search for minima along the direction $\mathbf{s}^j$. This modification, called also the *DSC method*, is described in the next section.

The same section contains, also, a description of a more economic and numerically superior method for generating sets of orthogonal directions. This last feature is important since the Gram-Schmidt orthogonalization procedure is highly susceptible to numerical errors. An alternative approach to Gram-Schmidt is the modified Gram-Schmidt method [Rice, 1966] or the Householder method [Golub and Saunders, 1970].

4-4 MODIFIED ROTATING COORDINATE METHOD

The method is very similar to the original method described above. The only difference is that instead of simple trials, a linear search to locate a minimum is performed along each orthogonal direction. So long as this search produces nonzero steps in each direction, the modified method works in the same way as the original method of Rosenbrock. If, however, the linear search fails in one or more directions, the Gram-Schmidt process breaks down. Modification of the orthogonalization procedure to avoid this difficulty is explained in this section.

The following notation is used:

1 $S = (\mathbf{s}^0, \mathbf{s}^1, \ldots, \mathbf{s}^{n-1})$ is a set of mutually orthogonal directions.

2 λ^j, $j = 0, 1, \ldots, n$ are steps taken in turn along the corresponding directions $\mathbf{s}^j$ to locate minima of the objective function. They are not necessarily all nonzero.

Iteration Steps

1 Given is $\mathbf{x}^0$ and the current set of orthogonal directions $S = (\mathbf{s}^0, \mathbf{s}^1, \ldots, \mathbf{s}^{n-1})$; initially, these are the coordinate axes. Find λ^j, satisfying $\min_{\lambda} M(\mathbf{x}^j + \lambda \mathbf{s}^j)$, and set $\mathbf{x}^{j+1} = \mathbf{x}^j + \lambda^j \mathbf{s}^j$, for $j = 0, 2, \ldots, n-1$. Let us assume that all $\lambda^j \neq 0$ except λ^p, which is zero.

2 Rearrange the order of the directions $\mathbf{s}^j$ as follows: $\mathbf{s}^0, \mathbf{s}^1, \ldots, \mathbf{s}^{p-1}, \ldots, \mathbf{s}^{n-1}, \mathbf{s}^p$, and call $\hat{S} = (\mathbf{s}^0, \mathbf{s}^1, \ldots, \mathbf{s}^{p-1}, \mathbf{s}^{p+1}, \ldots, \mathbf{s}^{n-1})$.

3 Compute the following set of independent directions:

$$Q = (\mathbf{q}^0, \mathbf{q}^1, \ldots, \mathbf{q}^{n-2}) = \hat{S}\Omega \tag{4-15}$$

where

$$\Omega = \begin{bmatrix} \lambda^0 & & & & \\ \lambda^1 & \lambda^1 & & & \\ \cdots & & & & \\ \lambda^{p-1} \cdots & & \lambda^{p-1} & & \\ \lambda^{p+1} \cdots & & & \lambda^{p+1} & \\ \cdots & & & & \\ \lambda^{n-1} \cdots & & & & \lambda^{n-1} \end{bmatrix}$$

4 Use the Gram-Schmidt orthogonalization procedure to produce a new set of orthogonal directions $\hat{\mathbf{s}}^0, \hat{\mathbf{s}}^1, \ldots, \hat{\mathbf{s}}^{n-2}$ and $\mathbf{v}^0 = \mathbf{q}^0$:

$$\mathbf{v}^i = \mathbf{q}^i - \sum_{j=0}^{i-1} [(\mathbf{q}^i)^T \hat{\mathbf{s}}^j]\hat{\mathbf{s}}^j \qquad i = 1, 2, \ldots, n-2 \tag{4-16}$$

and

$$\hat{\mathbf{s}}^i = \frac{\mathbf{v}^i}{\|\mathbf{v}^i\|} \qquad i = 0, 1, \ldots, n-2$$

5 Repeat from step 1 with $S = (\hat{\mathbf{s}}^0, \hat{\mathbf{s}}^1, \ldots, \hat{\mathbf{s}}^{n-2}, \mathbf{s}^p)$ and $\mathbf{x}^0 = \mathbf{x}^n$.

Remarks The Gram-Schmidt process is performed on the $n-1$ directions (with $\lambda^j \neq 0$), and only $n-1$ new $\hat{\mathbf{s}}^j$ are computed in steps 3 and 4. These new directions are mutually orthogonal, and they do not have components in the direction $\mathbf{s}^p$ (with $\lambda^p = 0$). Therefore, the direction $\mathbf{s}^p$ can be added to the set $\hat{\mathbf{s}}^0, \hat{\mathbf{s}}^1, \ldots, \hat{\mathbf{s}}^{n-2}$ as a complementary orthogonal direction. If $r > 1$ steplengths are zero, then only $(n-r)$ vectors are reorthogonalized by the Gram-Schmidt process. The remaining r vectors are unaltered and the new directions system remains orthogonal.

Box, et al. [1969] indicate that to find an approximation to λ^j it is sufficient to perform a single quadratic interpolation along each direction (Chap. 3).

Convergence is assumed and the process is terminated if $\max_j |\lambda^j| < \tau$, where τ is a prescribed small number.

The other aspects of the method are similar to those of the original rotating coordinate method.

Modifications and Acceleration Techniques

Powell [1968] suggested a more efficient method for computing a new orthonormal set of vectors than the Gram-Schmidt orthogonalization process. Powell's method requires $O(n^2)$ multiplications instead of $O(n^3)$ required by the Gram-Schmidt process, where O denotes "the order of." The method works if $n - 1$ of the vectors are already orthogonal. Our application satisfies this requirement.

Assume that the last orthonormal basis is $\mathbf{s}^0, \mathbf{s}^1, \ldots, \mathbf{s}^{n-1}$. Considered are the following two cases:

1 Steplengths $\lambda^j \neq 0$ for $j = 0, 1, \ldots, k$ and $\lambda^{k+1} = \lambda^{k+2} = \ldots = \lambda^{n-1} = 0$ (DSC method).

2 Steplengths $\lambda^j = \Lambda^j \neq 0$ for all $j = 0, 1, \ldots, n-1$ (Rosenbrock method).

The new orthonormal directions $\hat{\mathbf{s}}^1, \hat{\mathbf{s}}^2, \ldots, \hat{\mathbf{s}}^k$ (in the Rosenbrock method, $k = n - 1$) are computed by the same procedure for both cases:

1 Set

$$\begin{aligned} j &= k \\ \tau &= (\lambda^k)^2 \\ \boldsymbol{\sigma} &= \lambda^k \mathbf{s}^k \end{aligned} \tag{4-17}$$

2 If $j = 0$, terminate the process; otherwise ($j > 0$), compute

$$\hat{\mathbf{s}}^j = \frac{(\tau \mathbf{s}^{j-1} - \lambda^{j-1}\boldsymbol{\sigma})}{[\tau(\tau + \lambda^{j-1})^2]^{1/2}} \tag{4-18}$$

3 Set

$$\begin{aligned} j &:= j - 1 \\ \tau &:= \tau + (\lambda^j)^2 \\ \boldsymbol{\sigma} &:= \boldsymbol{\sigma} + \lambda^j \mathbf{s}^j \end{aligned} \tag{4-19}$$

and return to step 2.

The remaining vectors are computed as follows for cases 1 and 2, respectively:

Case 1:

$$\begin{aligned} &\hat{\mathbf{s}}^0 = \frac{\boldsymbol{\sigma}}{\sqrt{\tau}} \qquad \boldsymbol{\sigma} = \sum_{j=0}^{k} \lambda^j \mathbf{s}^j, \quad \tau = \sum_{j=0}^{k} (\lambda^j)^2 \\ &\hat{\mathbf{s}}^j = \mathbf{s}^j \qquad j = k+1, k+2, \ldots, n-1 \end{aligned} \tag{4-20}$$

and

Case 2:
$$\hat{\mathbf{s}}^0 = \frac{\sum_{j=0}^{n-1} \lambda^j \mathbf{s}^j}{\left\| \sum_{j=0}^{n-1} \lambda^j \mathbf{s}^j \right\|} \tag{4-21}$$

The method produces accurate orthogonal vectors. Powell has also shown that the method is numerically stable; that is, there is no dangerous accumulation of roundoff errors when the iteration is repeated many times. A derivation of Powell's method for calculating orthogonal vectors and discussion of its stability has been given by Osborne [1969].

Andrew [1969] conjectures, on the basis of his numerical experience, that the Powell vectors can also accelerate the convergence of the Rosenbrock and DSC methods, measured in terms of the number of function evaluations performed.

Linear Constraints

The DSC method can be extended to handle the linear constraints $A\mathbf{x} - \mathbf{b} = 0$ or $A\mathbf{x} - \mathbf{b} \geq 0$ [Box, et al., 1969]. This extension is similar to the gradient projection method (Chap. 7) and uses the projection matrices (4-13) to generate directions restricted to the manifold defined by the linear constraints.

4-5 THE SIMPLEX METHOD

The simplex method is based on the comparison of the objective function values at the $(n + 1)$ vertices of a general simplex* and moving this simplex towards the optimum point [Nelder and Mead, 1965]. This movement is achieved by three basic operations: reflection, contraction, and expansion (Fig. 4-4).

The following notation is used:

1 $\mathbf{x}^h$ is the vertex corresponding to the highest value of the objective function:

$$M(\mathbf{x}^h) = \max_i (\mathbf{x}^i) \qquad i = 1, 2, \ldots, (n + 1) \tag{4-22}$$

2 $\mathbf{x}^s$ is the vertex with the second highest value of M.

3 $\mathbf{x}^l$ is the vertex with the lowest value of M.

---

*A set of $(n + 1)$ equidistant points in n-dimensional space forms a *regular simplex*. A *general simplex* is a simplex without the property that the vertices are equally distanced. The simplex method described here should not be confused with Dantzig's simplex method for linear programming problems.

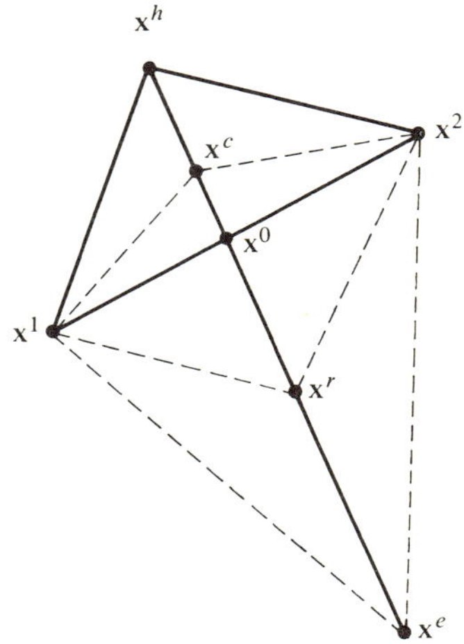

Figure 4-4 The reflection, expansion, and contraction operations in the simplex method

4 $\mathbf{x}^0$ is the centroid of all $\mathbf{x}^i$ except $i = h$, given by

$$\mathbf{x}^0 = \frac{1}{n} \sum_{\substack{i=1 \\ i \neq h}}^{n+1} \mathbf{x}^i \tag{4-23}$$

The three basic operations of the method are *reflection*, where $\mathbf{x}^h$ is replaced by

$$\mathbf{x}^r = (1 + \alpha)\mathbf{x}^0 - \alpha \mathbf{x}^h \qquad \alpha > 0 \tag{4-24}$$

expansion, where $\mathbf{x}^r$ is expanded in the direction along which a further improvement of the function value may be expected

$$\mathbf{x}^e = \gamma \mathbf{x}^r + (1 - \gamma)\mathbf{x}^0 \qquad \gamma > 1 \tag{4-25}$$

and *contraction*, by which we contract the simplex

$$\mathbf{x}^c = \beta \mathbf{x}^h + (1 - \beta)\mathbf{x}^0 \quad 0 < \beta < 1 \tag{4-26}$$

Nelder and Mead found that useful values for the coefficients α, β, and γ were: $\alpha = 1$, $\gamma = 2$, $\beta = 0.5$.

Given an initial estimate of the minimum $\mathbf{x}^1 = (x_1, x_2, \ldots, x_n)$, an initial simplex is formed using two auxiliary values

$$p_n = \frac{\sqrt{n+1} - 1 + n}{n\sqrt{2}} S \qquad \text{and} \qquad q_n = \frac{\sqrt{n+1} - 1}{n\sqrt{2}} S \tag{4-27}$$

where S is a scaling factor. The $n + 1$ vertices of a regular simplex with edge of length S are given by

$$
\begin{aligned}
\mathbf{x}^1 &= (x_1, x_2, x_3, \ldots, x_n)^T \\
\mathbf{x}^2 &= (p_n + x_1, q_n + x_2, q_n + x_3, \ldots, q_n + x_n)^T \\
\mathbf{x}^3 &= (q_n + x_1, p_n + x_2, q_n + x_3, \ldots, q_n + x_n)^T \\
&\vdots \\
\mathbf{x}^{n+1} &= (q_n + x_1, q_n + x_2, q_n + x_3, \ldots, p_n + x_n)^T
\end{aligned}
\tag{4-28}
$$

The construction of the initial simplex assures that its vertices span the full n-dimensional space.

Iteration Steps

The simplex procedure is shown in Fig. 4-5. Additional explanations follow.

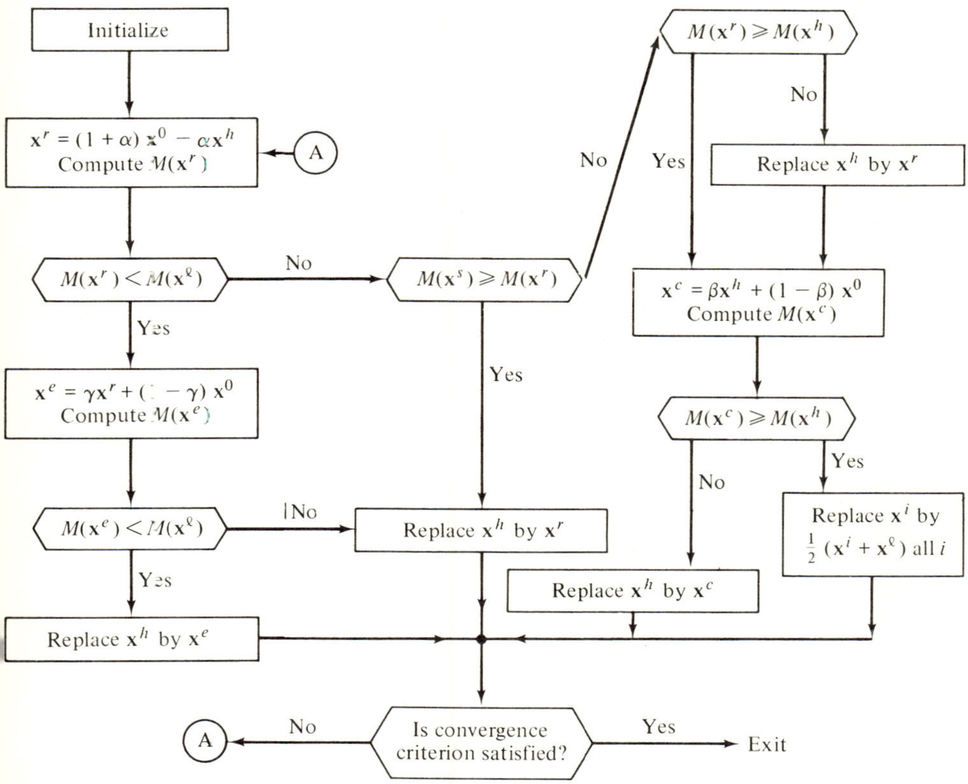

Figure 4-5 Flow diagram of the simplex method

1 The vertices $\mathbf{x}^h$, $\mathbf{x}^s$, $\mathbf{x}^l$ and the centroid $\mathbf{x}^0$ of the current simplex are determined, and a convergence test is performed as outlined in the remark following step 6.
2 $\mathbf{x}^h$ is reflected and the value of $M(\mathbf{x}^r)$ is computed.
3 If $M(\mathbf{x}^s) \geq M(\mathbf{x}^r) \geq M(\mathbf{x}^l)$ then $\mathbf{x}^h$ is replaced by $\mathbf{x}^r$, and the process is restarted from step 1.
4 If $M(\mathbf{x}^r) < M(\mathbf{x}^l)$, we expand the new simplex further in the direction $\mathbf{x}^r - \mathbf{x}^0$. The expansion is successful if $M(\mathbf{x}^e) < M(\mathbf{x}^l)$, and in this case $\mathbf{x}^h$ is replaced by $\mathbf{x}^e$. In the case of failure $\mathbf{x}^h$ is replaced by $\mathbf{x}^r$, and in either case we restart from step 1.
5 If the reflection (step 2) produces $\mathbf{x}^r$ such that $M(\mathbf{x}^h) > M(\mathbf{x}^r) > M(\mathbf{x}^s)$, we replace $\mathbf{x}^h$ by $\mathbf{x}^r$ and make the contraction move. The contraction is also applied if $M(\mathbf{x}^r) \geq M(\mathbf{x}^h)$.
6 If $M(\mathbf{x}^h) > M(\mathbf{x}^c)$, then $\mathbf{x}^h$ is replaced by $\mathbf{x}^c$ and the procedure is restarted from step 1. If $M(\mathbf{x}^h) \leq M(\mathbf{x}^c)$, the current simplex is shrunk about the point $\mathbf{x}^l$ as follows:

$$\mathbf{x}^i := \tfrac{1}{2}(\mathbf{x}^i + \mathbf{x}^l) \qquad i = 1, \ldots, n+1 \tag{4-29}$$

and we restart from step 1.

Remark To test for convergence, the standard deviation of the function at the $(n+1)$ vertices of the current simplex is compared with some preselected tolerance $\tau > 0$,

$$\left\{\sum_{i=1}^{n+1} \frac{[M(\mathbf{x}^i) - M(\mathbf{x}^0)]^2}{n}\right\}^{1/2} \leq \tau \tag{4-30}$$

The process is terminated if (4-30) is satisfied.

Modifications

Paviani and Himmelblau [1969] have suggested an extended version of the simplex method for solving the general nonlinear programming problem

$$\min_{\mathbf{x}} \{M(\mathbf{x}) \mid g_j(\mathbf{x}) \geq 0, \qquad j = 1, \ldots, m_1,$$
$$g_j(\mathbf{x}) = 0, \qquad j = m_1 + 1, \ldots, m\} \tag{4-31}$$

Formulation (4-31) is replaced by

$$\min_{\mathbf{x}} M(\mathbf{x}) \tag{4-32}$$

subject to

$$\phi - \left[\sum_{j=1}^{m_1} \{\min[g_j(\mathbf{x}), 0]\}^2 + \sum_{j=m_1+1}^{m} g_j^2(\mathbf{x})\right]^{1/2} \geq 0 \tag{4-33}$$

with the tolerance ϕ being a positive decreasing function of $\mathbf{x}$. This function serves as a tolerance criterion for constraint violation and also as a termination criterion. Paviani and Himmelblau have defined ϕ as:

$$\phi^k = \min \left\{ \phi^{k-1}, (r+1)^{-1} \sum_{i=1}^{r+1} \| \mathbf{x}^i - \mathbf{x}^0 \| \right\} \tag{4-34}$$

where k is the number of completed stages of the simplex method (reflection, expansion, or contraction), ϕ^0 is a suitable initial value of ϕ, and $r = n - (m - m_1)$ is a number of degrees of freedom in the presence of the equality constraints.

In general, the modified algorithm proceeds in the same manner as the simplex method does. However, after any change in $\mathbf{x}^i$ (reflection, expansion, or contraction), inequality (4-33) is checked. If this inequality is not satisfied, then we minimize the sum of squares of expression (4-33) to find a satisfactory new vertex before going to the next stage. In effect, the simplex method is used to improve $M(\mathbf{x})$ and at the same time we maintain near feasibility of the simplex vertices by satisfying (4-33).

Discussion

The simplex method has a wide application since it does not make any assumptions about the objective function except that the function is continuous. It may be useful when the locations of minima are needed with a limited accuracy and the final rapid convergence is not essential. Some disadvantages of the method are:

1. There is a possibility that the simplex may collapse into a subspace and will not find the solution in the desirable space of all variables.
2. It can also shrink drastically in a steep valley and stop the procedure prematurely.
3. Computational efficiency depends on the choice of the coefficients α, β, and γ, and on the size of the initial simplex. Criteria for an optimal choice of these values are not known. However, $\alpha = 1$, $\beta = 0.5$ and $\gamma = 2$ can be recommended. Examples of numerical tests are given in Sec. 4-9.
4. No direct use is made of the information acquired in previous steps.

4-6 RANDOM METHODS

Consider a problem of finding the minimum of $M(\mathbf{x})$ in the n-dimensional hypercube defined by

$$b_i \leq x_i \leq u_i \qquad i = 1, 2, \ldots, n \tag{4-35}$$

where b_i and u_i are given lower and upper bounds of the variables x_i. The simplest concept of random searching (random jumping) inside the hypercube is to generate sets of n random numbers $\mathbf{s} = (s_1, s_2, \ldots, s_n)^T$ that are uniformly distributed between 0 and 1, then to evaluate $M(\mathbf{x})$ at points $\{b_1 + (u_1 - b_1)s_1, \ldots, b_n + (u_n - b_n)s_n\}$. These points are inside the hypercube defined by (4-35). If a sufficient number of tests is performed, we take the least value of $M(\mathbf{x})$ as a desired minimum. Unfortunately, this simple method is not practical for problems with many variables.

Let us assume that the minimum of a continuous function is equally likely to lie anywhere in the n-dimensional hypercube, and that we wish to reduce the uncertainty of each variable to $(u_i - b_i)/R_i$, where R_i is a number of subintervals of variable x_i. Spang [1962] shows that in an approximate number of function evaluations, which are necessary to have 0.9 probability, that the minimum found and the true minimum lie in the same small hypercube (with sides equal $(u_i - b_i)/R_i$) is

$$N(0.9) \cong 2.3 \prod_{i=1}^{n} R_i \tag{4-36}$$

A simple grid test where $M(\mathbf{x})$ is evaluated at the nodes of a grid defined by R_i equal subintervals of the variable x_i requires

$$N_g = \prod_{i=1}^{n} (R_i + 1) \tag{4-37}$$

We see that even the grid test may require a smaller number of function evaluations. However, in both methods the number of function tests goes up exponentially with the number of dimensions. This feature makes these methods very inefficient for multidimensional problems.

A somewhat more refined concept is based on the assumption that the minimum is located close to a set of points with the lowest objective function values. After a certain number of test points we limit the search area to some neighborhood of the most successful points. We economize the number of function tests, but on the other hand we introduce a certain risk of missing the global solution.

Iteration Steps

Random Jumping. A temporary point $\hat{\mathbf{x}}^{k+1}$ is given by

$$\hat{x}_i^{k+1} = (u_i - b_i)s_i \qquad i = 1, 2, \ldots, n \tag{4-38}$$

where s_i are randomly generated in the (0, 1) interval. The new point $\mathbf{x}^{k+1}$ is determined by

$$\min\,\{M(\mathbf{x}^k), M(\hat{\mathbf{x}}^{k+1})\} \tag{4-39}$$

Random Stepping. Determine a temporary point $\hat{\mathbf{x}}^{k+1}$ by adding n random numbers to the components of $\mathbf{x}^k$. Different scales may be applied to the components of $\mathbf{x}$ and random numbers. The new point is determined as in the random jumping method.

Random Direction Stepping. A set of the n components of the direction vector $\mathbf{s}^k = (s_1^k, s_2^k, \ldots, s_n^k)^T$ is produced by the random number generator. The new point $\mathbf{x}^{k+1}$ is determined by

$$\mathbf{x}^{k+1} = \mathbf{x}^k + \lambda^k \mathbf{s}^k \tag{4-40}$$

where
$$M(\mathbf{x}^k + \lambda^k \mathbf{s}^k) = \min_{\lambda} M(\mathbf{x}^k + \lambda \mathbf{s}^k) \tag{4-41}$$

The latter problem is a one-dimensional optimization problem, for which methods are discussed in Chap. 3.

Notice that if the interval of the pseudorandom numbers is $[-1, 1]$ and we accept only such $\mathbf{s}^k$ that $\|\mathbf{s}^k\| \leq 1$, then the probability of locating $\mathbf{x}^k + \mathbf{s}^k$ in any element of solid angle centered at the origin is the same.

History Vector Method [*Wheeling, 1960*]. Determine a temporary new point $\hat{\mathbf{x}}^{k+1}$ as follows:

$$\hat{\mathbf{x}}^{k+1} = \mathbf{x}^k + \beta_1 \mathbf{h}^k + \beta_2 \mathbf{s}^k \tag{4-42}$$

where $\mathbf{s}^k$ is a random direction (see preceding method), $\mathbf{h}^k$ is a normalized history vector obtained from

$$\mathbf{h}^k = \gamma_1 \mathbf{h}^{k-1} + \gamma_2(\mathbf{x}^k - \mathbf{x}^{k-1}), \qquad \mathbf{h}^0 = 0 \tag{4-43}$$

and scalars β_1, β_2, γ_1, γ_2 can balance the influence and sensitivity of the history vector. Wheeling reports that he tried $\beta_1 = \beta_2 = 0.5$ and $\gamma_1 = \gamma_2$ such that $\|\mathbf{h}^k\| = 1$ in every step, but other strategies are also possible. The new point $\mathbf{x}^{k+1}$ is determined by (4-39).

Discussion

The random search methods are independent of the local behavior of the objective function (such as continuity and smoothness properties) and unbiased in the selection of search points. For this reason, they may be useful for the location of a global minimum when the objective function possesses multiple minima. In this case, such a procedure as the random jumping may be used in the early stage of optimization to detect the area where the global solution is likely to be found. These methods can also be applied if other deterministic methods fail to improve the objective function, due to local difficulties such as sharply varying functions or shallow regions. It is suggested that these methods be used as auxiliary techniques in combination algorithms. Another application of random methods is to slightly perturb solutions obtained by another method and seek a neighborhood in which the function value may be smaller.

4-7 MINIMIZATION BY A SEQUENCE OF QUADRATIC PROGRAMS

A method based on a sequence of quadratic programming problems has been proposed by Winfield [1967]. In general, this technique may be more economical in terms of function evaluations than techniques based on a sequence of one-dimensional searches. The disadvantage of the technique is in the computational effort required to fit a quadratic function $q(\mathbf{x})$. If $M(\mathbf{x})$ is strictly quadratic

$$M(\mathbf{x}) = q(\mathbf{x}) = a + \mathbf{b}^T\mathbf{x} + \tfrac{1}{2}\mathbf{x}^T A\mathbf{x} \tag{4-44}$$

the computation terminates at an exact minimum in one step; that is, the method is quadratically convergent. To determine a local quadratic approximation to $M(\mathbf{x})$, one needs $N = \frac{1}{2}(n+1)(n+2)$ distinct items of information [e.g., values of $M(\mathbf{x}^k)$, $k = 1, \ldots, N$], because the matrix A contains $\frac{1}{2}n(n+1)$ distinct elements, the vector b contains n, and the scalar a needs to be determined. The local quadratic function $q(\mathbf{x})$ is determined from this information (by surface fitting techniques) and used as the objective function in a quadratic programming problem solved in the neighborhood (hypercube) of the available approximation to the solution.

Iteration Steps

1 Select $N - 1$ points in the neighborhood of $\mathbf{x}^0$, the given starting point. Assuming λ is a given step size, pick points $\mathbf{x}^1, \ldots, \mathbf{x}^{N-1}$ according to the following scheme: $(x_1^0 \pm \lambda_2^0, x_2^0)$, $(x_1^0, x_2^0 \pm \lambda)$, $(x_1^0 + \lambda, x_2^0 + \lambda)$. Evaluate $M(\mathbf{x}^k)$, $k = 0, \ldots, N - 1$. This procedure has been explored only for two-dimensional problems, where $N = 6$, but it can be generalized.

2 Determine the basepoint, the point at which $\min_k M(\mathbf{x}^k)$, $k = 0, \ldots, N - 1$ occurs. Arrange data in a data table, the rows of which contain $\mathbf{x}^k$ and $M(\mathbf{x}^k)$, starting with the information associated with basepoint and then add other points in order of increasing distance from the basepoint. The number of rows in the table is arbitrarily limited to $\bar{N} > N$ (e.g., 10 for $n = 2$). In the first cycle $\bar{N} = N$, but then it can be increased. Only the first N rows participate in forming the local quadratic model. The remaining $\bar{N} - N$ rows constitute a reserve that may be called upon in case a basepoint is found near one of these $\bar{N} - N$ points. Shift the origin of the coordinate system to the basepoint, and set $a = M(\mathbf{x}^0)$ in (4-44).

3 Solve the hypersurface fitting problem by solving $N - 1$ equations in the variables $\mathbf{b}$ and A

$$\mathbf{b}^T\mathbf{x}^k + \tfrac{1}{2}[\mathbf{x}^k]^T A\mathbf{x}^k = M(\mathbf{x}^k) - a \qquad k = 1, \ldots, N - 1 \tag{4-45}$$

4 Let r be the distance between the base point $\mathbf{x}^0$ (which is now the origin) and the farthest point $\mathbf{x}^{N-1}$:

$$r = [\mathbf{x}^{N-1}]^T \mathbf{x}^{N-1} \tag{4-46}$$

Let R_{max} be a constant maximum region size (lower/upper bound on shifted variables). Find $\mathbf{x}_{min}$, the solution of the quadratic programming problem

$$\min q(\mathbf{x}) \tag{4-47}$$

subject to $\qquad -R \leq x_j \leq R \qquad j = 1, 2, \ldots, n$

where

$$R = \min\left(R_{max}, \frac{0.999\alpha r}{\sqrt{n}}\right) \tag{4-48}$$

and α is a constraint region reduction factor. Set $\alpha = 1$ for the first quadratic programming problem.

5 Compute $M(\mathbf{x}_{min})$ and compare it with $M(\mathbf{x}^0)$. If $M(\mathbf{x}^0) > M(\mathbf{x}_{min})$, set $\mathbf{x}^0 = \mathbf{x}_{min}$; reorder other points in the data table according to their distances from the new basepoint and continue from step 3 with $\alpha = 1$. If $M(\mathbf{x}^0) < M(\mathbf{x}_{min})$, set $\alpha := 0.95\alpha$ and enter $\mathbf{x}_{min}$, $M(\mathbf{x}_{min})$ in the appropriate row of the data table according to the scheme described in step 2. Note that (4-48) ensures that $\mathbf{x}_{min}$ is closer to the basepoint than the farthest of the $N - 1$ remaining points. Thus, the new point $\mathbf{x}_{min}$ will always participate in the forming of the next quadratic approximation. The computation then proceeds from step 3.

Termination Criteria and Convergence

Convergence is assumed and the process is terminated when the improvement in $M(\mathbf{x})$ is small, $|M(\mathbf{x}^0) - M(\mathbf{x}_{min})| < \tau_1$, and the hypercube in which the quadratic programming is solved is small, $R < \tau_2$, where τ_1 and τ_2 are given tolerances.

Modifications and Acceleration Techniques

When the iteration cycle proceeds from step 5 to step 3 only, one new row may be inserted into the data table, and only one of equations (4-45) may be changed. Appropriate shortcuts could be introduced into step 3 so that the hypersurface fitting problem is not solved every time from the beginning.

Discussion

The method reduces the nonlinear minimization problem to a sequence of simple quadratic hypersurface fitting and quadratic programming problems. Past information relevant to the progress of minimization is retained in the data table and influences the current iteration (old information is gradually discarded). For problems with a high number of variables the quadratic hypersurface fitting subproblem becomes time consuming (computing time is proportional to $N^3 \cong n^6/8$). The method may be useful for solving problems with few variables where $M(\mathbf{x})$ is nearly quadratic, but it is rather expensive to evaluate.

Another method, called the method of hypercubes, which minimizes a general objective function $M(\mathbf{x})$ subject to linear constraints by solving a sequence of quadratic programming subproblems, has been proposed by Fletcher [1970].

4-8 COMPUTING SYSTEMS

Computing systems applying search techniques are listed in Table. 4-1. Programs from VIM are listed in the 1971 *Program Library Catalog* of VIM, Users Organization for Control Data Corporation Computers.

Table 4-1 DIRECT SEARCH TECHNIQUES

Title and Reference	*Type, Machine and Approximate Storage*	*Experience*	*Remarks*
ALGORITHM 178 DIRECT SEARCH [Hooke and Jeeves, 1961] [Kaupe, 1966] [Bell and Pike, 1966]	ALGOL 60 program	Has been modified for BAC compiler and run successfully on Burroughs 205 and 220	See corrections [Bell and Pike, 1966] Program based on flow chart presented by original author of method [Hooke and Jeeves, 1961] Can be adapted to any machine with ALGOL 60 compiler
H2 SAND MIN VIM [Hooke and Jeeves, 1961] [Kaupe, 1966] [Beisinger and Bell, 1963]	FORTRAN IV program CDC 6600 $291 + 4n$ words plus storage for function evaluation	Success on Wood problem	Based on ALGORITHM 178 [Kaupe, 1966] Coded for $n = 50$, can be extended and adapted to any machine with FORTRAN compiler

n = number of variables.

Table 4-1 (*Cont.*)

Title and Reference	*Type, Machine and Approximate Storage*	*Experience*	*Remarks*
APMINS SDA 3214 SHARE [Lindamood, 1964]	FAP program IBM 7090–7094 $221 + 3n$ words plus storage for function evaluation	Tests reported [Lindamood, 1964]	Based on original algorithm [Hooke and Jeeves, 1961] Coded for $n = 144$
UYSEAR SDA 3376 SHARE [King, 1964]	MAP program IBM 7094, $191 + n$ words plus storage for function evaluation	Success on Wood problem	Based on original algorithm [Hooke and Jeeves, 1961] Coded for $n = 50$
CLIM—SHARE SDA 3466 [Rosenbrock, 1960] Method of Rotating Coordinates	FORTRAN IV program 7094, Operating System IBSYS, $1441 + n(n + 8) + 3m$ words plus user supplied subroutines	Success on Wood Problem. Several problems solved by the subroutine authors.	*Authors:* E. Davison K. Cowan Problem size limited by core size
SCOOP-Linear Surface Minimization [Shapiro, 1965]	FORTRAN II program IBM 7094 727 words plus storage required by subroutines	Tested by the author	*Author:* Melvin Sobol *Method:* Tests are made to determine behavior of the function as each variable is varied separately. Then all variables are varied simultaneously in directions such that each separately would decrease the function. Code limited to 24 variables
E4 BKY SIFMIN VIM Simplex Method	FORTRAN program, CDC 6600 CHIPPEWA Operating System		*Author:* E. Beals

n = number of variables
m = number of constraints

Combination Program AESOP

Program AESOP, originated by Hague and Glatt [1968*a*, *b*], combines nine methods for unconstrained minimization. The program user determines the sequence in which these are applied and the number of iterations taken in each. The methods are as follows:

1 *Sectioning.* A sequence of searches for local minima in a direction parallel to the coordinate axes. The golden section technique (Sec. 3-4) is used for these searches.

2 *Adaptive creeping.* The exploratory move of the pattern search method (Sec. 4-2).

3 *Magnification.* A search among points obtained by proportional changes in all variables.

4 *Steepest descent.* Different variable scaling options are available (Sec. 5-2). The golden section technique of Sec. 3-4 is used for minimizations along the search lines.

5 *Quadratic search.* This method is similar to the second-order gradient method (Sec. 5-3). Here a quadratic is fitted to a number of points generated as described in Sec. 4-7. Then the golden section method is used to search for a minimum along the line connecting the current point to a stationary point of the fitted quadratic.

6 *Variable metric.* Described in Sec. 5-6, this method is used to generate descent directions with the golden section method (Sec. 3-4) for minimization along the descent directions.

7 *Pattern search.* A search along the ("experience") direction of the previous search or search sequence; the golden section method is also used in this case.

8 *Random jumping.* As described in Sec. 4-6.

9 *Random direction stepping.* As described in Sec. 4-6.

The program is a FORTRAN IV program for the IBM 7094, CDC 6600, UNIVAC 1108, and IBM 360 requiring approximately 20,000 core locations. It can handle problems of up to 100 independent variables and up to 100 upper/lower bound constraints on the variables (the latter are required). Experience in solving a number of problems of practical size has been reported.

Combination Program GROPE

Program GROPE [Flood and Leon, 1964] combines four optimization methods which are selected automatically by the executive part of the program. The selection strategy is based on the progress and performance of each optimizing subroutine. The objective functions can be unconstrained and constrained.

The methods are as follows:

1 Modified version of the *pattern search* method of Hooke and Jeeves [1961].
2 *Simple random strategy.* Similar to the idea of random jumping.
3 *Random shrinkage.* The same as step 2, but the searching space is gradually reduced as the search continues.
4 *Random direction stepping.* The step size is also chosen randomly.

Other methods have also been considered by the authors of the code as potential components of GROPE (e.g., variable metric, steepest descent, and PARTAN). GROPE is available in FORTRAN IV for IBM 7090/7094 computers with IBJOB processors. To utilize the core storage efficiently, it has a flexible overlay structure, and several overlaying modes are possible. A simplified flowchart of GROPE and complete FORTRAN IV listing are given by Flood and Leon [1964].

4-9 NUMERICAL RESULTS

The simplex and Rosenbrock methods have been selected to illustrate a typical performance of the direct search methods. The test problem of Wood (Sec. 1-10) has been solved with different sets of parameters α, β, and γ. The work required to yield the solution is estimated by the number of function evaluations in the first column of Tables 4-2 through 4-5. From analysis

Table 4-2 SIMPLEX METHOD; [Wood] $\mathbf{x}^0 = (-3, -1, -3, -1)^T$; $\alpha = 1$, $\beta = 0.5$, $\gamma = 2$, $S = 0.5$.

Number of Function Evaluations	$M(\mathbf{x})$	x_1	x_2	x_3	x_4
25	19.707	0.6165	0.4555	0.4938	0.1868
50	1.301	1.0340	0.9865	0.9523	0.8463
75	5.501×10^{-2}	0.9990	0.9773	1.0039	0.9991
100	3.151×10^{-3}	0.9800	0.9594	1.0168	1.0383
150	1.082×10^{-3}	0.9826	0.9651	1.0163	1.0329
200	7.675×10^{-5}	0.9999	0.9996	0.9995	0.9985
250	1.232×10^{-6}	1.000	1.0000	0.9998	0.9997

of these tables we can make a general observation that the methods progress well at the beginning of the calculations and slow down in the vicinity of the solution. It may be desirable, therefore, to switch to the second-order or conjugate gradient methods when a good approximation to the solution is obtained.

The simplex method produced the best results for $\alpha = 1$, $\beta = 0.5$, $\gamma = 2$, and $S = 0.5$ (Table 4-2). Table 4-3 illustrates a quite inferior performance of the method with $\alpha = 1.5$, $\beta = 0.7$, and $\gamma = 2$. It is also essential to choose a sufficiently large initial simplex. Table 4-4 shows that the initial regular simplex with edge length $S = 0.2$ was too small at the beginning of the minimization process. Table 4-5 illustrates the performance of the method of rotating coordinates (Rosenbrock) with $\alpha = 3$, $\beta = 0.5$.

Table 4-3 SIMPLEX METHOD; [Wood] $\mathbf{x}^0 = (-3, -1, -3, -1)^T$; $\alpha = 1.5$, $\beta = 0.7$, $\gamma = 2$, $S = 0.5$.

Number of Function Evaluations	$M(\mathbf{x})$	x_1	x_2	x_3	x_4
25	65.114	−0.4847	0.3017	−0.9633	0.2752
50	4.566	1.0968	1.0513	1.1148	1.3339
75	3.516	1.0746	0.9819	1.1202	1.2298
100	3.952×10^{-1}	0.8887	0.8395	1.0683	1.1085
200	1.821×10^{-2}	0.9349	0.8726	1.0665	1.1401
250	1.607×10^{-2}	0.9309	0.8658	1.0639	1.1323
400	1.748×10^{-3}	1.0014	0.9991	0.9974	0.9952

Table 4-4 SIMPLEX METHOD; [Wood] $\mathbf{x}^0 = (-3, -1, -3, -1)^T$; $\alpha = 1$, $\beta = 0.5$, $\gamma = 2$, $S = 0.2$.

Number of Function Evaluations	$M(\mathbf{x})$	x_1	x_2	x_3	x_4
25	34.485	0.6751	0.1541	0.5441	0.2652
50	12.035	0.6378	0.4459	0.7696	0.5049
75	2.838×10^{-1}	1.0730	1.1134	1.0039	0.9956
100	4.803×10^{-3}	1.0176	1.0382	0.9812	0.9573
200	7.521×10^{-4}	1.0138	1.0272	0.9869	0.9746
260	7.349×10^{-7}	1.0002	1.0004	0.9999	0.9997

More extensive numerical studies [Fletcher, 1965; Box, 1966] of optimization methods, which do not use function derivatives, show that on the basis of function evaluations the most efficient is the method of Powell (Sec. 5-4). However, according to Fletcher [1965], as the number of variables increases, the comparison with the modified rotating coordinates method (DSC) becomes less favorable. Powell's method converges rapidly in the vicinity of a minimum and for nearly quadratic functions. The methods of

Table 4-5 METHOD OF ROTATING COORDINATES [Rosenbrock, 1960]; [Wood], $\mathbf{x}^0 = (-3, -1, -3, -1)^T$; $\alpha = 3$, $\beta = 0.5$.

Number of Function Evaluations	$M(\mathbf{x})$	x_1	x_2	x_3	x_4
19	112.700	1.0000	0.3000	1.0000	0.3000
46	16.944	0.0749	0.1748	0.8920	0.7085
112	2.314	0.5583	0.3144	1.1041	1.2789
202	0.264	0.7725	0.5989	1.1648	1.3511
317	4.009×10^{-5}	1.0002	1.0008	0.9988	0.9979
426	2.717×10^{-6}	1.0006	1.0013	0.9993	0.9987
496	1.6621×10^{-6}	1.0005	1.0010	0.9995	0.9991

rotating coordinates attempt to align the direction of search along the axes of the valleys, which probably explains their good performance even if they operate far from local solutions.

Kowalik and Osborne [1968] experiments have shown that the efficiency of the simplex method is comparable with that of Powell's conjugate direction method. However, its behavior depends strongly on the choice of the coefficients α, β, and γ, as illustrated by our numerical results. Box [1966] also remarked that the superior performance of the simplex method does not continue as the dimensionality of the problem increases.

The pattern search method shows good efficiency, comparable to that of the simplex or rotating coordinate methods, but only for problems with few variables.

References

ANDREW, A. M.: "The Calculation of Orthogonal Vectors," *Computer Journal*, Vol. 12, 411, 1969.

BEISINGER, Z. E. and S. BELL: Program H2 SAND MIN, *VIM—CDC Users Organization*, 1963.

BELL, M. and M. PIKE: "Remarks on Algorithm 178 [E4] Direct Search," *Communications of the ACM*, Vol. 9, 684, 1966.

BOX, G. E. P.: "Evolutionary Operation: A Method for Increasing Industrial Productivity," *Applied Statistics*, Vol. 2, 81, 1957.

BOX, M. J.: "A Comparison of Several Current Optimization Methods, and the Use of Transformations in Constrained Problems," *Computer Journal*. Vol. 9, 67, 1966.

———, D. DAVIES and W. H. SWANN: *Nonlinear Optimization Techniques*, ICI Ltd., Monograph No. 5, Oliver and Boyd, Ltd., Edinburgh, 1969.

FLETCHER, R.: "Function Minimization Without Evaluating Derivatives—A Review," *Computer Journal*, Vol. 8, 33, 1965.

———: *An Efficient Globally Convergent Algorithm for Unconstrained and Linearly Constrained Optimization Problems*, T. P. 431, Atomic Energy Res. Establishment, Harwell, England, 1970.

Flood, M. M. and A. Leon: *A Universal Adaptive Code for Optimization*, Working Paper No. 19, Space Sciences Laboratory, University of California, Berkeley, 1964.

Golub, G. H. and M. A. Saunders: "Linear Least Squares and Quadratic Programming," in J. Abadie (ed.), *Integer and Nonlinear Programming*, North-Holland Publishing Company, Amsterdam, Ch. 10, 229, 1970.

Hague, D. S. and C. R. Glatt: *An Introduction to Multivariable Search Techniques for Parameter Optimization (and Program AESOP)*, NASA CR-73200, 1968*a*.

———: *A Guide to the Automated Engineering and Scientific Optimization Program—AESOP*, NASA CR-73201, 1968*b*.

Hooke, R. and T. A. Jeeves: "Direct Search Solution of Numerical and Statistical Problems," *Journal of the ACM*, Vol. 8, 212, 1961.

Kaupe, A. F., Jr.: "Algorithm 178 [E4] Direct Search," *Communications of the ACM*, Vol. 7, 498, 1964.

King, J. C.: *SHARE Program UYSEAR—SDA 3376*, SHARE—IBM Users Organization, 1964.

Kowalik, J. and M. R. Osborne: *Methods for Unconstrained Optimization Problems*, American Elsevier, New York, 1968.

Lindamood, G. E.: *SHARE Program APMINS Direct Search Minimization—SDA 3214*, SHARE—IBM Users Organization, 1964.

Nelder, J. A. and R. Mead: "A Simplex Method for Function Minimization," *Computer Journal*, Vol. 7, 308, 1965.

Osborne, M. R.: "A Note on Powell's Method for Calculating Orthogonal Vectors," *The Australian Computer Journal*, Vol. 1, 216, 1969.

Paviani, D. A. and D. M. Himmelblau: "Constrained Nonlinear Optimization by Heuristic Programming," *Journal of ORSA*, Vol. 17, 872, 1969.

Powell, M. J. D.: "On the Calculation of Orthogonal Vectors," *Computer Journal*, Vol. 11, 302, 1968.

Rice, J.: "Experiments on Gram-Schmidt Orthogonalization," *Math. of Computation*, Vol. 20, 325, 1966.

Rosenbrock, H. H.: "An Automatic Method for Finding the Greatest or Least Value of a Function," *Computer Journal*, Vol. 3, 175, 1960.

Schechter, S.: "Relaxation Methods for Convex Problems," *SIAM Journal Numer. Anal.*, Vol. 5, 601, 1968.

Shapiro, M. S. and M. Goldstein: *A Collection of Mathematical Computer Routines*, AEC Computing and Applied Mathematics Center, Courant Institute of Math. Sciences, New York University, New York, 1965.

Spang, H. A.: "A Review of Minimization Techniques for Nonlinear Functions," *SIAM Review*, Vol. 4, 343, 1962.

Spendley, W., G. R. Hext and F. R. Himsworth: "Sequential Application of Simplex Designs in Optimization and Evolutionary Operation," *Technometrics*, Vol. 4, 441, 1962.

Sugie, N.: "An Extension of Fibonaccian Searching to Multidimensional Cases," *IEEE Transactions on Automatic Control*, AC-9, 105, 1964.

vim, Users Organization for Control Data Corporation Computers, *Program Library Catalog*, 1971.

Wheeling, R. F.: "Optimizers: Their Structure," *Communications of the ACM*, Vol. 3, 632, 1960.

Winfield, D. H.: *Function Minimization Without Derivatives by a Sequence of Quadratic Programming Problems*, Technical Report No. 537, Division of Engineering and Applied Physics, Harvard University, 1967.

5 DESCENT TECHNIQUES

This chapter deals with descent techniques for the solution of unconstrained minimization problems. The unconstrained minimization problem is min $M(\mathbf{x})$. Here a value of the variable vector $\mathbf{x} = (x_1, \ldots, x_n)^T$ is sought that minimizes the objective function $M(\mathbf{x})$. The problem min $M(\mathbf{x})$ is a special case of the general nonlinear programming or optimization problem stated in Sec. 1-3. Its special feature is that no other conditions or constraints need to be satisfied by the solution.

The descent techniques dealt with in this chapter differ from search techniques (Chap. 4) in a number of respects. One of these is the amount of information carried over from iteration to iteration. Some of the search techniques frequently do not use information from previous iterations, while most of the descent techniques carry over and use this information. Furthermore, new directions of descent are generated in each iteration of most descent techniques. Most search techniques, on the other hand, just proceed to probe the objective function along a set of predetermined directions in some systematic fashion. The descent iterations usually involve more computational work than the search steps. The latter require only evaluations of the objective function $M(\mathbf{x})$. The descent iteration may involve in addition the evaluation of first- and possibly higher-order derivatives of the objective function. On the other hand, as a result of more computation, each step of the descent process frequently yields considerable improvement of the objective function. Thus, overall computational efficiency of these techniques is usually high in comparison to that of search techniques, and convergence to a solution in relatively fewer steps can be expected.

5-1 THEORETICAL BACKGROUND OF DESCENT TECHNIQUES

The descent techniques involve iterations which consist essentially of three parts. First, a direction of descent $\mathbf{s}^k$ is found, then a descent steplength λ^k is determined and finally the descent step $\mathbf{x}^{k+1} = \mathbf{x}^k + \lambda^k \mathbf{s}^k$ is performed.

The Descent Direction

The descent direction is an n-dimensional vector $\mathbf{s} = (s_1, \ldots, s_n)^T$. We assume here that $\mathbf{s}$ has been normalized so that $\|\mathbf{s}\| = 1$; where convenient, however, we shall use descent directions that have not been normalized. At the kth iteration the direction vector $\mathbf{s}^k$ originates at the current point $\mathbf{x}^k$. It points in a descent or "downhill" direction. That is, the value of the objective function decreases from $\mathbf{x}^k$ to a point at some distance in that direction. A unit vector $\mathbf{s}^k$ is said to be a *descent direction*, with respect to the objective function $M(\mathbf{x})$ at $\mathbf{x}^k$, if there is a $\bar{\lambda} > 0$, such that for all λ satisfying $\bar{\lambda} \geq \lambda > 0$ we have

$$M(\mathbf{x}^{k+1}) = M(\mathbf{x}^k + \lambda \mathbf{s}^k) < M(\mathbf{x}^k) \tag{5-1}$$

If $M(\mathbf{x})$ is differentiable, $\mathbf{s}^k$ is a descent direction if

$$\begin{aligned} \lim_{\lambda \to 0} \frac{M(\mathbf{x}^k + \lambda \mathbf{s}^k) - M(\mathbf{x}^k)}{\lambda} &= \left. \frac{dM(\mathbf{x}^k + \lambda \mathbf{s}^k)}{d\lambda} \right|_{\lambda = 0} \\ &= (\mathbf{s}^k)^T \nabla M(\mathbf{x}^k) < 0 \end{aligned} \tag{5-2}$$

where $\nabla M(\mathbf{x}^k)$ denotes the gradient of the objective function $M(\mathbf{x})$ evaluated at the point $\mathbf{x}^k$. If $M(\mathbf{x})$ is differentiable, the product $(\mathbf{s}^k)^T \nabla M(\mathbf{x}^k)$ of the direction $\mathbf{s}^k$ and the gradient $\nabla M(\mathbf{x}^k)$ is, by definition, the directional derivative of $M(\mathbf{x})$ in the direction $\mathbf{s}^k$ evaluated at $\mathbf{x}^k$. If this directional derivative exists and is negative, then $\mathbf{s}^k$ is a descent direction. Figure 5-1 depicts the sector of descent directions at a point $\mathbf{x}^k$ in two dimensions.

The Descent Steplength

The descent steplength λ is a scalar. It is a measure of the distance along the descent direction $\mathbf{s}^k$ between two successive iteration points $\mathbf{x}^k$ and $\mathbf{x}^{k+1}$. In other words, at the kth iteration a step of length λ^k is taken from point $\mathbf{x}^k$ in the direction $\mathbf{s}^k$ to the point $\mathbf{x}^{k+1}$. The scalar λ entering the definitions (5-1) and (5-2) is a descent steplength.

The Descent Iteration

With the notation introduced so far we can summarize the typical descent iteration as follows:

1 Compute a descent direction $\mathbf{s}^k = (s_1^k, \ldots, s_n^k)^T$.

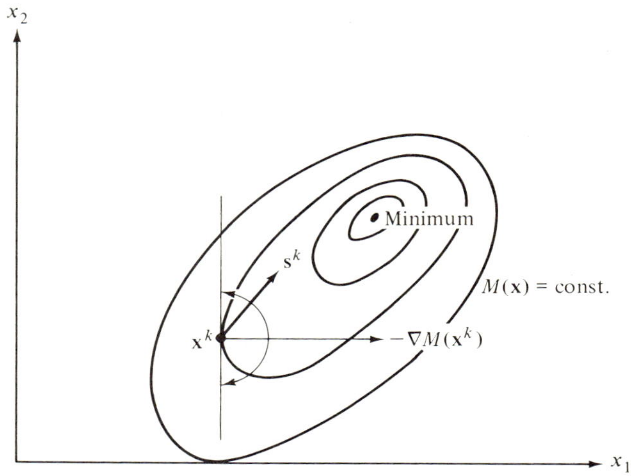

Figure 5-1 Sector of descent directions at $\mathbf{x}^k$ in two dimensions

2 Compute a descent steplength λ^k.
3 Perform a descent step to obtain a new point

$$\mathbf{x}^{k+1} = \mathbf{x}^k + \lambda^k \mathbf{s}^k \tag{5-3}$$

For reference below we define the kth descent step as

$$\Delta \mathbf{x}^k = \mathbf{x}^{k+1} - \mathbf{x}^k = \lambda^k \mathbf{s}^k \tag{5-4}$$

A sequence of k descent steps leads from a starting point $\mathbf{x}^0$ to a point $\mathbf{x}^k$ given by

$$\mathbf{x}^k = \mathbf{x}^0 + \sum_{l=0}^{k-1} \lambda^l \mathbf{s}^l = \mathbf{x}^0 + \sum_{l=0}^{k-1} \Delta \mathbf{x}^l \tag{5-5}$$

At the kth iteration we define a matrix $\Delta \bar{X}_k$ by

$$\Delta \bar{X}_k = [\Delta \mathbf{x}^0, \Delta \mathbf{x}^1, \ldots, \Delta \mathbf{x}^{k-1}] \tag{5-6}$$

That is, the columns of $\Delta \bar{X}_k$ are the k descent steps $\Delta \mathbf{x}^0, \Delta \mathbf{x}^1, \ldots, \Delta \mathbf{x}^{k-1}$ preceeding $\Delta \mathbf{x}^k$.

Types of Descent Techniques

The descent techniques differ from each other in the type of descent directions and in the strategy used to determine the lengths of the descent steps. Descent directions can be classified in a number of ways and considered accordingly: *Steepest descent methods* are based on the premise that the overall efficiency of the descent process is best if each of the iterations is optimal. Optimal descent iterations are those resulting in maximum improvements (reductions) in the value of the objective function $M(\mathbf{x})$. To achieve such optimal steps, it may be argued that steepest descent directions should be used. In practice, of course, these are only "locally steepest." There are different descent directions, depending on how distances are measured, and the formulas for measuring distances and determining descent directions are related, as we shall see. In all cases the descent direction depends on the gradient of the objective function. The descent techniques also differ in the order of partial derivatives of $M(\mathbf{x})$ involved in the computation, and in other computational details.

An alternative argument is that better overall computational efficiency may be obtained if a sequence of descent iterations is considered, rather than individual iterations. Information, or in its absence a reasonable assumption, concerning the behavior of the objective function $M(\mathbf{x})$ is often involved in the determination of these sequences of iterations. One class of such techniques uses a sequence of *conjugate directions.* A conjugate direction technique which does not involve differentiation of the objective function is described in Sec. 5-4. It is also possible to combine the arguments leading to steepest descent and conjugate direction techniques. If first-order gradient descent is involved, this leads to *conjugate gradient techniques. Variable metric techniques* involve conjugate directions and under appropriate conditions these directions approach, in a sense to be defined, second-order gradient directions. They are, therefore, frequently referred to as *quasi-second-order* or *quasi-Newton techniques.*

In this chapter, the descent techniques are grouped in two classes: steepest descent and conjugate gradient. In each group, the techniques are arranged in the order of highest derivative of the objective function involved in determining the descent direction.

Gradient Descent

The choice of the locally steepest direction as a descent direction leads to steepest descent techniques. A locally steepest direction is obtained if the descent step $\Delta\mathbf{x}^k$ defined by (5-4) minimizes

$$\Delta M = \sum_{i=1}^{n} \frac{\partial M(\mathbf{x}^k)}{\partial x_i} \Delta x_i^k$$

However, before any meaning can be attached to this statement, we have to specify a method of measuring the distance between two points in n-dimensional space. The *distance* between two points $\mathbf{x}^1$ and $\mathbf{x}^2$ in the n-dimensional space is defined as

$$d(\mathbf{x}^1, \mathbf{x}^2) = [(\mathbf{x}^1 - \mathbf{x}^2)^T A(\mathbf{x}^1 - \mathbf{x}^2)]^{1/2} \tag{5-7}$$

where A is a positive definite $n \times n$ symmetric *metric matrix.* The definition of positive definiteness assures that $d(\mathbf{x}^1, \mathbf{x}^2) > 0$ for any nonzero $\mathbf{x}^1$ and $\mathbf{x}^2$. Notice that, in general, it is not necessary to use the same unit of distance along the different coordinate axes. The choice of I, the $n \times n$ identity matrix, as a metric matrix, i.e., $A = I$, leads to the first-order steepest descent technique (Sec. 5-2). Any rescaling of the variables, e.g., $\hat{\mathbf{x}} = B\mathbf{x}$, is equivalent to introducing a new metric matrix relative to the old coordinate system $\mathbf{x}$. In this example, the new metric matrix is B^TB. It is not necessary to use the same metric matrix throughout the whole iterative process of a descent technique. Changing this matrix would be tantamount to rescaling the variables as the solution process progresses. The variable metric descent techniques described in Sec. 5-6 have that feature.

In view of the definition (5-7) of distance, the locus of points at distance δ from $\mathbf{x}^k$ is given by the n-dimensional ellipsoid

$$\delta^2 = [(\mathbf{x} - \mathbf{x}^k)^T A(\mathbf{x} - \mathbf{x}^k)]$$

with center $\mathbf{x}^k$. Consider the step $\Delta\mathbf{x}^k$ from $\mathbf{x}^k$ to that point on the above ellipsoid at which the value of the objective function $M(\mathbf{x})$ is the least. That would be the point given by $\mathbf{x}^k + \Delta\mathbf{x}^k$ satisfying

$$\min_{\mathbf{x}} \{M(\mathbf{x}^k + \Delta\mathbf{x}^k) \,|\, (\Delta\mathbf{x}^k)^T A\, \Delta\mathbf{x}^k = \delta^2\}$$

Approximating $M(\mathbf{x})$ by the first-order Taylor series about $\mathbf{x}^k$, we have

$$M(\mathbf{x}^k + \Delta\mathbf{x}^k) \cong M(\mathbf{x}^k) + (\Delta\mathbf{x}^k)^T \nabla M(\mathbf{x}^k)$$

To minimize $M(\mathbf{x}^k + \Delta\mathbf{x}^k)$, we must minimize $(\Delta\mathbf{x}^k)^T\nabla M(\mathbf{x}^k)$ because $M(\mathbf{x}^k)$ is fixed. Therefore, our problem reduces to

$$\min_{\mathbf{x}} \{(\Delta\mathbf{x}^k)^T \nabla M(\mathbf{x}^k) \,|\, (\Delta\mathbf{x}^k)^T A\, \Delta\mathbf{x}^k = \delta^2\}$$

A solution can be obtained by forming the corresponding Lagrangian and seeking its saddlepoint (Sec. 1-7). The Lagrangian is

$$L(\mathbf{x}, u) = (\Delta\mathbf{x}^k)^T \nabla M(\mathbf{x}^k) - u[(\Delta\mathbf{x}^k)^T A\, \Delta\mathbf{x}^k - \delta^2]$$

and at the saddlepoint

$$\frac{\partial L}{\partial \mathbf{x}} = \nabla M(\mathbf{x}^k) - 2uA\, \Delta\mathbf{x}^k = 0$$

$$\frac{\partial L}{\partial u} = (\Delta\mathbf{x}^k)^T A\, \Delta\mathbf{x}^k - \delta^2 = 0$$

The first condition gives

$$\Delta\mathbf{x}^k = \frac{1}{2u} A^{-1}\, \nabla M(\mathbf{x}^k)$$

and we see that step $\Delta\mathbf{x}^k$ is taken in the direction $A^{-1}\nabla M(\mathbf{x}^k)$. The constant of proportionality involves the Lagrange multiplier u, and can be found by combining the first and second saddlepoint conditions.

The direction of (locally) steepest descent thus is given by

$$\mathbf{s}^k = -(A_k)^{-1} \nabla M(\mathbf{x}^k) \tag{5-8}$$

Recall that A_k is positive definite and symmetric. Therefore, if $\nabla M(\mathbf{x}^k) \neq 0$,

$$(\mathbf{s}^k)^T \nabla M(\mathbf{x}^k) = -[\nabla M(\mathbf{x}^k)]^T [A_k^{-1}]^T \nabla M(\mathbf{x}^k) < 0$$

because the inverse of a positive definite symmetric matrix is also positive definite. Consequently, the directions $\mathbf{s}^k$ as defined by (5-8) are descent directions. The metric matrix was labeled A_k in this definition to indicate that it may vary from step to step. Gradient descent techniques differ in their choice of A_k. These techniques are arranged, for discussion, in increasing order of the highest derivative of the objective functions $M(\mathbf{x})$ involved in the computation of the descent directions by means of (5-8).

Conjugate Gradient Descent

The consideration of a sequence of descent steps, rather than individual steps, may lead to the use of conjugate descent directions in the sequence. This is justified on the basis that the objective function $M(\mathbf{x})$ may frequently be quadratic with a positive definite matrix, or it may be sufficiently smooth and therefore approximated, at least locally, by a quadratic. It is assumed then that

$$M(\mathbf{x}) = a + \mathbf{b}^T\mathbf{x} + \tfrac{1}{2}\mathbf{x}^T A \mathbf{x} \tag{5-9}$$

where A is a positive definite symmetric $n \times n$ matrix, $\mathbf{b}$ is an n vector, and a is a constant. We may argue that methods that can efficiently minimize a quadratic of this form will also efficiently minimize a general function, at least if minimization starts sufficiently close to a minimum. Furthermore, it may be conjectured that theory applicable to quadratic functions applies essentially also to sufficiently smooth general functions.

In any case, it may be interesting to explore the argument motivated by the assumption that (5-9) is, at least locally, valid. Under these circumstances the necessary and sufficient condition for $M(\hat{\mathbf{x}})$ to be a minimum is that $\hat{\mathbf{x}}$ satisfies

$$\nabla M(\hat{\mathbf{x}}) = \mathbf{b} + A\hat{\mathbf{x}} = 0$$

This problem has a unique solution: $-A^{-1}\mathbf{b}$. The computation of this solution is fairly laborious and it is not always practical to attack the problem min $M(\mathbf{x})$ by attempting to find $\mathbf{x}$ satisfying the latter expression directly. The indirect approach is more justified, particularly when (5-9) is only a local approximation and a, $\mathbf{b}$, as well as A, vary with $\mathbf{x}$.

A set of n nonzero directions $\mathbf{s}^0, \mathbf{s}^1, \ldots, \mathbf{s}^{n-1}$ are said to be *conjugate* to each other with respect to a given positive definite symmetric matrix A if

$$(\mathbf{s}^k)^T A \mathbf{s}^l \begin{cases} = 0 & \text{for } 0 \leq k \neq l \leq n-1 \\ > 0 & \text{for } k = l \end{cases} \tag{5-10}$$

The second part of the definition, for $k = l$, is an obvious consequence of the positive definiteness of the matrix A. There exists at least one set of n directions satisfying definition (5-10). This is the set of eigenvectors of the matrix A. The n conjugate directions satisfying (5-10) are also linearly independent. A result of the linear independence of the n conjugate directions is that any vector conjugate to all n of them must be the zero vector.

If we set $k = n$, the sequence of n descent steps defined by (5-5) can be considered [Pierre, 1969] as the dot product of an n vector of steplengths $\boldsymbol{\lambda} = [\lambda^0, \lambda^1, \ldots, \lambda^{n-1}]^T$ and a matrix S the columns of which are the descent directions $S = [\mathbf{s}^0, \mathbf{s}^1, \ldots, \mathbf{s}^{n-1}]$. If these directions are conjugate, the matrix S is nonsingular because of the linear independence of the direction vectors forming its columns. We have

$$\sum_{l=0}^{n-1} \lambda^l \mathbf{s}^l = [\mathbf{s}^0, \mathbf{s}^1, \ldots, \mathbf{s}^{n-1}][\lambda^0, \lambda^1, \ldots, \lambda^{n-1}]^T = S\boldsymbol{\lambda}$$

Now set $\mathbf{x}^0 = 0$ for convenience and suppose $\mathbf{x}^n$ has been obtained by means of this sequence of descent steps. Then $\mathbf{x}^n = S\boldsymbol{\lambda}$, and consider the corresponding quadratic term, $(\mathbf{x}^n)^T A \mathbf{x}^n = (S\boldsymbol{\lambda})^T A S\boldsymbol{\lambda} = \boldsymbol{\lambda}^T(S^T A S)\boldsymbol{\lambda}$. The matrix $S^T A S$ is diagonal because the columns of S—that is, the directions ($\mathbf{s}^0$, $\mathbf{s}^1$, $\ldots$, $\mathbf{s}^{n-1}$) are mutually conjugate with respect to A. We can say, therefore, that the conjugate directions are used to diagonalize the positive definite matrix A. Notice, however, that neither the matrix $S^T A S$, nor the set of conjugate directions $\mathbf{s}^0, \mathbf{s}^1, \ldots, \mathbf{s}^{n-1}$ is uniquely determined by the aforementioned argument.

The use of conjugate gradients was introduced by Hestenes and Stiefel [1952] as a method for solving sets of simultaneous linear equations having positive definite, symmetric coefficient matrices. This problem of finding $\hat{\mathbf{x}}$ satisfying $A\mathbf{x} + \mathbf{b} = 0$ is equivalent (Sec. 2-4) to the problem of minimizing the quadratic function (5-9). We shall examine now the process of solving the latter problem by a descent technique using conjugate directions (not necessarily gradients). Suppose then that $M(\mathbf{x})$ is of the form (5-9) and that a descent process consisting of n steps defined by (5-5) with $k = n$ is performed. Suppose that the directions used are mutually conjugate, satisfying the definition (5-10) and that the descent steplengths are determined by minimization along these descent directions. Such steplengths are referred to as "optimal." These n optimal steps in conjugate descent directions lead from an arbitrary starting point $\mathbf{x}^0$ to the point $\mathbf{x}^n$ given by (5-5) with $k = n$. We have

$$M(\mathbf{x}^n) = M(\mathbf{x}^0 + \sum_{l=0}^{n-1} \lambda^l \mathbf{s}^l)$$

and using (5-9) and (5-10)

$$\begin{aligned} M(\mathbf{x}^n) &= a + \mathbf{b}^T\mathbf{x}^n + \tfrac{1}{2}(\mathbf{x}^n)^T A \mathbf{x}^n \\ &= a + \mathbf{b}^T\mathbf{x}^0 + \tfrac{1}{2}(\mathbf{x}^0)^T A \mathbf{x}^0 \\ &\quad + \sum_{l=0}^{n-1} [\tfrac{1}{2}(\lambda^l)^2(\mathbf{s}^l)^T A \mathbf{s}^l + (\mathbf{b} + A\mathbf{x}^0)^T \lambda^l \mathbf{s}^l] \end{aligned} \tag{5-11}$$

Notice that each of the summands in (5-11) involves only one value of the index l.

Following Pearson [1969], the conjugate directions descent process leading to (5-11) is considered now from three points of view.

1 Consideration of the descent process as a sequence of n minimizations along the descent lines: The descent steps taken were optimal—that is, each of the steplengths λ^l, $l = 0, \ldots, n-1$ is a solution of a (one-dimensional) minimization along the corresponding descent line

$$M(\mathbf{x}^l + \lambda^l \mathbf{s}^l) = \min_{\lambda} M(\mathbf{x}^l + \lambda \mathbf{s}^l),\ l = 0, \ldots, n-1 \qquad (5\text{-}12)$$

The descent line at $\mathbf{x}^l$ is a line $\{\mathbf{x}^l + \lambda \mathbf{s}^l \mid \lambda > 0\}$ through $\mathbf{x}^l$ pointing in the direction of descent $\mathbf{s}^l$. Distance along this line is measured in terms of λ, a positive number. (Finding λ^l that satisfy (5-12) is discussed in Chap. 3.) Because of (5-12), the derivative of $M(\mathbf{x})$ in the direction $\mathbf{s}^l$ at $\mathbf{x}^{l+1} = \mathbf{x}^l + \lambda^l \mathbf{s}^l$ is zero. This holds for any $l = 0, \ldots, n-1$ and is illustrated in Fig. 5-2. Hence,

$$(\mathbf{s}^l)^T \nabla M(\mathbf{x}^{l+1}) = 0 \qquad l = 0, \ldots, n-1 \qquad (5\text{-}13)$$

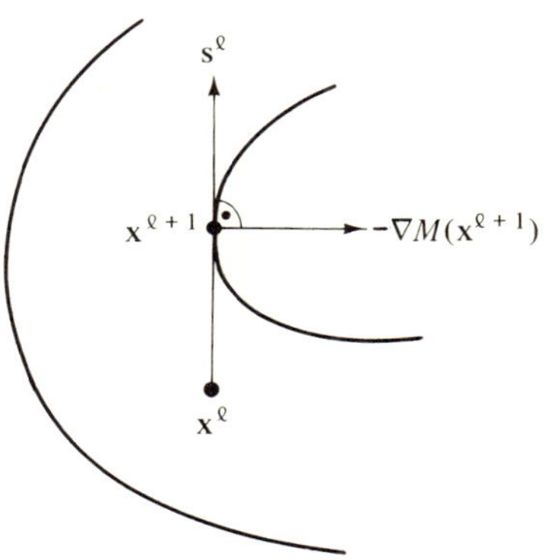

Figure 5-2 Orthogonality of the vectors $\mathbf{s}^l$ and $\nabla M(\mathbf{x}^{l+1})$ at $\mathbf{x}^{l+1}$

In other words, the gradient $\nabla M(\mathbf{x})$ at $\mathbf{x}^{l+1}$ is orthogonal to the direction $\mathbf{s}^l$ for $l = 0, \ldots, n-1$. An expression for $\nabla M(\mathbf{x}^{l+1})$ can be obtained from (5-9) using (5-5) and (5-10)

$$\nabla M(\mathbf{x}^{l+1}) = \mathbf{b} + A\mathbf{x}^0 + \sum_{j=0}^{l} \lambda^j A \mathbf{s}^j \qquad l = 0, \ldots, n-1 \qquad (5\text{-}14)$$

then combining (5-13) and (5-14) and making use of the conjugacy of the directions defined by (5-10), we obtain

$$\lambda^l = -(\mathbf{s}^l)^T(\mathbf{b} + A\mathbf{x}^0)/[(\mathbf{s}^l)^T A \mathbf{s}^l] \qquad l = 0, \ldots, n-1 \qquad (5\text{-}15)$$

for the descent steplengths. In this formula, the steplength λ^l depends on the starting point $\mathbf{x}^0$ and on the direction $\mathbf{s}^l$. It does not depend on the descent steps performed between $\mathbf{x}^0$ and $\mathbf{x}^l$. It follows that $M(\mathbf{x})$ as given by (5-11) does not depend on the order in which the directions $\mathbf{s}^l$, $l = 0, \ldots, n-1$ were used in the descent process that iterated $\mathbf{x}^0$ to $\mathbf{x}^n$. Therefore, this order is arbitrary, except that each direction is used only once.

2 Consideration of the descent process by regarding the function at the terminal point, $M(\mathbf{x}^n)$ as a function of the n steplengths λ^l, $l = 0, \ldots, n-1$: Suppose that $M(\mathbf{x}^n)$ as given by (5-11) is a function of the n steplengths λ^l, $l = 0, \ldots, n-1$. When this function assumes its minimum, $\nabla M(\mathbf{x}^n) = 0$; consequently, the steplengths satisfy

$$\frac{\partial M(\mathbf{x}^n)}{\partial \lambda^l} = (\mathbf{s}^l)^T \nabla M(\mathbf{x}^n) = 0 \qquad l = 0, \ldots, n-1 \tag{5-16}$$

This expression simply states that the directional derivative of the function $M(\mathbf{x})$ in each of the conjugate descent directions, evaluated at $\mathbf{x}^n$, is zero. We can obtain another expression for this derivative by differentiating $M(\mathbf{x}^n)$ as given by (5-11) with respect to the steplengths λ^l. This yields

$$\frac{\partial M(\mathbf{x}^n)}{\partial \lambda^l} = \lambda^l (\mathbf{s}^l)^T A \mathbf{s}^l + (\mathbf{b} + A\mathbf{x}^0)^T \mathbf{s}^l = 0 \qquad l = 0, \ldots, n-1$$

It is easy to see that this expression is satisfied if the steplengths λ^l satisfy (5-15). This means that the process of at most n minimizations along conjugate directions is equivalent to minimizing $M(\mathbf{x}^n)$ as a function of the descent steplengths. Fewer than n steps may actually be required because one or more of the steplengths may be zero. Of course, this would not save much computational work even if it occurred.

3 Consideration of the descent process by examining whether $M(\mathbf{x}^n)$ as defined by (5-11) is a minimum of $M(\mathbf{x})$: a general result can be obtained as an answer to this question. It pertains to any strictly convex differentiable function $M(\mathbf{x})$ having an unconstrained minimum (Sec. 1-5). The quadratic $M(\mathbf{x})$ defined by (5-9) is such a function and is included in Lemma 5-1.

Lemma 5-1 Given is $M(\mathbf{x})$, a strictly convex, differentiable function of $\mathbf{x}$ which has an unconstrained minimum. A necessary and sufficient condition that $M(\hat{\mathbf{x}})$ is the unique minimum of $M(\mathbf{x})$ is that

$$(\mathbf{s}^l)^T \nabla M(\hat{\mathbf{x}}) = 0 \qquad l = 0, \ldots, n-1 \tag{5-17}$$

where $\mathbf{s}^l$, $l = 0, \ldots, n-1$ is any set of linearly independent vectors.

The proof of this lemma is immediate since $\nabla M(\hat{\mathbf{x}})$ is, by (5-16), orthogonal to n linearly independent vectors, therefore $\nabla M(\hat{\mathbf{x}}) = 0$. The latter is a necessary condition for $M(\hat{\mathbf{x}})$ to be a minimum. Since $M(\mathbf{x})$ is strictly convex this condition is also sufficient.

The n conjugate directions $\mathbf{s}^l$, $l = 0, \ldots, n-1$ that satisfy the definition (5-10) are, as stated, linearly independent. The quadratic function $M(\mathbf{x})$ defined by (5-9) is a strictly convex, differentiable function of $\mathbf{x}$ with an unconstrained minimum. Consider now point $\mathbf{x}^n$, which was obtained from starting point $\mathbf{x}^0$ by the process of n optimal steps along the conjugate descent directions $\mathbf{s}^l$, $l = 0, \ldots, n-1$. The lengths of these steps are given by (5-15); hence, from (5-14) and (5-10)

$$(\mathbf{s}^l)^T\nabla M(\mathbf{x}^n) = (\mathbf{s}^l)^T(\mathbf{b} + A\mathbf{x}^0) + \lambda^l(\mathbf{s}^l)^T A\mathbf{s}^l = 0 \quad \text{for } l = 0, \ldots, n-1.$$

This means that by Lemma 5-1 $M(\mathbf{x}^n)$ is the unique minimum of the quadratic function (5-9). Furthermore, the point $\mathbf{x}^n$ was obtained by a process of n descent iterations (along n linearly independent conjugate directions). Consequently, this descent process is quadratically convergent. In practice, we may minimize a function that is not quadratic; minimizations along the descent lines may be approximate; and computation generally involves errors. Hence, usually several cycles of the n-step conjugate direction descent process may be required. Notice that Lemma 5-1 deals with general functions and sets of n linearly independent vectors. The lemma does not determine how $\hat{\mathbf{x}}$ is obtained, provided that (5-17) is satisfied. It is important, however, to understand that conjugate directions are defined only for quadratic functions, and when conjugate direction methods are used for general functions the directions may not satisfy the conditions of the lemma.

We return to the analogy of our process to that of solving $A\mathbf{x} + \mathbf{b} = 0$. Suppose that

$$\mathbf{x}^n = \mathbf{x}^0 - \sum_{l=0}^{n-1} \frac{(\mathbf{s}^l)^T(\mathbf{b} + A\mathbf{x}^0)}{(\mathbf{s}^l)^T A\mathbf{s}^l}\mathbf{s}^l$$

was obtained starting with $\mathbf{x}^0 = 0$; then

$$\mathbf{x}^n = -\sum_{l=0}^{n-1} \frac{(\mathbf{s}^l)^T\mathbf{b}}{(\mathbf{s}^l)^T A\mathbf{s}^l}\mathbf{s}^l = -\sum_{l=0}^{n-1} \frac{\mathbf{s}^l(\mathbf{s}^l)^T}{(\mathbf{s}^l)^T A\mathbf{s}^l}\mathbf{b}$$

and if $A\mathbf{x}^n + \mathbf{b} = 0$, then $\mathbf{x}^n = -A^{-1}\mathbf{b}$, and

$$A^{-1} = \sum_{l=0}^{n-1} \frac{\mathbf{s}^l(\mathbf{s}^l)^T}{(\mathbf{s}^l)^T A\mathbf{s}^l}$$

A geometric interpretation may shed more light on the conjugate directions descent process: We start from $\mathbf{x}^0$, move along $\mathbf{s}^0$ to $\mathbf{x}^1$ and then along $\mathbf{s}^1$ to $\mathbf{x}^2$ and so on. The point $\mathbf{x}^1$ minimizes $M(\mathbf{x})$ along the line $\{\mathbf{x}^0 + \lambda\mathbf{s}^0 \mid \lambda > 0\}$, by construction. The point $\mathbf{x}^2$ minimizes $M(\mathbf{x})$ in the plane through $\mathbf{x}^0$ containing $\mathbf{s}^0$ and $\mathbf{s}^1$; that is, $\{\mathbf{x}^0 + \lambda^0\mathbf{s}^0 + \lambda^1\mathbf{s}^1 \mid \lambda^0, \lambda^1 > 0\}$. This can be shown to be true because the steplengths λ^0 and λ^1 satisfy (5-15).

In general, it can be shown in the same manner that at the kth step of this process, a minimum of $M(\mathbf{x})$ is produced on a k-dimensional manifold through $\mathbf{x}^0$ and spanned by the k directions $\mathbf{s}^0, \mathbf{s}^1, \ldots, \mathbf{s}^{k-1}$.

Theorem 5-1 summarizes the main features of the conjugate directions descent process applied to the quadratic function given by (5-9).

Theorem 5-1 Given is $M(\mathbf{x})$, a quadratic function

$$M(\mathbf{x}) = a + \mathbf{b}^T\mathbf{x} + \tfrac{1}{2}\mathbf{x}^T A\mathbf{x}$$

with a symmetric positive definite matrix A. Suppose that an arbitrary starting point $\mathbf{x}^0$ has been selected, and that n optimal descent steps have been performed to obtain a terminal point $\mathbf{x}^n$:

$$\mathbf{x}^n = \mathbf{x}^0 + \sum_{l=0}^{n-1} \lambda^l \mathbf{s}^l$$

where λ^l minimizes $M(\mathbf{x}^l + \lambda\mathbf{s}^l)$, $l = 0, \ldots, n-1$.

Suppose further that the descent directions used were mutually conjugate with respect to the matrix A:

$$(\mathbf{s}^k)^T A\mathbf{s}^l \begin{cases} = 0 & \text{for } 0 \leq k \neq l \leq n-1 \\ > 0 & \text{for } k = l \end{cases}$$

The directions $[\mathbf{s}^0, \ldots, \mathbf{s}^{n-1}]$ may be used in arbitrary order, provided that each direction is used only once. Then

$$M(\mathbf{x}^n) = M(\mathbf{x}^0) + \sum_{l=0}^{n-1} [\tfrac{1}{2}(\lambda^l)^2(\mathbf{s}^l)^T A\mathbf{s}^l + (\mathbf{b} + A\mathbf{x}^0)^T \lambda^l \mathbf{s}^l]$$

where the steplengths satisfy

$$\lambda^l = -(\mathbf{s}^l)^T \frac{\mathbf{b} + A\mathbf{x}^0}{(\mathbf{s}^l)^T A\mathbf{s}^l} \qquad l = 0, \ldots, n-1$$

Furthermore,

$$(\mathbf{s}^l)^T \nabla M(\mathbf{x}^n) = 0 \qquad l = 0, \ldots, n-1$$

and therefore:

1 This process provides the necessary condition for minimizing $M(\mathbf{x}^n)$, when considered as a function of the n steplengths λ^l, $l = 0, \ldots, n-1$.

2 This process provides the necessary and sufficient condition for $M(\mathbf{x}^n)$ to be the unique minimum of $M(\mathbf{x})$.

3 This process is quadratically convergent.

The Length of the Descent Steps

Different approaches are possible in determining the length λ^k of the individual descent steps $\Delta\mathbf{x}^k$ defined by (5-4). Each of these can be (and probably has been) used in combination with different descent direction strategies. Consequently, the descent techniques can also be classified according to the type of steplength used:

Optimal Descent Steps. A descent step λ^k is said to be an *optimal step* when it is determined as a solution to the problem of minimizing $M(\mathbf{x})$ along the line $\{\mathbf{x}^k + \lambda\mathbf{s}^k \mid \lambda > 0\}$.

$$M(\mathbf{x}^k + \lambda^k\mathbf{s}^k) = \min_{\lambda} M(\mathbf{x}^k + \lambda\mathbf{s}^k) \tag{5-18}$$

The theoretical justification of the conjugate direction descent techniques is based on optimal descent steps. In addition, the use of optimal descent steps may be motivated by efficiency considerations. These are mainly that the computational effort involved in finding the directions of descent may be so considerable that the improvement in the objective function in each step should be as great as possible. If an optimal step strategy is used, the descent technique is referred to as an *optimal step descent technique.* Methods for actually solving the problem stated in (5-18) are discussed in Chap. 3.

Nonoptimal Descent Steps. A descent step λ^k is said to be a *nonoptimal step* if it satisfies

$$M(\mathbf{x}^k + \lambda^k\mathbf{s}^k) \leq M(\mathbf{x}^k) \qquad \lambda^k > 0, \tag{5-19}$$

but not (5-18). Contrary to the logic leading to the use of optimal descent steps, it is often argued that to increase computational efficiency one should not use optimal steps. This is correct in the sense that usually more computation would be required to find λ^k satisfying (5-18) than just (5-19). It is also true, of course, that less improvement in the objective function might result in the latter case. Experience indicates, however, that overall computational efficiency is frequently improved when nonoptimal steps are used. Methods for determining the stepsize by means of (5-19) are discussed in Chap. 3.

Descent Steps with Theoretically Justified Length. In certain cases the theoretical considerations leading to the determination of the descent direction will also provide a step length. The second-order gradient descent method (Newton's method), for example, suggests a fixed steplength of $\lambda^k = 1$, for all k with the not-normalized direction. Similarly, the theory of conjugate directions is based on optimal descent steps. There are also situations in which the steplength argument is related to the argument leading to the descent directions—for example, in being based on the same assumptions. Suppose that a function can be expanded in a second-order Taylor series.

Then a scheme may be derived to determine the stepsize from the eigenvalues of the metric matrix. An extension of these ideas may lead to even more sophisticated methods [Stiefel, 1958]. The common drawback of many of these theories is that they imply a computational effort that is usually unwarranted by the objective function improvement that is produced. Therefore, these theories are often ignored in practice, or they are at most used to justify a nonoptimal step strategy. Furthermore, in many cases when a general nonquadratic function is minimized the considerations leading to the theoretical steplength are only locally true. If a step of theoretically justified length is actually taken, it may "overshoot." This means that at $\mathbf{x}^{k+1}$, the new point, considerations leading to $\mathbf{s}^k$ as a descent direction and λ^k as a steplength are no longer valid. Therefore, smaller steps are frequently used.

Convergence of Descent Techniques and Termination Criteria

Before discussing convergence and termination, we have to determine how to identify solution points. Ideally we would like sufficient conditions for a minimum to be satisfied at such a point. This is often not practical, as is briefly explained in the discussion of "practical solutions" in Sec. 1-8. One can, however, determine whether necessary conditions for a minimum are satisfied at a point, and these are frequently used to identify solutions. If the objective function $M(\mathbf{x})$ is differentiable, a necessary condition for $\mathbf{x}^k$ to be a solution point is

$$\nabla M(\mathbf{x}^k) = 0 \tag{5-20}$$

because if $\nabla M(\mathbf{x}^k) \neq 0$, one could find a descent direction—for example, $\mathbf{s}^k = -\nabla M(\mathbf{x}^k)$ —and perform a descent step to a point $\mathbf{x}^{k+1}$ where $M(\mathbf{x}^{k+1}) < M(\mathbf{x}^k)$.

If the objective function is not differentiable, one could identify the solution point $\mathbf{x}^k$ after actually performing another descent step to $\mathbf{x}^{k+1}$ and finding that

$$M(\mathbf{x}^k) - M(\mathbf{x}^{k+1}) \cong 0 \tag{5-21}$$

If the objective function $M(\mathbf{x})$ is twice continuously differentiable, it can be expanded in a second-order Taylor series about $\mathbf{x}^k$:

$$M(\mathbf{x}^k + \lambda\mathbf{s}) \cong M(\mathbf{x}^k) + \lambda(\mathbf{s})^T\nabla M(\mathbf{x}^k) + \tfrac{1}{2}\lambda^2(\mathbf{s})^T \operatorname{Hess}(\mathbf{x}^k + \theta\lambda\mathbf{s})\mathbf{s} \tag{5-22}$$

for some $\lambda > 0$ and θ such that $0 \leq \theta \leq 1$. Following Zangwill [1969], suppose that the Hessian is not positive semidefinite at $\mathbf{x}^k$. Then, by definition, there must be a vector $\mathbf{s} \neq 0$ such that

$$(\mathbf{s})^T \operatorname{Hess}(\mathbf{x}^k)\mathbf{s} < 0$$

A descent direction $\mathbf{s}^k$ may be chosen to be this vector $\mathbf{s}$ multiplied by unity with a sign opposite to that of the term $(\mathbf{s})^T\nabla M(\mathbf{x}^k)$. (If this term is zero, the sign of $\mathbf{s}$ does not matter.) Then $(\mathbf{s}^k)^T\nabla M(\mathbf{x}^k) \leq 0$, and because the Hessian is continuous, there must be a steplength $\bar{\lambda} > 0$ small enough such that $(\mathbf{s}^k)^T \operatorname{Hess}(\mathbf{x}^k + \theta\bar{\lambda}\mathbf{s}^k)\,\mathbf{s}^k < 0$. Using $\bar{\lambda}$ and $\mathbf{s}^k$ in (5-22), we can perform a descent step and obtain $M(\mathbf{x}^{k+1}) = M(\mathbf{x}^k + \bar{\lambda}\mathbf{s}^k) < M(\mathbf{x}^k)$.

In other words, if the objective function is twice continuously differentiable and the Hessian is not positive semidefinite, the descent process could be continued. This motivates the addition of another condition to (5-20), namely that $\operatorname{Hess}(\mathbf{x}^k)$ be positive semidefinite. This additional condition may prevent the erroneous identification of nonoptimal points as optimal points if they satisfy the necessary condition (5-20).

A very general approach to convergence has been taken by Zangwill [1969]. He shows that the first- and second-order gradient techniques, for instance, are convergent, provided that optimal steps are used in the first and the theoretical steplength of unity in the second. In computational practice, the steps of these algorithms frequently are not carried out exactly. Even if attempted, this may not be achieved because of computational error. Therefore, it is important to understand how much an algorithm can be "relaxed" without adversely affecting its convergence properties.

In a recent article, Wolfe [1969] tried to find the weakest possible conditions on the individual descent steps so that when a process with k steps satisfying these conditions is applied to $M(\mathbf{x})$, as $k \longrightarrow \infty$, either $M(\mathbf{x}^k) \longrightarrow -\infty$ or $(\mathbf{s}^k)^T\nabla M(\mathbf{x}^k) \longrightarrow 0$. A descent process having this property will converge as $k \longrightarrow \infty$. It is of interest to outline a part of Wolfe's argument because it yields the weakest convergence conditions provided so far, as well as additional criteria for termination of the computational process. In the derivation of these convergence conditions it is assumed that the derivative of the objective function $M(\mathbf{x})$ in the direction of descent $\mathbf{s}^k$ exists for any k and all λ such that $0 \leq \lambda \leq \lambda^k$ along the descent line $\mathbf{x}^k + \lambda\mathbf{s}^k$. These derivatives are given by

$$\frac{dM(\mathbf{x}^k - \lambda\mathbf{s}^k)}{d\lambda} = (\mathbf{s}^k)^T\nabla M(\mathbf{x}^k + \lambda\mathbf{s}^k) \qquad 0 \leq \lambda \leq \lambda^k \qquad k = 0, 1, \ldots$$

It is assumed that these derivatives possess certain continuity properties. Wolfe [1969] lists five properties that a descent step may have:

1 $M(\mathbf{x}^k) - M(\mathbf{x}^{k+1}) \geq \epsilon$; that is, the step results in an improvement of at least ϵ in the objective function value.

2 $\|\Delta\mathbf{x}^k\| \geq \epsilon \min\{1, -(\mathbf{s}^k)^T\nabla M(\mathbf{x}^k)\}$; that is, the stepsize is reasonably large in relation to the directional derivative.

3 $M(\mathbf{x}^k) - M(\mathbf{x}^{k+1}) \geq \epsilon\|\Delta\mathbf{x}^k\|[-(\mathbf{s}^k)^T\nabla M(\mathbf{x}^k)]$; that is, the step results in a relatively reasonable improvement in the value of the objective function.

4 $(\mathbf{s}^k)^T\nabla M(\mathbf{x}^k + \bar{\lambda}\mathbf{s}^k) \geq (1 - \epsilon)(\mathbf{s}^k)^T\nabla M(\mathbf{x}^k)$ for some $\bar{\lambda}$ where $0 \leq \bar{\lambda} \leq \lambda^k$ such that $M(\mathbf{x})$ is nonincreasing from $\mathbf{x}^k$ to $(\mathbf{x}^k + \lambda\mathbf{s}^k)$.

5 With $\bar{\lambda}$ as in step 4, $M(\mathbf{x}^{k+1}) \leq M(\mathbf{x}^k + \bar{\lambda}\mathbf{s}^k)$.

A different ϵ, $0 < \epsilon < 1$, may be used in each of the above properties and also for each step. A descent step is said to be a *serious step* if it has a combination of properties from among (1) through (5) such that if it produces a small (less than ϵ) reduction in the objective function $M(\mathbf{x})$, then the directional derivative $(\mathbf{s}^k)^T\nabla M(\mathbf{x}^k)$ must be close to zero. A step is called a *serious step* if it has property (1), or (2) and (3), or (3) and (4), or (4) and (5). Notice that a step can be serious without being optimal in the previously defined sense of (5-18). However, the case of an optimal step is included in (4) and (5) if $(\mathbf{s}^k)^T\nabla M(\mathbf{x}^k + \bar{\lambda}\mathbf{s}^k) = 0$ and $\bar{\lambda} = \lambda^k$. Further, if the Hessian Hess($\mathbf{x}$) is positive definite in the region of interest, it can be shown, using a second-order Taylor series expansion (5-22) of $M(\mathbf{x})$, that serious steps can be generated for which properties (2) and (3) hold. The analysis is similar to that leading to the second-order termination criterion mentioned earlier, except that, of course, we may have $\nabla M(\mathbf{x}) \neq 0$ in this case. A step cannot be serious and arbitrarily short because when the directional derivative $(\mathbf{s}^k)^T\nabla M(\mathbf{x}^k) \neq 0$ properties (1), (2), and (4) will not be satisfied for arbitrarily small λ^k. On the other hand, so long as the descent process has not converged by Wolfe's definition [the directional derivative $(\mathbf{s}^k)^T\nabla M(\mathbf{x}^k) \neq 0$], one can always take serious descent steps. These can be of two types: $\epsilon < 1$, then λ^k may be chosen so that the step has either property (1), or (4) and (5); and $\epsilon < 1/2$, then λ^k can be chosen so that the step has either properties (2) and (3), or (3) and (4). Wolfe [1969] summarizes a comparison of his definition of serious steps with attempts made by others to determine properties required of descent steps. It is shown that his definition constitutes a considerable weakening of these other requirements.

Wolfe distinguishes between two exhaustive but not exclusive situations that may arise when a sequence of serious steps is taken:

1 $||\Delta\mathbf{x}^k|| \geq \epsilon > 0$ for all k on some subsequence. Then if $M(\mathbf{x}^k)$ does not tend to $-\infty$, the directional derivatives $(\mathbf{s}^k)^T\nabla M(\mathbf{x}^k)$ must tend to zero.

2 $||\Delta\mathbf{x}^k|| \longrightarrow 0$ on the subsequence. Then the method for determining the steplength λ^k must be such that the directional derivative $(\mathbf{s}^k)^T\nabla M(\mathbf{x}^k + \lambda\mathbf{s}^k)$ tends to zero. This is assured by property (4), which states that there must be a relative increase in the directional derivative as one moves from $\mathbf{x}^k$ to $\mathbf{x}^k + \lambda\mathbf{s}^k$. So if $\lambda^k \longrightarrow 0$ also, the directional derivative tends to zero, provided it satisfies appropriate continuity conditions.

If a descent sequence has a subsequence of serious steps and the directional derivative satisfies appropriate continuity conditions, then either the function $M(\mathbf{x}^k) \longrightarrow -\infty$ or the directional derivative $(\mathbf{s}^k)^T\nabla M(\mathbf{x}^k) \longrightarrow 0$ as $k \longrightarrow \infty$. Suppose that, in addition, the descent directions $\mathbf{s}^k$ are chosen so that the angle they form with the gradient $\nabla M(\mathbf{x}^k)$ is uniformly bounded away from 90° (Fig. 5-1). That is, for some $\delta > 0$ (recall that $\mathbf{s}^k$ is a unit vector), $(\mathbf{s}^k)^T\nabla M(\mathbf{x}^k) \leq -\delta||\nabla M(\mathbf{x}^k)||$. If the subsequence of serious steps satisfies also this condition, then either the function $M(\mathbf{x}^k) \longrightarrow -\infty$ or its gradient $\nabla M(\mathbf{x}^k) \longrightarrow 0$ as $k \longrightarrow \infty$. In the latter case, a limit point of the subsequence is a stationary point of $M(\mathbf{x})$. This is Wolfe's main result.

In practice, the descent iterations are continued until a predetermined termination criterion has been satisfied. Either a point is identified as a solution point, or the algorithm is incapable of generating successor points where the value of the objective function is any smaller. The strategy outlined here for terminating the descent techniques is essentially motivated by the aforementioned theoretical considerations.

All techniques discussed in this chapter with the exception of the conjugate directions method (Sec. 5-4) involve at least first derivatives of the objective functions. If it is desired to determine the value of the objective function, but not the values of the variables, with high accuracy, these techniques should be terminated at $\mathbf{x}^k$ such that

$$||\nabla M(\mathbf{x}^k)|| < \tau \tag{5-23}$$

where $\tau > 0$ is a predetermined tolerance. If the objective function is not differentiable, then $[M(\mathbf{x}^k) - M(\mathbf{x}^{k+1})]$ should be used instead of $||\nabla M(\mathbf{x}^k)||$ in (5-23). This may be suitable for the conjugate directions technique. If the objective function is twice differentiable, (5-23) may be supplemented by a test concerning the positive definiteness of the Hessian Hess($\mathbf{x}^k$). This may be suitable for the second-order gradient method (Sec. 5-3) and for the variable metric method (Sec. 5-6).

If an accurate value of the variable vector $\mathbf{x}$ at the termination point is desired, the kth step should be completed even if (5-23) is satisfied. Then the process should be terminated only if

$$M(\mathbf{x}^k) - M(\mathbf{x}^{k+1}) < \epsilon_1 \qquad \text{and} \qquad ||\Delta \mathbf{x}^k|| < \epsilon_2 \tag{5-24}$$

where $\epsilon_1 > 0$ and $\epsilon_2 > 0$ are predetermined problem-dependent parameters. For example, one could set ϵ_1 proportional to $M(\mathbf{x}^0)$ and ϵ_2 proportional to $\mathbf{x}^0$.

If a nonoptimal step strategy is used, the process can be terminated if

$$M(\mathbf{x}^k) - M(\mathbf{x}^{k+1}) < \epsilon_3 \qquad \text{and} \qquad ||\Delta \mathbf{x}^k|| < \epsilon_4 \tag{5-25}$$

where $\epsilon_3 > 0$ and $\epsilon_4 > 0$ are predetermined problem-dependent parameters, but some care should be exercised to ensure that (5-25) does not terminate the process due to unreasonably short nonoptimal steps. The parameters ϵ_3 and ϵ_4 could, for example, be determined in the same way as ϵ_1 and ϵ_2 in (5-24). Processes using nonoptimal steps should be terminated if criterion (5-25) is satisfied even if criterion (5-23) is not.

Computational experience indicates that it is advisable, after termination, to restart the computation from a point obtained by a slight perturbation of the termination point. Termination of the new process at a point considerably different from the previous one is a possible indication of multiple local minima. Such a problem should be examined carefully. Notice that if $(\mathbf{s}^k)^T\nabla M(\mathbf{x}^k)$, the directional derivative along the direction $\mathbf{s}^k$, is nonnegative, this direction is a direction of ascent and $-\mathbf{s}^k$ should be used in the process. Certain of the following techniques have to be recycled (but not terminated) under these circumstances.

5-2 FIRST-ORDER GRADIENT (STEEPEST DESCENT) TECHNIQUES

The choice of locally steepest directions as directions of descent leads to different gradient techniques. As explained in the preceding section, these techniques differ in the metric matrix A_k which enters the definitions (5-7) of distances and (5-8) of descent directions. The simplest choice for A_k is I, the $n \times n$ identity matrix, that is, $A_k = I$, and this remains fixed for all k (throughout the iterative process). Because $I = I^{-1}$ we obtain

$$\mathbf{s}^k = -\nabla M(\mathbf{x}^k) \tag{5-26}$$

This formula provides descent directions because $(\mathbf{s}^k)^T\nabla M(\mathbf{x}^k) = -[\nabla M(\mathbf{x}^k)]^T\nabla M(\mathbf{x}^k) < 0$. This choice of direction was first suggested by Cauchy [1847] and is termed the *steepest descent direction.* These directions are simply gradients of the objective function $M(\mathbf{x})$, and their computation involves only first-order differentiation of this function.

Note that the first-order gradient technique that uses steepest descent directions is applicable on analog computers. The gradient directions can be used with infinitesimal steplengths and this leads to a continuous scheme. The descent path (trajectory) $\mathbf{x}(t)$ is considered as a function of time t, and the descent steps are taken as steps in time: $[d\mathbf{x}(t)]/dt = -\nabla M[\mathbf{x}(t)]$. This is a set of n first-order differential equations that can be solved on an analog computer. The solution $\lim_{t\to\infty} \mathbf{x}(t)$ corresponds to a stationary point of $M(\mathbf{x})$.

Iterative techniques are, however, our main theme and we shall proceed to show how one may establish an appropriate steplength λ^k to be used with the steepest descent direction (5-26). If the objective function $M(\mathbf{x})$ is twice differentiable, the argument can be based on a second-order Taylor series expansion of $M(\mathbf{x})$ about the point $\mathbf{x}^{k+1}$. Assuming that $\mathbf{s}^k = -\nabla M(\mathbf{x}^k)$, we have

$$\begin{aligned} M(\mathbf{x}^k + \lambda^k\mathbf{s}^k) = M(\mathbf{x}^{k+1}) \cong M(\mathbf{x}^k) &- \lambda^k[\nabla M(\mathbf{x}^k)]^T\nabla M(\mathbf{x}^k) \\ &+ \tfrac{1}{2}(\lambda^k)^2[\nabla M(\mathbf{x}^k)]^T \operatorname{Hess}(\mathbf{x}^k)\nabla M(\mathbf{x}^k) \end{aligned}$$

We wish to maximize $[M(\mathbf{x}^k) - M(\mathbf{x}^{k+1})]$ in the kth descent step, and we shall pick λ^k so as to achieve this. Regarding the difference $[M(\mathbf{x}^k) - M(\mathbf{x}^{k+1})]$ as a function of λ^k, differentiating the Taylor series approximation of this difference with respect to λ^k and setting the derivative equal to zero, we obtain

$$\lambda^k = \frac{[\nabla M(\mathbf{x}^k)]^T \nabla M(\mathbf{x}^k)}{[\nabla M(\mathbf{x}^k)]^T \operatorname{Hess}(\mathbf{x}^k) \nabla M(\mathbf{x}^k)} \tag{5-27}$$

To ensure that this steplength yields the desired maximum improvement in $M(\mathbf{x})$, it is necessary that the second derivative of $[M(\mathbf{x}^{k+1}) - M(\mathbf{x}^k)]$ with respect to λ^k be positive: $[\nabla M(\mathbf{x}^k)]^T \operatorname{Hess}(\mathbf{x}^k)\nabla M(\mathbf{x}^k) > 0$, which is guaranteed if the Hessian $\operatorname{Hess}(\mathbf{x}^k)$ is positive definite. In practice, the computation of the steplength by means of (5-27) may be too cumbersome. It is also questionable whether, if computed, the second-order information may not be better utilized in the determination of the descent direction. There are some indications that this may indeed be advisable, particularly in the vicinity of local minima. Furthermore, the objective function may not be twice differentiable. Consequently, the steepest descent technique is frequently used with optimal or nonoptimal steps (Sec. 5-1). Figure 5-3 depicts minimization of a quadratic function by the steepest descent method using optimal steps.

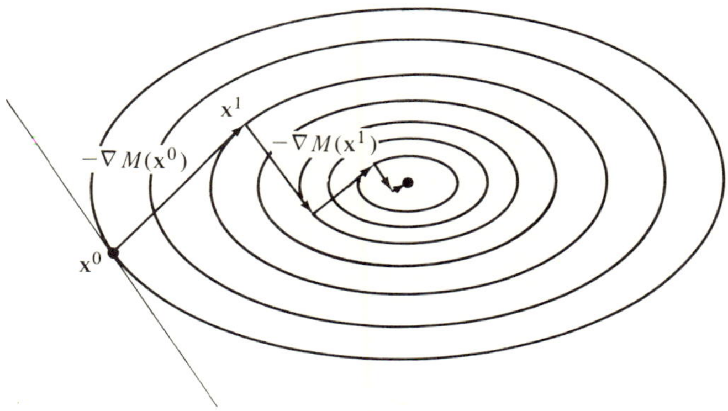

Figure 5-3 Quadratic function $M(\mathbf{x}) = a + b^T\mathbf{x} + \frac{1}{2}\mathbf{x}^T A\mathbf{x}$ minimized by the steepest descent method

Iteration Steps

1 Initialize, set $k = 0$, evaluate $M(\mathbf{x}^0)$; $\mathbf{x}^0$ is given.
2 Evaluate the gradient $\nabla M(\mathbf{x}^k)$ at the current point.
3 If $\|\nabla M(\mathbf{x}^k)\| < \tau$, and a very accurate value of the solution vector is not required, terminate the iteration; otherwise, proceed to step 4 ($\tau > 0$ is a predetermined tolerance).
4 Set the current descent direction

$$\mathbf{s}^k = -\nabla M(\mathbf{x}^k) \tag{5-28}$$

Normalize $\mathbf{s}^k$ so that $\|\mathbf{s}^k\| = 1$.

5 Compute the current descent steplength λ^k [Chap. 3 gives methods of optimization along the line $(\mathbf{x}^k + \lambda \mathbf{s}^k)$].

a. Optimal step alternative: find λ^k which satisfies

$$M(\mathbf{x}^k + \lambda^k \mathbf{s}^k) = \min_{\lambda} M(\mathbf{x}^k + \lambda \mathbf{s}^k) \tag{5-29}$$

b. Nonoptimal step alternative: find λ^k which satisfies

$$M(\mathbf{x}^k + \lambda^k \mathbf{s}^k) \leq M(\mathbf{x}^k) \tag{5-30}$$

6 Compute the current descent step

$$\Delta \mathbf{x}^k = \lambda^k \mathbf{s}^k \tag{5-31}$$

7 Perform the current descent step

$$\mathbf{x}^{k+1} = \mathbf{x}^k + \Delta \mathbf{x}^k \tag{5-32}$$

8 Evaluate the objective function $M(\mathbf{x}^{k+1})$ at the new point $\mathbf{x}^{k+1}$.

9 If $M(\mathbf{x}^k) - M(\mathbf{x}^{k+1}) < \epsilon_1$ and $||\Delta \mathbf{x}^k|| < \epsilon_2$, terminate the iteration; otherwise, proceed to step 10 ($\epsilon_1 > 0$ is a predetermined minimal function improvement and $\epsilon_2 > 0$ a predetermined minimal stepsize).

10 Accept the new point $\mathbf{x}^{k+1}$, set $k : k + 1$ and repeat from step 2.

Termination Criteria and Convergence

The steepest descent process is terminated as indicated above; convergence theory for this algorithm is outlined in Sec. 5-1. Note that Forsythe [1968] shows that a more general class of methods, of which the steepest descent technique may be regarded as a special case, has similar convergence properties. These methods are referred to as "*S*-dimensional optimum gradient methods," and they can be described as follows: Consider a quadratic objective function given by

$$M(\mathbf{x}) = \tfrac{1}{2}\mathbf{x}^T A \mathbf{x} \tag{5-33}$$

with a positive definite symmetric matrix A. Then $\nabla M(\mathbf{x}) = A\mathbf{x}$, and steepest descent proceeds by moving from $\mathbf{x}^k$ to $\mathbf{x}^{k+1}$ according to $\mathbf{x}^{k+1} = \mathbf{x}^k + \lambda^k A \mathbf{x}^k$, where λ^k minimizes $\{M(\mathbf{x}^k + \lambda A \mathbf{x}^k) \mid -\infty < \lambda < \infty\}$. Consider the line of steepest descent through $\mathbf{x}^k$, given by

$$L_1^k = \{\mathbf{x}^k + \lambda A \mathbf{x}^k \mid -\infty < \lambda < \infty\} \tag{5-34}$$

The gradient $M(\mathbf{x})$ along this line is given by $\nabla M(\mathbf{x}) = \{A\mathbf{x}^k + \lambda A^2 \mathbf{x}^k \mid \mathbf{x} \in L_1^k\}$. This motivates the formulation of the two-dimensional plane of steepest descent through $\mathbf{x}^k$ as

$$L_2^k = \{\mathbf{x}^k + \lambda^1 A \mathbf{x}^k + \lambda^2 A^2 \mathbf{x}^k \mid -\infty < \lambda^j < \infty, \qquad j = 1, 2\}$$

and, in general, the *S*-dimensional plane of steepest descent is

$$L_S^k = \{\mathbf{x}^k + \sum_{j=1}^{S} \lambda^j (A)^j \mathbf{x}^k \mid -\infty < \lambda^j < \infty, \qquad j = 1, \ldots, S\} \tag{5-35}$$

The optimum S gradient method considers points $\mathbf{x}^{k+1}$ generated from points $\mathbf{x}^k$ so that $\mathbf{x}^{k+1}$ is the unique point in L_S^k for which $M(\mathbf{y})$ with $\mathbf{y} \in L_S^k$ is a minimum. These steps may actually be executed in a number of ways. The steepest descent technique can be regarded as a special case with $S = 1$. For $S = 1$ it was conjectured by Forsythe and Motzkin [1951] and proved by Akaike [1959] that the directions generated in the optimization process are ultimately asymptotic to just two directions, so that the method essentially operates in a two-dimensional subspace—the "cage" of Stiefel [1952]. Forsythe [1968] has extended some of the results from $S = 1$ to arbitrary $S > 1$.

Modifications and Acceleration Techniques

Different schemes have been proposed to speed up the steepest descent method. One of these has been extensively tested and found to be successful. It is based on an idea of Forsythe and Motzkin [1951]. They found that the solution of linear equation sets by steepest descent can be accelerated if the direction

$$\mathbf{s}^k = \mathbf{x}^k - \mathbf{x}^{k-2} \qquad k \geq 2 \tag{5-36}$$

is used from time to time (Fig. 5-4). Note that the method converges to the minimum in such a way that eventually the successive approximations alternate between two lines. The directions defined by (5-36) may coincide with one of these lines, and it is therefore reasonable to search along these directions from time to time.

The method of gradient PARTAN ("parallel tangents") is a generalization of

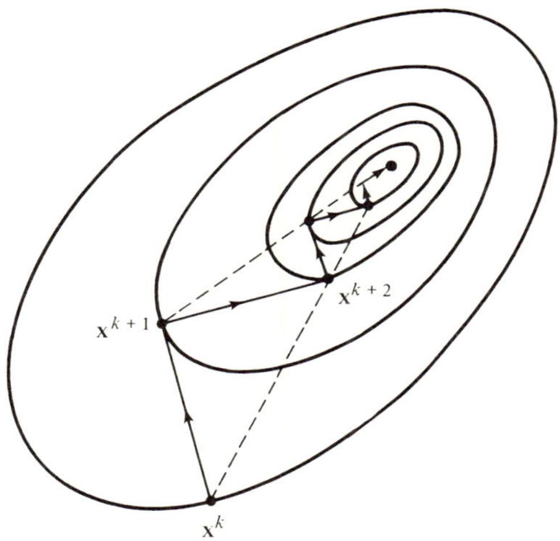

Figure 5-4 Forsythe-Motzkin acceleration

the Forsythe-Motzkin idea, [Shah, et al., 1964]. It proceeds by alternating between steepest descent direction and directions given by a formula like (5-36). Specifically,

$$\mathbf{s}^0 = -\nabla M(\mathbf{x}^0)$$
$$\mathbf{s}^1 = -\nabla M(\mathbf{x}^1)$$
$$\mathbf{s}^2 = -(\mathbf{x}^2 - \mathbf{x}^0) \tag{5-37}$$

and

$$\mathbf{s}^k \begin{cases} = -\nabla M(\mathbf{x}^k) & k = 3, 5, 7, \ldots, 2l - 1 \\ = -(\mathbf{x}^k - \mathbf{x}^{k-3}) & k = 4, 6, 8, \ldots, 2l \end{cases}$$

The optimal steplength strategy (5-29) for λ^k is used with both types of directions.

Pierre [1969] shows that this method is a type of a conjugate direction method. Sorenson [1969] shows that if a quadratic function is minimized, the vectors $(\mathbf{x}^1 - \mathbf{x}^0), (\mathbf{x}^3 - \mathbf{x}^1), (\mathbf{x}^5 - \mathbf{x}^3), \ldots, (\mathbf{x}^{2l-1} - \mathbf{x}^{2l-3})$ are mutually conjugate, but the vectors generated by (5-37) are not, and that the gradient vectors $\nabla M(\mathbf{x}^0), \nabla M(\mathbf{x}^1), \nabla M(\mathbf{x}^3), \ldots, \nabla M(\mathbf{x}^{2l-1})$ are orthogonal.

Sorenson [1969] has also shown that when a quadratic function is minimized PARTAN-accelerated steepest descent is considerably less efficient than the conjugate gradient method (Sec. 5-5). In particular, the former method is shown to go through the same sequence of points as the latter, and through an equal number of other points. The PARTAN-accelerated steepest descent thus requires twice as many minimizations along the line as are necessary.

Discussion

The steepest descent technique is a simple technique involving relatively little computational work per step. Each iteration is independent of the previous iterations; hence, the method is relatively economical in computer storage. The method may converge to a solution even from a poor initial approximation. Unfortunately, the method has several shortcomings. These essentially result in slow convergence due to the deterioration of the process in a global sense, even though the individual iterations are locally optimal. Very short steps may be produced, accompanied by sharp changes in the gradient $\nabla M(\mathbf{x})$. This is frequently referred to as "zigzagging" and implies, of course, slow convergence and low computational efficiency in spite of the small amount of work per step. Furthermore, the steepest descent technique depends on the variable scaling (Sec. 1-9).

It has also been shown [Kowalik and Osborne, 1968] that the method is basically unstable with respect to small perturbations. Such perturbations may be the result of rounding error or the inaccuracy in determining the optimal steplengths. In the extreme case, these perturbations may cause the descent directions to be orthogonal to those that might have been computed if no errors had been involved. In general, the descent directions may not be linearly independent. Consequently, the descent process may be unable to "penetrate" the whole n-dimensional space in the search for a minimum.

The rate of convergence of a steepest descent process applied to a quadratic form $M(\mathbf{x})$ can be estimated by

$$M(\mathbf{x}^{k+1}) \leq M(\mathbf{x}^k)\left(\frac{\gamma_{\max} - \gamma_{\min}}{\gamma_{\max} + \gamma_{\min}}\right)^2 \tag{5-38}$$

where $\gamma_{\max}$ and $\gamma_{\min}$ are the largest and smallest eigenvalues of the matrix involved in the quadratic form. (These are positive because of the positive definiteness of the matrix.) Furthermore, it has been demonstrated by Akaike [1959] that approximate equality holds in (5-38). This indicates that the method would converge hopelessly slowly, particularly if $\gamma_{\max} \gg \gamma_{\min}$. The latter is, of course, the case if the matrix is ill-conditioned, and this in turn is related to scaling of the variables (Sec. 5-1). Consequently, we can say that the steepest descent method is highly scale dependent. Relationship (5-38) can also serve as an estimate of the ultimate rate of convergence of the method for a general nonlinear function.

In conclusion, it is recommended that the steepest descent method be used only as an auxiliary technique in combination with other minimization techniques.

5-3 SECOND-ORDER GRADIENT (NEWTON) TECHNIQUES

The choice of locally steepest directions as directions of descent leads to gradient techniques that differ in the metric matrix A_k, which enters the definitions (5-7) of distances and (5-8) of descent directions. The second-order gradient technique is frequently referred to as Newton's method. It uses the Hessian matrix $\text{Hess}(\mathbf{x}^k)$ as the metric matrix A_k, which means that second-order derivatives of the objective function $M(\mathbf{x})$ are involved in computing the *second-order gradient directions* that are given, in view of (5-8), by

$$\mathbf{s}^k = -[\text{Hess}(\mathbf{x}^k)]^{-1}\nabla M(\mathbf{x}^k) \tag{5-39}$$

This formula provides descent directions if the Hessian $\text{Hess}(\mathbf{x}^k)$ is positive definite and symmetric. Then

$$(\mathbf{s}^k)^T\nabla M(\mathbf{x}^k) = -[\nabla M(\mathbf{x}^k)]^T[\text{Hess}(\mathbf{x}^k)]^{-1}\nabla M(\mathbf{x}^k) < 0,$$

because the inverse of a positive definite symmetric matrix is also positive definite.

One way to look at the second-order gradient technique is to consider a second-order Taylor series expansion of $M(\mathbf{x})$ about $\mathbf{x}^k$. This expansion is possible if the objective function $M(\mathbf{x})$ is twice differentiable. We obtain

$$M(\mathbf{x}) \cong M(\mathbf{x}^k) + (\mathbf{x} - \mathbf{x}^k)^T\nabla M(\mathbf{x}^k) + \tfrac{1}{2}(\mathbf{x} - \mathbf{x}^k)^T \,\text{Hess}(\mathbf{x}^k)(\mathbf{x} - \mathbf{x}^k)$$

If this expansion is differentiated with respect to $\mathbf{x}$, we obtain

$$\nabla M(\mathbf{x}) \cong \nabla M(\mathbf{x}^k) + \text{Hess}(\mathbf{x}^k)(\mathbf{x} - \mathbf{x}^k).$$

Suppose that the expansion of $\nabla M(\mathbf{x})$ is computed at $\mathbf{x} = \mathbf{x}^{k+1}$. Our objective is to maximize $[M(\mathbf{x}^k) - M(\mathbf{x}^{k+1})]$ at the kth iteration. One may attempt to achieve this by picking $\mathbf{x}^{k+1}$ so that $\nabla M(\mathbf{x}^{k+1}) = 0$, which implies then that

$$\mathbf{x}^{k+1} - \mathbf{x}^k = -[\text{Hess}(\mathbf{x}^k)]^{-1}\nabla M(\mathbf{x}^k) \tag{5-40}$$

This provides the second-order gradient direction $\mathbf{s}^k$ as defined by (5-39). It follows that a steplength of $\lambda^k = 1$ for all k is appropriate.

If $M(\mathbf{x})$ is a quadratic as defined by (5-9), then a single step of unit length along the second-order gradient direction leads to a solution, because the minimum of $M(\mathbf{x}) = a + \mathbf{b}^T\mathbf{x} + \frac{1}{2}\mathbf{x}^T A\mathbf{x}$ is $\hat{\mathbf{x}} = -A^{-1}\mathbf{b}$, and one step along the second-order gradient direction gives

$$\mathbf{x}^{k+1} = \mathbf{x}^k - A^{-1}[\mathbf{b} + A\mathbf{x}^k] = -A^{-1}\mathbf{b} = \hat{\mathbf{x}}$$

regardless of k. Figure 5-5 illustrates this process. One is, of course, tempted

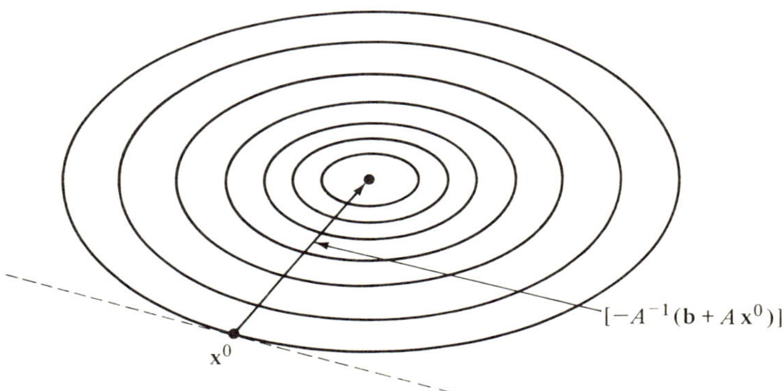

Figure 5-5 Quadratic funtion $M(\mathbf{x}) = a + b^T\mathbf{x} + \frac{1}{2}\mathbf{x}^T A\mathbf{x}$ minimized by the second-order gradient technique.

to use the steplength $\lambda^k = 1$, $k = 0, 1, \ldots$ because, in addition to the aforementioned justification, it involves no added computation. In practice, this is not done because in many cases $M\{\mathbf{x}^k - [\text{Hess}(\mathbf{x}^k)]^{-1}\nabla M(\mathbf{x}^k)\} > M(\mathbf{x}^k)$, which is referred to as "overshooting" or "undershooting" and is basically a result of the fact that considerations leading to $\mathbf{x}^{k+1} = \mathbf{x}^k - [\text{Hess}(\mathbf{x}^k)]^{-1}\nabla M(\mathbf{x}^k)$ are not valid at $\mathbf{x}^{k+1}$. It is recommended that the method be used in conjunction with an optimal or nonoptimal step strategy (Sec. 5-1).

Iteration Steps

1 Initialize, set $k = 0$, evaluate $M(\mathbf{x}^0)$; $\mathbf{x}^0$ is given.
2 Evaluate the gradient $\nabla M(\mathbf{x}^k)$ at the current point.

3 Compute the current Hessian matrix Hess($\mathbf{x}^k$). This matrix is symmetric and only $(\frac{1}{2})n(n+1)$ distinct terms need to be evaluated and stored.

4 Solve the system of linear equations

$$\text{Hess}(\mathbf{x}^k)\mathbf{s}^k = -\nabla M(\mathbf{x}^k) \tag{5-41}$$

for the direction vector $\mathbf{s}^k$. Normalize $\mathbf{s}^k$ so that $\|\mathbf{s}^k\| = 1$.

5 Compute the normalized current derivative of the objective function in the descent direction

$$\delta^k = \frac{|(\mathbf{s}^k)^T \nabla M(\mathbf{x}^k)|}{\|\nabla M(\mathbf{x}^k)\|} \tag{5-42}$$

6 If $\delta^k < \tau_1$, $\|\nabla M(\mathbf{x}^k)\| < \tau_2$, and a very accurate value of the solution vector is not required, terminate the iteration ($\tau_1 > 0$ and $\tau_2 > 0$ are predetermined tolerances). Otherwise, if $(\mathbf{s}^k)^T \nabla M(\mathbf{x}^k) \geq 0$, set $\mathbf{s}^k := -\mathbf{s}^k$. Proceed to step 7.

7 Compute the current descent steplength λ^k [Chap. 3 gives methods of optimization along the line $(\mathbf{x}^k + \lambda \mathbf{s}^k)$].

a. Optimal step alternative: find λ^k which satisfies

$$M(\mathbf{x}^k + \lambda^k \mathbf{s}^k) = \min_{\lambda} M(\mathbf{x}^k + \lambda \mathbf{s}^k) \tag{5-43}$$

b. Nonoptimal step alternative: find λ^k which satisfies

$$M(\mathbf{x}^k + \lambda^k \mathbf{s}^k) \leq M(\mathbf{x}^k) \tag{5-44}$$

8 Compute the current descent step

$$\Delta \mathbf{x}^k = \lambda^k \mathbf{s}^k \tag{5-45}$$

9 Perform the current descent step

$$\mathbf{x}^{k+1} = \mathbf{x}^k + \Delta \mathbf{x}^k \tag{5-46}$$

10 Evaluate the objective function $M(\mathbf{x}^{k+1})$ at the new point.

11 If $M(\mathbf{x}^k) - M(\mathbf{x}^{k+1}) < \epsilon_1$ and $\|\Delta \mathbf{x}^k\| < \epsilon_2$, terminate the iteration; otherwise, proceed to step 12 ($\epsilon_1 > 0$ is a predetermined minimal function improvement and $\epsilon_2 > 0$ a predetermined minimal stepsize).

12 Accept the new point $\mathbf{x}^{k+1}$, set $k := k + 1$ and repeat from step 2.

Termination Criteria and Convergence

The second-order gradient process is terminated as indicated above; convergence theory for this algorithm is outlined in Sec. 5-1. According to this theory the second-order algorithm permits an additional second-order convergence test. It consists of an examination (in step 6) that essentially indicates whether the Hessian matrix Hess($\mathbf{x}^k$) is positive definite. In principle, the iteration may be continued if the Hessian fails this test, because $\nabla M(\mathbf{x}^k) = 0$ indicates that $\mathbf{x}^k$ is a stationary, but not a minimum, point of $M(\mathbf{x})$. However, the process should be continued from a different point, because at $\mathbf{x}^k$ the termination criteria are satisfied and a direction of descent cannot be generated.

Modifications and Acceleration Techniques

Acceleration by Updating the Hessian Once in Several Iterations. To save computational work, the Hessian matrix Hess($\mathbf{x}$) may be updated less frequently than once in every iteration. For example, it may be updated once in every l iterations and then $[\text{Hess}(\mathbf{x}^k)]^{-1}$ is used to compute the direction of descent in relation (5-41) instead of $[\text{Hess}(\mathbf{x}^j)]^{-1}$ for $j = k, k + 1, \ldots, k + l - 1$. It is conceivable that the objective function improvement per step may be reduced if this shortcut is adopted, because the Hessian is used in a neighborhood in which it may not be strictly valid. Nevertheless, the reduction in computational work per step may result in increased overall efficiency of the modified method. In practice, this modification is frequently successful. The parameter l varies from case to case and is best determined by experimentation with each particular problem.

Modifications to Ensure Generation of Descent Directions. If the Hessian matrix Hess($\mathbf{x}$) is not positive definite, relation (5-41) may not generate descent directions. Furthermore, the Hessian may not be invertible and then the technique may break down altogether. In this sense, the second-order gradient technique may be considered unstable. This is a serious disadvantage of the technique, and modifications for its correction should be implemented. These modifications are of four types.

1 In the solution of (5-41) the Hessian matrix may have negative eigenvalues. In this case, Pearson [1969] proposes using a direction $\mathbf{s}^k$ "roughly" along an eigenvector corresponding to a negative eigenvalue. Thus, a region may be located where the Hessian is positive definite. The reader may recall that one of the factors ensuring positive definiteness of the Hessian is the positivity of its eigenvalues (Sec. 1-5). This scheme is based on the computation of the eigenvalues and certain eigenvectors of the Hessian matrix. In a slightly different scheme, also based on eigenvalue computation, Greenstadt [1967] simply replaces any negative eigenvalues of the Hessian matrix by their absolute values and zero eigenvalues by units. Suppose then, that $\{\gamma_1^k, \gamma_2^k, \ldots, \gamma_n^k\}$ are the eigenvalues of Hess($\mathbf{x}^k$) and $\{\xi_1^k, \ldots, \xi_n^k\}$ are the corresponding eigenvectors. Because Hess($\mathbf{x}^k$) is symmetric we have

$$\text{Hess}(\mathbf{x}^k) = \sum_{j=1}^{n} \gamma_j^k \xi_j^k (\xi_j^k)^T$$

and Greenstadt recommends the use of

$$\overline{\text{H}}\text{ess}(\mathbf{x}^k) = \sum_{j=1}^{n} |\gamma_j^k| \xi_j^k (\xi_j^k)^T \tag{5-47}$$

in relation (5-41). The remainder of the technique is unchanged.

2 Another approach, proposed by Goldfeld, et al. [1966, 1968], combines the Hessian with a diagonal matrix αI and uses

$$[\text{Hess}(\mathbf{x}^k) + \alpha I] \tag{5-48}$$

instead of $[\text{Hess}(\mathbf{x}^k)]$ in (5-41). The matrix αI is such that the matrix (5-48) is positive definite. A similar scheme was proposed earlier by Marquardt [1963] and others for the solution of nonlinear equation sets. We may observe that as $\alpha \longrightarrow 0$ the scheme reduces to a second-order gradient technique. On the other hand, as $\alpha \longrightarrow \infty$ the relative weight of the Hessian in (5-48) decreases and the scheme tends to the steepest descent scheme. Thus, depending on the value of α, the iterative procedure behaves like the steepest descent or the second-order gradient descent. Furthermore, as α increases from zero, the step when regarded as a function of α—that is $\|\Delta\mathbf{x}^k(\alpha)\| = \|[\text{Hess}(\mathbf{x}^k) + \alpha I]^{-1}\nabla M(\mathbf{x}^k)\|$ decreases monotonically and is, in general, automatically of the appropriate magnitude. As α increases, the angle between $\Delta\mathbf{x}^k(\alpha)$ and the steepest descent direction decreases monotonically. Goldfeld, et al. [1966] proposed to determine α in a manner dependent upon the smallest eigenvalue of the Hessian $\text{Hess}(\mathbf{x}^k)$ and a parameter adjusted, on the basis of experience, from step to step.

In a more general scheme, Goldfeld, et al. [1968] introduced a metric matrix B instead of I, and used

$$[\text{Hess}(\mathbf{x}^k) + \alpha B] \tag{5-49}$$

instead of $[\text{Hess}(\mathbf{x}^k)]$ in Eq. [5-41]. The matrix αB is chosen so that the matrix (5-49) is positive definite. Here α and B are updated from iteration to iteration, offering somewhat more maneuverability than the previous scheme.

These two schemes were compared to the conjugate direction and variable metric techniques in numerical experiments. The Goldfeld, et al. schemes were somewhat more economical than the other techniques in terms of function evaluations. Furthermore, they were superior in reliability of convergence. Their slightly lower overall computational efficiency for the type of problem tested might have been expected. The evaluation of the tested functions was inexpensive in comparsion to the added computation in determining α or αB. The latter involves finding the eigenvalues of the Hessian matrix. A stepsize adjustment scheme was superimposed on the method, in effect resulting in nonoptimal steps. This makes it hard to assess which part of the result is due to the use of (5-48) or (5-49) in relation (5-41), and which is due to the nonoptimal step strategy. Shanno [1970] investigates the selection of the parameter α in (5-48). He shows that it is desirable to pick α such that $dM/d\alpha = 0$ or nearly so, minimizing $M(\alpha)$. In practice, the zero point of this derivative is not found to high accuracy, for reasons of economy in function evaluations. If α is thus chosen with precision, it frequently may be possible to save function evaluations in determining a proper steplength λ; instead, the step is performed with $\lambda = 1$ (not normalized direction). Further experimentation with this scheme seems desirable.

From our own computational experience, it is recommended that the determination of α or αB from the eigenvalues of the Hessian matrix be bypassed and, instead, that α (not αB) be used and updated on the basis of computational progress. An initial value of α is increased (decreased) by a factor of 5 to 10 depending on failure (success) with the resulting direction. This variation should be lower and upper bounded in some reasonable manner (for example: $10^{-10} \leq \alpha \leq 10^5$). The use of the matrix (5-48) is recommended, but (5-49) can also be used with B determined so that problem variables are properly scaled (Sec. 1-9).

3 Another modification of the method was proposed by Goldstein and Price [1967]. It avoids the Hessian and is particularly suitable if this matrix is ill-behaved. Instead of step 3 of the unmodified algorithm, the jth column of the kth metric matrix A_k (not the inverse) is computed by means of

$$\text{col}_j\,(A_k) = \frac{\nabla M(\mathbf{x}^k + \theta_k I_j) - \nabla M(\mathbf{x}^k)}{\theta_k} \tag{5-50}$$

where $\theta_0 = r > 0$ a predetermined number and $\theta_k = r\|\mathbf{s}^{k-1}\|$ for $k = 1, 2, \ldots$ and where I_j is the jth column of the $n \times n$ identity matrix, and $0 < r$. Then a direction $\mathbf{s}^k$ is computed as follows (compare with step 4 of the unmodified technique): If $k = 0$ or if A_k is singular or if it would generate a direction of ascent—that is,

$$[\nabla M(\mathbf{x}^k)]^T(A_k)^{-1}\nabla M(\mathbf{x}^k) \leq 0 \tag{5-51}$$

set $\mathbf{s}^k = -\nabla M(\mathbf{x}^k)$; otherwise, set $\mathbf{s}^k = -(A_k)^{-1}\nabla M(\mathbf{x}^k)$. Next a steplength is determined (compare with step 7 of the unmodified technique). A temporary new point is computed

$$\hat{\mathbf{x}}^{k+1} = \mathbf{x}^k + \lambda\mathbf{s}^k \tag{5-52}$$

where $\lambda = 1$ is a trial value (in general, $\lambda > 0$) and the auxiliary function

$$g(\mathbf{x}^k, \lambda) = \frac{M(\mathbf{x}^k) - M(\mathbf{x}^{k+1})}{\lambda(\mathbf{s}^k)^T\nabla M(\mathbf{x}^k)} \tag{5-53}$$

If $g(\mathbf{x}^k, \lambda) < \delta$, where δ is a predetermined number so that $0 < \delta < 1/2$, we adjust λ so that $\delta \leq g(\mathbf{x}^k, \lambda^k) \leq 1 - \delta$; otherwise, we set $\lambda^k = \lambda$.

Finally, the descent step is calculated and performed with this steplength as in steps 8 and 9 of the unmodified technique. The remainder of the technique is also unchanged. For certain $M(\mathbf{x})$, the convergence has been proved to be superlinear and possibly of second order [Goldstein and Price, 1967]. Numerical experiments with $n = 2$ and $n = 6$ show that the modified method compares favorably with the unmodified method. A variation of this method has been proposed by Vetters [1970]. It is shown to have higher-order convergence without requiring significantly more computation.

4 Techniques for forcing the direction $\mathbf{s}^k$ to be descent by modifying the LU decomposition of the Hessian matrix Hess($\mathbf{x}^k$) have been suggested by Fiacco and McCormick [1968] and by Matthews and Davies [1971].

Discussion

The second-order gradient technique has good overall convergence and is especially efficient in the vicinity of the minimum, or in general for convex problems. The method will yield the minimum of a positive definite quadratic form in a single step of unit length. It is invariant under linear transformations of the variables (Sec. 1-9). The method involves a relatively large amount of computational work per step, because second derivatives of the objective function $M(\mathbf{x})$ have to be computed and n equations in n unknowns must be solved. In some cases, this shortcoming outweighs the faster convergence of the method. Computer codes of the method are fairly complex and require considerable storage. When the Hessian matrix is not positive definite the method may become unstable with respect to the generation of descent directions. To avoid this, additional computational work may be required as noted in the preceding section.

The second-order gradient method is recommended under the following circumstances:

1 The objective function $M(\mathbf{x})$ is nearly quadratic, or a good initial approximation $\mathbf{x}^0$ to the minimum is available, in which case the iteration will probably be in a region where the function can be reasonably represented by a quadratic.

2 Analytical expressions are available for first and second derivatives of the objective function $M(\mathbf{x})$, thus eliminating the need for much of the computational labor in numerical differentiation.

The method should be used with the acceleration modification whereby the Hessian matrix is updated less frequently than once in every iteration. Furthermore, it should be used either modified to ensure positive definiteness of the inverse of the metric matrix or with automatic termination when directions generated are no longer directions of descent. Note, however, that such termination could be premature.

5-4 CONJUGATE DIRECTIONS TECHNIQUES

This section deals with Powell's [1964] simple method using conjugate descent directions that satisfy the definition (5-10). Because conjugate directions are used, the technique will minimize a quadratic in n steps. The conjugate directions descent technique does not involve derivatives of the objective function $M(\mathbf{x})$, and is particularly useful if one is interested in avoiding the computational work involved in differentiation.

The basic cycle of the method starts with a point $\mathbf{x}^0$ and generates n points $\mathbf{x}^1, \ldots, \mathbf{x}^n$ by n minimizations along n given linearly independent directions $\mathbf{s}^0, \ldots, \mathbf{s}^{n-1}$ as follows:

1 Initialize, set $k = 0$.

2 Solve the problem of minimization along the line (λ may be positive or negative, because the direction may be descent or ascent—the negative of descent); that is, find λ^k that minimizes $M(\mathbf{x}^k + \lambda\mathbf{s}^k)$.

3 Compute the current descent step $\Delta\mathbf{x}^k = \lambda^k\mathbf{s}^k$.

4 Perform the current descent step $\mathbf{x}^{k+1} = \mathbf{x}^k + \Delta\mathbf{x}^k$.

5 If $k < n - 1$, set $k := k + 1$ and repeat from step 2; otherwise, terminate the cycle.

The coordinate axes are used as an initial set of directions. Consequently, the initial cycle is the (Southwell) relaxation process (Chap. 4). Subsequent cycles use new directions determined so that the last l of the n directions chosen for the $(l + 1)$st cycle are mutually conjugate. After n cycles all the directions are mutually conjugate with respect to the positive definite matrix A, if the quadratic function (5-9) is minimized. As a result, in theory the minimum of the quadratic function can be found in the $(n + 1)$st cycle. In this sense, the conjugate directions descent technique of Powell can be thought of as an improvement of the method of relaxation or of rotating coordinates, because it uses mutually conjugate, rather than simply orthogonal, search directions.

The set of new directions for use in a new iteration cycle is obtained from the previous set by $\mathbf{s}^k := \mathbf{s}^{k+1}$, $k = 0, \ldots, n - 2$; and $\mathbf{s}^{n-1} = \mathbf{x}^n - \mathbf{x}^0$. The starting point for the new cycle is the point at which $\min_{\lambda} M[\mathbf{x}^n + \lambda(\mathbf{x}^n - \mathbf{x}^0)]$ occurs. Consider now the direction $\mathbf{s}^{n-1}$ being reused (as $\mathbf{s}^{n-2}$) in the new iteration cycle. In the current cycle the minimum found in this direction was at $\mathbf{x}^n$. Hence, using the current cycle superscripts we have for the quadratic $M(\mathbf{x})$:

$$(\mathbf{s}^{n-1})^T\nabla M(\mathbf{x}^n) = (\mathbf{s}^{n-1})^T(A\mathbf{x}^n + \mathbf{b}) = 0 \tag{5-54}$$

Unless the current cycle was the first, its starting point $\mathbf{x}^0$ was also found as the minimum along a line in the direction $\mathbf{s}^{n-1}$ (in view of the definition of $\mathbf{s}^{n-1}$). Hence,

$$(\mathbf{s}^{n-1})^T\nabla M(\mathbf{x}^0) = (\mathbf{s}^{n-1})^T(A\mathbf{x}^0 + \mathbf{b}) = 0 \tag{5-55}$$

Note that the minimization leading to $\mathbf{x}^n$ was along $\{\mathbf{x}^{n-1} + \lambda\mathbf{s}^{n-1} \mid -\infty < \lambda < \infty\}$; minimization leading to $\mathbf{x}^0$ was along $\{\hat{\mathbf{x}}^n + \lambda\mathbf{s}^{n-1} \mid -\infty < \lambda < \infty\}$, where $\hat{\mathbf{x}}^n$ denotes $\mathbf{x}^n$ of the previous cycle. Normally $\mathbf{x}^{n-1}$ and $\hat{\mathbf{x}}^n$ and consequently also $\mathbf{x}^0$ and $\mathbf{x}^n$ are distinct points, provided at least one λ^k, $k = 0, \ldots, n-1$ is not zero. This condition is assumed to be satisfied. Subtracting (5-55) from (5-54) yields $(\mathbf{s}^{n-1})^T A(\mathbf{x}^n - \mathbf{x}^0) = 0$, which shows that the newly introduced direction $\mathbf{x}^n - \mathbf{x}^0$ is conjugate to the direction $\mathbf{s}^{n-1}$ introduced in the previous cycle. In general, the last l of the n directions used in the $(l+1)$st cycle are mutually conjugate with respect to the positive definite matrix A.

Unfortunately, Powell [1964] found that this simple procedure tends to generate directions that are linearly dependent, or very nearly so. This situation may arise if one or more of the steplengths turn out to be zero or very small. To avoid this difficulty, Powell suggested an improved version of the procedure in which the newly introduced direction $(\mathbf{x}^n - \mathbf{x}^0)$ is not accepted if its acceptance would result in destruction of the linear independence of the directions. A criterion was developed for determining whether to accept the new direction $(\mathbf{x}^n - \mathbf{x}^0)$ in the set or reuse old directions in the new search cycle. The criterion also indicates which direction from the old set should be replaced by $(\mathbf{x}^n - \mathbf{x}^0)$. Thus, one of the previously chosen conjugate directions may be discarded or the same set of directions may be reused in several cycles, and therefore more than n cycles may be needed to find the minimum of a quadratic.

Iteration Steps

1 Initialize, set $k = 0$, and set the n initial directions equal to the coordinate axes:

$$\mathbf{s}^l = \mathbf{e}_{l+1} \qquad l = 0, \ldots, n-1 \tag{5-56}$$

where $\mathbf{e}_j$ is the jth unit vector. A starting point $\mathbf{x}^0$ is given.

2 Evaluate $M(\mathbf{x}^k)$, and save this information for $k = 0, \ldots, n-1$ for use in step 8.

3 Solve the problem of minimization along the line (λ may be positive or negative); that is, find λ^k satisfying

$$M(\mathbf{x}^k + \lambda^k \mathbf{s}^k) = \min_{\lambda} M(\mathbf{x}^k + \lambda \mathbf{s}^k) \tag{5-57}$$

4 Compute the current descent step

$$\Delta \mathbf{x}^k = \lambda^k \mathbf{s}^k \tag{5-58}$$

5 Perform the current descent step

$$\mathbf{x}^{k+1} = \mathbf{x}^k + \Delta \mathbf{x}^k \tag{5-59}$$

6 If $k < n-1$, set $k := k+1$ and repeat from step 2; otherwise, terminate the cycle and proceed to step 7.

7 If $|x_i^n - x_i^0| < \epsilon_i$, $i = 1, \ldots, n$, terminate the iteration; otherwise, proceed to step 8 ($\epsilon_i > 0$, $i = 1, \ldots, n$ are predetermined minimum steplength tolerances).

8 Find the integer j, $1 \leq j \leq n$ so that

$$\delta = [M(\mathbf{x}^{j-1}) - M(\mathbf{x}^{j})] = \max_{1 \leq i \leq n} [M(\mathbf{x}^{i-1}) - M(\mathbf{x}^{i})] \tag{5-60}$$

9 Let $M_1 = M(\mathbf{x}^0)$ and $M_2 = M(\mathbf{x}^n)$, and compute

$$M_3 = M(2\mathbf{x}^n - \mathbf{x}^0) \tag{5-61}$$

10 If either $M_3 \geq M_1$ or

$$(M_1 - 2M_2 + M_3)(M_1 - M_2 - \delta)^2 \geq \tfrac{1}{2}\delta(M_1 - M_3)^2 \tag{5-62}$$

reuse the current set of n directions $\mathbf{s}^0, \ldots, \mathbf{s}^{n-1}$ in the next cycle, set $\mathbf{x}^0 := \mathbf{x}^n$, $k := 0$ and repeat from step 2; otherwise, proceed to step 11.

11 Compute $\mathbf{s} = (\mathbf{x}^n - \mathbf{x}^0)$ and $\bar{\lambda}$, which minimizes $M(\mathbf{x}^n + \lambda\mathbf{s})$, and set

$$\mathbf{x}^0 = \mathbf{x}^n + \bar{\lambda}\mathbf{s} \tag{5-63}$$

12 Determine a new set of directions: $\{\mathbf{s}^0, \ldots, \mathbf{s}^{j-1}, \mathbf{s}^{j+1}, \ldots, \mathbf{s}^{n-1}, \mathbf{s}\}$ (that is, the current direction $\mathbf{s}^j$ is eliminated, j is as determined in step 8, and the new direction $\mathbf{s}$ is added). Set $k := 0$ and repeat from step 2.

Termination Criteria and Convergence

The iteration is terminated if in a cycle each of the variables has been changed by less than a preset tolerance:

$$|x_i^n - x_i^0| < \epsilon_i \qquad i = 1, \ldots, n \tag{5-64}$$

where $\mathbf{x}^0$ is the starting point and $\mathbf{x}^n$ is the terminal point of the cycle. The $\epsilon_i > 0$, $i = 1, \ldots, m$ are predetermined minimum steplength tolerances. This simple criterion is usually satisfactory. There are cases, however, in which it may terminate the process prematurely. Therefore, it may be advisable to investigate the behavior of the objective function $M(\mathbf{x})$ in the vicinity of this point. Powell suggests the following "safe" termination procedure for any method not using gradients:

1 The conjugate direction method is used until a point $\mathbf{x}^n$ is found that satisfies the termination criterion (5-64) with ϵ_i set equal to one tenth of the accuracy required of the variable x_i, for all $i = 1, \ldots, n$.

2 Each variable x_i is increased by $100\epsilon_i$.

3 Step 1 is repeated; let the resulting point be $\hat{\mathbf{x}}^n$.

4 Find $\hat{\hat{\mathbf{x}}}^n$, the minimum along the line through $\mathbf{x}^n$, and $\hat{\mathbf{x}}^n$; that is, $M(\hat{\hat{\mathbf{x}}}^n) = \min_{\lambda} \{M(\mathbf{x}) \mid \mathbf{x} = \mathbf{x}^n + \lambda(\hat{\mathbf{x}}^n - \mathbf{x}^n)\}$.

5 If

$$|x_i^n - \hat{\hat{x}}_i^n| < \epsilon_i \qquad \text{and} \qquad |\hat{\hat{x}}_i^n - \hat{x}_i^n| < \epsilon_i \tag{5-65}$$

the process is terminated.

6 Otherwise, the direction $\mathbf{s}^1$ is replaced by $(\mathbf{x}^n - \hat{\hat{\mathbf{x}}}^n)$, and the process is repeated from $\hat{\hat{\mathbf{x}}}^n$.

Convergence theory for this algorithm is provided by Theorem 5-1.

Modifications and Acceleration Techniques

A simplified version of Powell's conjugate directions descent technique has been developed by Zangwill [1967]: Given an initial point $\mathbf{x}_0^1$, the normalized coordinate system $\mathbf{s}_0^1, \ldots, \mathbf{s}_{n-1}^1$, and a scalar ϵ, $0 < \epsilon \leq 1$, the normalization is such that $\|\mathbf{s}_j^1\| = 1$ for $j = 0, \ldots, n-1$. Set $k = 1$ and $\delta^1 = 1$, and proceed to iteration cycle k:

1 Compute λ_j^k as the solution to the problem of minimization along the line

$$\min_{\lambda} M(\mathbf{x}_j^k + \lambda \mathbf{s}_j^k), \tag{5-66}$$

and let

$$\mathbf{x}_{j+1}^k = \mathbf{x}_j^k + \lambda_j^k \mathbf{s}_j^k \qquad j = 0, \ldots, n-1 \tag{5-67}$$

2 If $\|\mathbf{x}_n^k - \mathbf{x}_0^k\| < \tau$, a predetermined tolerance, terminate the process; otherwise, define $\mathbf{s}_n^k = (\mathbf{x}_n^k - \mathbf{x}_0^k)/\|\mathbf{x}_n^k - \mathbf{x}_0^k\|$. Compute λ_n^k as the solution to the problem of minimization along the line

$$\min_{\lambda} M(\mathbf{x}_n^k + \lambda \mathbf{s}_n^k) \tag{5-68}$$

and let

$$\mathbf{x}_0^{k+1} = \mathbf{x}_n^k + \lambda_n^k \mathbf{s}_n^k \tag{5-69}$$

3 Find $\lambda_m^k = \max_j \{\lambda_j^k,\ j = 0, 1, \ldots, n-1\}$.

4 If $\lambda_m^k \delta^k/\|\mathbf{x}_n^k - \mathbf{x}_0^k\| \geq \epsilon$, then set $\mathbf{s}_j^{k+1} = \mathbf{s}_j^k$ for all $j \neq m$, $j = 0, \ldots, n-1$, and set $\mathbf{s}_m^{k+1} = \mathbf{s}_n^k$. Set $\delta^{k+1} = \lambda_m^k \delta^k/\|\mathbf{x}_n^k - \mathbf{x}_0^k\|$ and $k := k + 1$; then repeat from step 1.

5 Otherwise, set $\mathbf{s}_j^{k+1} := \mathbf{s}_j^k$ for $j = 0, \ldots, n-1$, set $\delta^{k+1} := \delta^k$ and $k := k + 1$; then repeat from step 1.

Zangwill [1967] has shown that steps 3, 4, and 5 ensure the linear independence of the directions. These steps appear to be a rather simple scheme for generating, and constitute a criterion for selecting, new directions of descent. The scheme is aimed at overcoming the major shortcoming of the unmodified technique, which is the linear dependence of the directions. Computational experience should be gathered to see how it performs in practice.

Zangwill also suggested a new conjugate technique, which is essentially a combination of the original Powell idea and the relaxation technique. (Recall that in the relaxation technique only one variable is changed at a time.) The search directions generated by this method are mutually conjugate and linearly independent. Consequently, Theorem 5-1 applies. Convergence of the technique has been established, in theory, for strictly convex continuously differentiable functions. The technique should be subjected to numerical experimentation so that it can be properly assessed.

Discussion

The conjugate directions techniques have essentially quadratic convergence. Furthermore, no differentiation of the objective function $M(\mathbf{x})$ is required. These two properties make the technique very useful. It should be noted that n cycles are required to find the minimum of a quadratic; each cycle requires approximately the same amount of computational work as one steepest descent step involving first-order differentiation. The method and its modified versions are rather complicated in their computational logic and do not require less storage than other methods with first-order differentiation (for example, steepest descent or conjugate gradient). The method can detect directions along which the objective function varies very slowly. Such directions are included in the set of search directions, and further minimization thus becomes possible. It has been found, however, that the termination criterion is too stringent, and when problems with more variables are solved, the new directions are chosen less often. This may cause an increase in the number of iteration cycles required for convergence.

The conjugate directions technique is one of the best among those that do not use derivatives. It has been reported by Box [1966] that the technique has performed in several cases as well as the variable metric method (Sec. 5-6). It can be regarded as one of the most efficient techniques involving function evaluations only and should be tried before any other method of this group is used, with the possible exception of the modified variable metric method, which accepts difference approximations (Sec. 5-6).

5-5 GENERALIZED CONJUGATE GRADIENT METHODS

These are first-order gradient techniques in which the descent directions are mutually conjugate with respect to a positive definite matrix A corresponding to the quadratic function

$$M(\mathbf{x}) = a + \mathbf{b}^T\mathbf{x} + \tfrac{1}{2}\mathbf{x}^T A\mathbf{x}$$

The first is that of Fletcher and Reeves [1964], which has been tested thoroughly and used extensively. A second method in this class is the supermemory gradient method (actually, a group of methods), and a third is the projected gradient method, which also has a number of variants. These two methods are more recent than the conjugate gradient method and have been less exposed to testing and applications. They do appear to be very efficient and reliable, however.

The Conjugate Gradient Method [Fletcher and Reeves, 1964]

This method uses a general form of the Gram-Schmidt orthogonalization process to generate the conjugate descent directions. It has been utilized by Hestenes and Stiefel [1952] to solve sets of linear equations by minimizing the corresponding quadratic functions as discussed in Sec. 2-4 [Beckman, 1960]. The method hinges on (5-16)—that is, $(\mathbf{s}^l)^T \nabla M(\mathbf{x}^k) = 0, \quad l = 0, \ldots, k-1$—and the theory preceding it. It follows from this equation that the current gradient $\nabla M(\mathbf{x}^k)$ is linearly independent of all the previous conjugate descent directions $\mathbf{s}^0, \mathbf{s}^1, \ldots, \mathbf{s}^{k-1}$, and therefore it can be used to generate a subsequent conjugate direction. This suggests the following recurrence formula for the descent directions $\mathbf{s}^k$

$$\mathbf{s}^k = -\nabla M(\mathbf{x}^k) + \sum_{i=1}^{k} \alpha^i \mathbf{s}^{i-1} \tag{5-70}$$

where α^i are scalars chosen to satisfy the condition that $\mathbf{s}^k$ should be conjugate to all previously used descent directions $\mathbf{s}^0, \mathbf{s}^1, \ldots, \mathbf{s}^{k-1}$. The process starts by setting $\mathbf{s}^0 = -\nabla M(\mathbf{x}^0)$, where $\mathbf{x}^0$ is a given starting point. The parameters α^i are determined from the condition that $\mathbf{s}^k$ must satisfy the definition (5-10) of conjugacy $(\mathbf{s}^k)^T A \mathbf{s}^l = 0$ for $l = 0, \ldots, k-1$.

It is assumed that the formula for $\mathbf{s}^1$ is like (5-70): $\mathbf{s}^1 = -\nabla M(\mathbf{x}^1) + \alpha^1 \mathbf{s}^0$, and for conjugacy to hold, α^1 must satisfy

$$(\mathbf{s}^1)^T A \mathbf{s}^0 = [-\nabla M(\mathbf{x}^1) + \alpha^1 \mathbf{s}^0]^T A \mathbf{s}^0 = 0$$

Using (5-4) and (5-9), however,

$$A\mathbf{s}^0 = \frac{1}{\lambda^0} A(\mathbf{x}^1 - \mathbf{x}^0) = \frac{1}{\lambda^0}[\nabla M(\mathbf{x}^1) - \nabla M(\mathbf{x}^0)]$$

Hence, if $\lambda^0 \neq 0$

$$[-\nabla M(\mathbf{x}^1) - \alpha^1 \nabla M(\mathbf{x}^0)]^T [\nabla M(\mathbf{x}^1) - \nabla M(\mathbf{x}^0)] = 0$$

from which one can determine α^1 as

$$\alpha^1 = \frac{[\nabla M(\mathbf{x}^1)]^T \nabla M(\mathbf{x}^1)}{[\nabla M(\mathbf{x}^0)]^T \nabla M(\mathbf{x}^0)}$$

because $[\nabla M(\mathbf{x}^1)]^T \mathbf{s}^0 = -[\nabla M(\mathbf{x}^1)]^T \nabla M(\mathbf{x}^0) = 0$.

In a similar fashion, a formula can be developed for $\mathbf{s}^2$: $\mathbf{s}^2 = -\nabla M(\mathbf{x}^2) + \alpha^2 \mathbf{s}^1 + \alpha^1 \mathbf{s}^0$, with α^2 and α^1 determined so that $\mathbf{s}^2$ is conjugate to $\mathbf{s}^0$ and $\mathbf{s}^1$. It turns out that

$$\alpha^2 = \frac{[\nabla M(\mathbf{x}^2)]^T \nabla M(\mathbf{x}^2)}{[\nabla M(\mathbf{x}^1)]^T \nabla M(\mathbf{x}^1)} \qquad \text{and} \qquad \alpha^1 = 0$$

The general formula (5-70) for $\mathbf{s}^k$ can be established inductively, resulting in $\alpha^1 = \ldots = \alpha^{k-1} = 0$. It has been established for $k = 1, 2$ and using (5-16) and (5-10) it can be shown to hold for any k, if it holds for $k - 1$. The general formula for α^k corresponding to (5-70) is

$$\alpha^k = \frac{[\nabla M(\mathbf{x}^k)]^T \nabla M(\mathbf{x}^k)}{[\nabla M(\mathbf{x}^{k-1})]^T \nabla M(\mathbf{x}^{k-1})} \tag{5-71}$$

Because $\mathbf{s}^0$ is a descent direction it follows from (5-70) and (5-71) that $\mathbf{s}^k$, $k = 1, \ldots$ are also descent directions. These formulas can be used to determine the descent directions also when $M(\mathbf{x})$ is a general nonlinear function, not necessarily quadratic. In this case, however, the concept of conjugacy is not defined. Practical experience with the conjugate gradient method suggests that its performance can be improved if it is restarted from time to time. The only information carried over from one step to another is $\alpha^k\mathbf{s}^{k-1}$, and this may be progressively contaminated by computational error. The minimizations along the descent directions to determine the optimal steplengths are also inaccurate. We propose, therefore, to recycle the method every $n + 1$ iterations.

Iteration Steps. Iteration of the Fletcher-Reeves [1964] conjugate gradient method proceeds as follows:

1 Initialize, set $k = 0$, evaluate $M(\mathbf{x}^0)$; $\mathbf{x}^0$ is given.

2 Evaluate the gradient $\nabla M(\mathbf{x}^k)$ at the current point.

3 If $\|\nabla M(\mathbf{x}^k)\| < \tau$, and accuracy of the terminal point is unimportant, terminate the iteration; otherwise, proceed to step 4 ($\tau > 0$ is a predetermined tolerance).

4 Compute the current descent direction $\mathbf{s}^k$

$$\mathbf{s}^k = -\nabla M(\mathbf{x}^k) + \alpha^k\mathbf{s}^{k-1}$$

where $\alpha^0 = 0$ and

$$\alpha^k = \frac{[\nabla M(\mathbf{x}^k)]^T \nabla M(\mathbf{x}^k)}{[\nabla M(\mathbf{x}^{k-1})]^T \nabla M(\mathbf{x}^{k-1})} \qquad k = 1, \ldots, n \tag{5-72}$$

Then normalize $\mathbf{s}^k$ so that $\|\mathbf{s}^k\| = 1$.

5 Compute the current descent steplength λ^k that satisfies

$$M(\mathbf{x}^k + \lambda^k\mathbf{s}^k) = \min_{\lambda} M(\mathbf{x}^k + \lambda\mathbf{s}^k) \tag{5-73}$$

[Chap. 3 gives methods of optimization along the line $(\mathbf{x}^k + \lambda\mathbf{s}^k)$.]

6 Compute the current descent step

$$\Delta\mathbf{x}^k = \lambda^k\mathbf{s}^k \tag{5-74}$$

7 Perform the current descent step

$$\mathbf{x}^{k+1} = \mathbf{x}^k + \Delta\mathbf{x}^k \tag{5-75}$$

8 Evaluate the objective function $M(\mathbf{x}^{k+1})$ at the new point $\mathbf{x}^{k+1}$.

9 If $M(\mathbf{x}^k) - M(\mathbf{x}^{k+1}) < \epsilon_1$ and $\|\Delta\mathbf{x}^k\| < \epsilon_2$, terminate the iteration; otherwise, proceed to step 10 ($\epsilon_1 > 0$ is a predetermined minimal function improvement and $\epsilon_2 > 0$ a predetermined minimal stepsize).

10 Accept the new point $\mathbf{x}^{k+1}$ and if $k < n$, set $k := k + 1$; otherwise, set $\mathbf{x}^0 := \mathbf{x}^{k+1}$, $k := 0$, return to step 2.

Termination Criteria and Convergence. Convergence theory for this algorithm is provided by Theorem 5-1. Convergence theory outlined in Sec. 5-1 also applies. Termination criteria are the same as those used in the steepest descent technique.

Supermemory Gradient Methods

The *memory gradient* and *supermemory gradient* methods, recently proposed by Miele and Cantrell [1969] and investigated by Cragg and Levy [1969], reduce to the conjugate gradient method of Fletcher and Reeves [1964] when a quadratic function is minimized, and therefore can be regarded as a modification or a generalization of this method. The memory gradient method in turn is a special case of the supermemory gradient method.

The supermemory gradient method can be defined as follows:

$$\mathbf{x}^{k+1} = \mathbf{x}^k + \Delta\mathbf{x}^k \tag{5-76}$$

$$\Delta\mathbf{x}^k = -\beta_0 \nabla M(\mathbf{x}^k) + \sum_{i=1}^{r} \beta_i \Delta\mathbf{x}^{k-i} \tag{5-77}$$

where $\boldsymbol{\beta} = (\beta_0, \beta_1, \ldots, \beta_r)^T$ are computed at each step so as to give the greatest decrease of the function $M(\mathbf{x})$—that is, $\min_{\boldsymbol{\beta}} [M(\mathbf{x}^{k+1}) - M(\mathbf{x}^k)]$—and the indices i indicate the ith iteration previous to the kth. Recall that a relationship similar to (5-77) is used to derive the generalized conjugate gradient method where we assume that

$$\mathbf{s}^k = -\nabla M(\mathbf{x}^k) + \sum_{i=1}^{k} \beta_i \mathbf{s}^{k-i}$$

and require that $\mathbf{s}^k$ be conjugate to all previous directions $\mathbf{s}^0, \ldots, \mathbf{s}^{k-1}$. This requirement gives

$$\mathbf{s}^k = -\nabla M(\mathbf{x}^k) + \frac{[\nabla M(\mathbf{x}^k)]^T \nabla M(\mathbf{x}^k)}{[\nabla M(\mathbf{x}^{k-1})]^T \nabla M(\mathbf{x}^{k-1})} \mathbf{s}^{k-1}$$

and
$$\mathbf{x}^{k+1} = \mathbf{x}^k + \Delta\mathbf{x}^k = \mathbf{x}^k + \lambda^k \mathbf{s}^k$$

where λ^k is determined from $\min_{\lambda} M(\mathbf{x}^k + \lambda\mathbf{s}^k)$. Thus, $\mathbf{s}^k$ is a combination of $-\nabla M(\mathbf{x}^k)$ and $\mathbf{s}^{k-1}$ only. The supermemory gradient method gives also $\beta_2 = \cdots = \beta_r = 0$ if $M(\mathbf{x})$ is quadratic, but for a nonquadratic function it retains all components of (5-77) and, therefore, has extra degrees of freedom available. This feature explains the good performance of the method in computational experiments.

The supermemory gradient method is formally derived as follows:

A first-order Taylor series expansion of $M(\mathbf{x})$ about $\mathbf{x}^{k+1}$ gives

$$M(\mathbf{x}^{k+1}) \cong M(\mathbf{x}^k) + [\nabla M(\mathbf{x}^k)]^T \Delta\mathbf{x}^k$$

At the kth iteration we wish to minimize $M(\mathbf{x}^{k+1}) - M(\mathbf{x}^k)$; or equivalently, $\min\{[\nabla M(\mathbf{x}^k)]^T\Delta\mathbf{x}^k\}$. The supermemory gradient methods use descent steps $\Delta\mathbf{x}$ satisfying the constraint

$$K = (\Delta\mathbf{x}^k - \sum_{i=1}^{r} \beta_i \Delta\mathbf{x}^{k-i})^T(\Delta\mathbf{x}^k - \sum_{i=1}^{r} \beta_i \Delta\mathbf{x}^{k-i}) \tag{5-78}$$

where K, β_i, $i = 1, \ldots, r$, and r are prescribed. The index i indicates the ith iteration previous to the kth. The solution of the constrained minimization problem defined above may be found by forming the Lagrangian (Sec. 1-7) and seeking its saddlepoint. This condition yields (5-77) and

$$K = \beta_0^2[\nabla M(\mathbf{x}^k)]^T\nabla M(\mathbf{x}^k) \tag{5-79}$$

where $(2\beta_0)^{-1}$ is the Lagrange multiplier that has been introduced. Because of the relation (5-79) between K and β_0, it is sufficient to specify β_0 instead of K. If $\beta_i = 0$, $i = 1, \ldots, r$, then

$$\Delta\mathbf{x}^k = -\beta_0 \nabla M(\mathbf{x}^k)$$

and the method reduces to steepest descent. If, on the other hand, $r = 1$ and

$$\beta_1 = \frac{\beta_0^{(k)}[\nabla M(\mathbf{x}^k)]^T \nabla M(\mathbf{x}^k)}{\beta_0^{(k-1)}[\nabla M(\mathbf{x}^{k-1})]^T \nabla M(x^{k-1})}$$

where $\beta_0^{(l)}$ denotes β_0 at step l, then the method reduces to the generalized conjugate gradient method [Cantrell, 1969]. Furthermore, Cragg and Levy [1969] claim that the method reduces to the conjugate gradient method whenever a quadratic function is minimized, regardless of the value of r.

If we define

$$\mathbf{u}^0 = -\nabla M(\mathbf{x}) \quad \text{and} \quad \mathbf{u}^{k-i} = \Delta\mathbf{x}^{k-i} \qquad i = 1, \ldots, r \tag{5-80}$$

then the descent step can be rewritten as

$$\Delta\mathbf{x}^k = \sum_{i=0}^{r} \beta_i \mathbf{u}^{k-i} \tag{5-81}$$

The kth step of the algorithm depends on r previous steps $i = 1, \ldots, r$, and the first step is a steepest descent step. This motivates the name supermemory gradient method of order r for the method. The memory gradient method for which $r = 1$ is a special case. In general, we may have $r > 1$.

In the first r iterations (for $k = 0, \ldots, r - 1$), it is necessary to modify the technique and use

$$\beta_i = 0 \qquad i = k + 1, \ldots, r \tag{5-82}$$

The first iteration is a steepest descent step, the second ($k = 1$) is a memory gradient step and, in general, for $0 \leq k \leq r - 1$ the kth iteration is a supermemory gradient step of order k. It is therefore sufficient to describe the typical iteration of a supermemory gradient method of order r.

In the typical iteration step $r + 1$ parameters $\beta_0, \beta_1, \ldots, \beta_r$ have to be determined (we assume $k \geq r$). We can consider

$$M(\mathbf{x}^{k+1}) = M(\mathbf{x}^k + \sum_{i=0}^{r} \beta_i \mathbf{u}^{k-i}) = M(\beta_0, \beta_1, \ldots, \beta_r) \tag{5-83}$$

These parameters are found by an $(r + 1)$-dimensional search procedure which is intended to minimize $M(\beta_0, \ldots, \beta_r)$. In this search one attempts to satisfy (approximately) the necessary conditions $\partial M/\partial \beta_i = 0$, $i = 0, \ldots, r$, for min $M(\boldsymbol{\beta})$, where $\boldsymbol{\beta} = (\beta_0, \ldots, \beta_r)^T$. This is achieved by computing corrections $\delta\beta_i$ to values of β_i used at the previous iteration for $i = 0, \ldots, r$. Denoting the latter by β_i^*, we have

$$\beta_i = \beta_i^* + \delta\beta_i \qquad i = 0, \ldots, r \tag{5-84}$$

Initially—that is, for $k = 0$—we have $\beta_i^* = 0 \qquad i = 0, \ldots, r$.

The Newton method with a step bisection is used to compute the corrections $\delta\beta_i$, $i = 0, \ldots, r$. The general iteration for $k \geq r$ proceeds as follows:

1 Determine a temporary new point

$$\hat{\mathbf{x}}^{k+1} = \mathbf{x}^k + \sum_{i=0}^{r} \beta_i^* \mathbf{u}^{k-i} \tag{5-85}$$

2 Compute elements f_i of a vector $\mathbf{f}$ as follows:

$$f_i = [\nabla M(\mathbf{x}^{k+1})]^T \mathbf{u}^{k-i} \qquad i = 0, \ldots, r \tag{5-86}$$

if $\|\mathbf{f}\| \leq \epsilon_1$, (a small positive number), terminate the process; otherwise, proceed and compute elements f_{ij} of a matrix F as follows:

$$f_{ij} = (\mathbf{u}^{k-i})^T \operatorname{Hess}(\hat{\mathbf{x}}^{k+1}) \mathbf{u}^{k-j} \qquad i, j = 0, \ldots, r \tag{5-87}$$

3 Solve the system of linear equations $F\Delta\boldsymbol{\beta} + \mathbf{f} = 0$ for the vector $\Delta\boldsymbol{\beta} = (\Delta\beta_0, \ldots, \Delta\beta_r)^T$.

4 Compute

$$\rho = -\operatorname{sign}(\mathbf{f}^T \Delta\boldsymbol{\beta}) \tag{5-88}$$

and set $\mu = 1$.

5 Compute

$$\begin{aligned} \delta\beta_i &= \rho\mu\Delta\beta_i \\ \beta_i &= \beta_i^* + \alpha\beta_i \end{aligned} \qquad i = 0, \ldots, r \tag{5-89}$$

6 If $M(\beta_0, \ldots, \beta_r) \leq M(\beta_0^*, \ldots, \beta_r^*)$, proceed to step 7; otherwise, set $\mu := \mu/2$ and repeat from step 5.

7 Compute the current descent step

$$\Delta\mathbf{x}^k = \sum_{i=0}^{r} \beta_i \mathbf{u}^{k-i} \tag{5-90}$$

8 Perform the current descent step

$$\mathbf{x}^{k+1} = \mathbf{x}^k + \Delta\mathbf{x}^k \tag{5-91}$$

9 Set $\beta_i^* := \beta_i$, $i = 0, \ldots, r$, and $k := k + 1$, and repeat from step 1.

Cragg and Levy [1969] propose to terminate the procedure when $\|\mathbf{f}\| \leq \epsilon_1$, where $\mathbf{f}$ is evaluated at the current β_i, or one may use $\|\mathbf{f}\| \leq \epsilon_2 \|\mathbf{f}^0\|$, where $\mathbf{f}$ on the right side is evaluated for the $\beta_i = 0$, $i = 0, \ldots, r$. A third termination criterion that may be used is $|\delta\beta_i| \leq \epsilon_3 |\beta_i|$, $i = 0, \ldots, r$. In these expressions, ϵ_1, ϵ_2, and ϵ_3 are small positive numbers.

Cragg and Levy [1969] report their comparison of the supermemory gradient method with $r = 1$ and $r = 3$ to a number of methods, including the steepest descent and conjugate gradient methods. Two problems were solved, including that of Wood (Sec. 1-10). In both cases, the supermemory gradient method with $r = 1$ and $r = 3$ performed better, in terms of computer time, than the steepest descent and conjugate gradient methods.

Projected Gradient Methods

A method proposed by Myers [1968] and later in a somewhat modified form by Pearson [1969] can also be regarded as a conjugate gradient method. Conjugate descent directions are generated and used, and only first-order differentiation of the objective function is required. In principle, the descent directions $\mathbf{s}^k$ are chosen as projections of the gradient $\nabla M(\mathbf{x}^k)$ orthogonal to $[\mathbf{y}^0, \mathbf{y}^1, \ldots, \mathbf{y}^{k-1}]$, where we define

$$\mathbf{y}^l = \nabla M(\mathbf{x}^{l+1}) - \nabla M(\mathbf{x}^l) \qquad l = 0, \ldots, k-1 \tag{5-92}$$

Define also a matrix Y_k as

$$Y_k = [\mathbf{y}^0, \mathbf{y}^1, \ldots, \mathbf{y}^{k-1}] \tag{5-93}$$

Then the directions $\mathbf{s}^k$ are chosen to satisfy $Y_k^T\mathbf{s}^k = 0$, $k = 0, 1, \ldots$, and, consequently,

$$Y_k^T\Delta\mathbf{x}^k = 0 \qquad k = 0, 1, \ldots \tag{5-94}$$

For the quadratic $M(\mathbf{x})$ given by (5-9) we have

$$Y_k = [A\Delta\mathbf{x}^0, A\Delta\mathbf{x}^1, \ldots, A\Delta\mathbf{x}^{k-1}] = A\Delta\bar{X}_k \tag{5-95}$$

therefore, (5-94) implies that the descent directions $\mathbf{s}^0, \mathbf{s}^1, \ldots, \mathbf{s}^{k-1}$ are mutually conjugate with respect to the positive definite matrix A. This is so, provided that the steplengths $\lambda^0, \lambda^1, \ldots, \lambda^{k-1}$ are nonzero.

Recall that $\Delta\mathbf{x}^l = \lambda^l\mathbf{s}^l$, $l = 0, \ldots, k-1$. Pearson [1969] shows that equation (5-94) is satisfied if the algorithm proceeds as follows:

$$\mathbf{s}^k = -H_k^T\nabla M(\mathbf{x}^k) \tag{5-96}$$

where $H_0 = R$ and

$$H_k = R - RY_k(Y_k^TRY_k)^{-1}Y_k^TR \tag{5-97}$$

with R an arbitrary, positive definite matrix. Equation (5-97) gives rise to a number of algorithms; one of these, which has been tested, is $H_0 = R_0 = I$ and, when k is a multiple of n, reset $H_k = I$. We have

$$H_k = H_{k-1} - \frac{(H_{k-1}\mathbf{y}^{k-1})(H_{k-1}\mathbf{y}^{k-1})^T}{(\mathbf{y}^{k-1})^T H_{k-1}\mathbf{y}^{k-1}}, \text{ for } k = 1, \ldots, n-1 \tag{5-98}$$

Myers [1968] proposes formula (5-98) as a modification of the variable metric method (Sec. 5-6). It is shown that the H_k computed by means of (5-98) is symmetric and positive semidefinite and, if used in (5-96) for a quadratic, generates directions identical to those generated by the variable metric method. It is not a quasi-Newton method because H_k does not tend to the inverse of the Hessian of the quadratic form in n steps. Since the quasi-Newton property is desirable, Myers [1968] proposes to reset the algorithm when k is a multiple of n to

$$H_k = R_k = \sum_{l=0}^{k-1} \frac{\Delta\mathbf{x}^l(\Delta\mathbf{x}^l)^T}{(\Delta\mathbf{x}^l)^T\mathbf{y}^l} \tag{5-99}$$

which approximates the inverse of the Hessian. (This resetting is motivated by the form of H_k used in the variable metric method, Sec. 5-6.) Kelley, et al. [1970] did numerical experiments with a technique using formulas (5-98) and (5-99). In these experiments the accuracy of the searches for a minimum along the descent lines varied. With a high-accuracy search, the variable metric technique of Sec. 5-6 was found to outperform this algorithm.

Another formula for resetting the algorithms when k is a multiple of n has been proposed by Pearson [1969]. This is

$$H_k = R_k = R_{k-1} + \frac{(\Delta\mathbf{x}^{k-1} - R_{k-1}\mathbf{y}^{k-1})(H_{k-1}\mathbf{y}^{k-1})^T}{(\mathbf{y}^{k-1})^T H_{k-1}\mathbf{y}^{k-1}} \tag{5-100}$$

It can be shown that R_k given by (5-100) satisfies $R_k Y_k = \Delta\bar{X}_k$; hence, (5-95) is satisfied with $R_k = A^{-1}$. Thus, the direction of search $\mathbf{s}^k$ is reset to a Newton direction when k is a multiple of n and a quadratic form is minimized.

The projected gradient method simply uses equations (5-96) and (5-98) to compute the descent directions, instead of step 4 of the iteration process of the generalized conjugate gradient method. The only other change of the iteration as described for the generalized method is in step 10 where one sets $k := k + 1$ even if $k \geq n$. Resetting then takes place in step 4, formula (5-98). Numerical tests reported by Pearson [1969] indicate that in certain cases the projected gradient method is less efficient than the generalized conjugate gradient method; however, it has provided a solution where the latter technique failed, as in problems that arose from a sequential transformation of a constrained problem. Because of the importance of reliability, the projected gradient method may be preferred to the conjugate gradient method for problems resulting from sequential transformations of constrained problems. The projected gradient method with resetting by means of formula (5-99) was tested by Myers [1968]. The results obtained for a quadratic function were almost identical to those obtained by the variable metric method, and more accurate than those obtained by conjugate gradient techniques. When applied to a nonquadratic function, the projected gradient method was superior in convergence to conjugate gradient techniques, but it was more susceptible to error propagation. Numerical experiments reported by Pearson [1969] indicate that formula (5-100) for resetting the algorithm did not produce sufficiently good results to make it worthwhile.

Sorenson [1969] proposes to replace (5-71) for α^k by

$$\alpha^k = \frac{(\mathbf{y}^{k-1})^T \nabla M(\mathbf{x}^k)}{(\mathbf{y}^{k-1})^T \mathbf{s}^{k-1}} \tag{5-101}$$

In view of (5-95), this formula is the same for a quadratic function as

$$\alpha^k = \frac{(\mathbf{s}^{k-1})^T A \nabla M(\mathbf{x}^k)}{(\mathbf{s}^{k-1})^T A \mathbf{s}^{k-1}} \tag{5-102}$$

The latter had been recommended by Hestenes and Stiefel [1952] as being less sensitive to round-off error than (5-71). It is also interesting to note that the Sorenson formula for α^k will generate direction vectors $\mathbf{s}^k$, by means of (5-70), which are orthogonal to $\mathbf{y}^{k-1}$. This is suggested by Sorenson [1969] as the main factor in the good performance of the variable metric method described in the next section. Of course, for general nonquadratic functions, even the modified conjugate gradient directions do not approach the second-order gradient (Newton) directions as the variable metric directions do; that is, this is not a quasi-Newton method. Sorenson further shows that the modified conjugate gradient procedure, in the quadratic case, implicitly constructs direction vectors with a projection matrix. With

$$\mathbf{s}^0 = -\nabla M(\mathbf{x}^0) \qquad \text{and} \qquad \mathbf{s}^k = -H_k^T \nabla M(\mathbf{x}^k) \tag{5-103}$$

one has

$$H_k^T = H_{k-1}^T - \frac{[H_{k-1}^T \nabla M(\mathbf{x}^k)](\mathbf{y}^{k-1})^T}{(\mathbf{y}^{k-1})^T H_{k-1}^T \nabla M(\mathbf{x}^k)} \tag{5-104}$$

and, furthermore, these directions are (for all k) scalar multiples of those generated by the variable metric method. This result has also been obtained in a slightly different form by Myers [1968] and Huang [1970].

Sorenson [1969] shows that the PARTAN-accelerated steepest descent process (Sec. 5-2) goes through the same points for a quadratic objective function as the conjugate gradient method, but it also goes through an equal number of other points, which are generated by the gradient steps. Thus, PARTAN-accelerated steepest descent requires twice as many minimizations along the line as this version of the conjugate gradient method.

Discussion

The various conjugate gradient methods have the properties of first-order gradient combined with quadratic convergence due to the conjugacy of the descent directions. They have overall good convergence properties and the Fletcher-Reeves method has also a relatively modest requirement for computer storage. The computer codes corresponding to the methods are simple. Normally, more than one cycle of n steps is required, and in resetting the computation, all information collected in previous iterations is discarded. The use of those methods in preference to the variable metric technique is recommended only if computer storage economy is more important than computational efficiency. They should be used, however, in preference to the first-order gradient method (Sec. 5-2). The supermemory gradient method and the projected gradient method appear very promising, and their applicability should be further explored.

Luenberger [1970] has provided an extension of the conjugate gradient method to handle problems with linear constraints. He shows how to reduce the problem of minimizing a quadratic function subject to linear constraints to an equivalent problem of solving a set of linear equations. This problem in turn can be reduced to an unconstrained minimization problem. The difficulty is that the coefficient matrix of the set of equations is not positive definite; to overcome this, a new algorithm is proposed, which has many of the advantages of the conjugate gradient technique. Furthermore, if the matrix is positive definite, the new algorithm essentially reduces to the latter technique.

5-6 VARIABLE METRIC METHODS

A significant development in the area of descent techniques was made by Davidon [1959] and extended by Fletcher and Powell [1963]. This work led to the so-called "variable metric" methods of minimization. Excellent discussions of these methods, including their theoretical background as well as numerical experimentation, have been recently published by Pearson [1969], Huang [1970], Huang and Levy [1970], and Adachi [1971]. Recall from Sec. 5-1 that the gradient techniques differ in the choice of the metric matrix A_k This matrix enters the definitions of distances and directions of search. The variable metric methods start with an arbitrary metric matrix. In most practical applications $A_0 = I$, the $n \times n$ identity matrix, just as in the first-order steepest descent technique (Sec. 5-2). Of course, also $A_0^{-1} = I$, and consequently the first descent direction is the gradient direction defined by (5-26). In each of the subsequent steps the inverse A_k^{-1} of the metric matrix is updated.

The Features of the Variable Metric Method

The variable metric method typically has the following algorithmic structure:

1 Given $\mathbf{x}^0$ and a positive definite metric matrix H_0 (for example, $H_0 = I$, the n-dimensional identity matrix), set $k = 0$ and compute $\nabla M(\mathbf{x}^0)$.
2 Perform a descent step to the new point $\mathbf{x}^{k+1}$ by means of $\mathbf{x}^{k+1} = \mathbf{x}^k - \lambda^k H_k^T \nabla M(\mathbf{x}^k)$ where λ^k minimizes $M[\mathbf{x}^k - \lambda H_k^T \nabla M(\mathbf{x}^k)]$.
3 Compute $\nabla M(\mathbf{x}^{k+1})$.
4 Update inverse of the metric matrix to obtain H_{k+1}. In general, H_{k+1} is a function of H_k, $\Delta \mathbf{x}^{k-1} = \mathbf{x}^k - \mathbf{x}^{k-1}$, and $\mathbf{y}^{k-1} = \nabla M(\mathbf{x}^k) - \nabla M\ (\mathbf{x}^{k-1})$.
5 Set $k := k + 1$ and repeat from step 2.

The updating in step 4 is done in such a manner that if a quadratic function $M(\mathbf{x}) = a + \mathbf{b}^T\mathbf{x} + \frac{1}{2}\mathbf{x}^T A \mathbf{x}$ is minimized, we have the following properties:

1 All descent directions generated are mutually conjugate with respect to the positive definite matrix A, which is associated with the quadratic function. Consequently, Theorem 5-1 applies, and a quadratic function will be minimized in at most n steps.
2 The updated matrices H_k are all positive definite and converge to the inverse of the matrix A—the Hessian of the quadratic function—when the minimum is approached. Therefore, in the vicinity of the minimum the directions generated by the variable metric method are approximately second-order gradient directions as defined by (5-39).

If a general nonlinear function is minimized, these properties are not generally obtained. The second property of the variable metric method makes it a quasi-second-order gradient (or quasi-Newton) method and distinguishes it from the conjugate gradient (Sec. 5-5). The descent directions $\mathbf{s}^k$ used by the variable metric technique are computed by means of

$$\mathbf{s}^k = -H_k^T \nabla M(\mathbf{x}^k) \tag{5-105}$$

where H_k is the inverse of the current (kth) metric matrix A_k. As H_k is updated, A_k varies since $A_k^{-1} = H_k$. The variation of this matrix in the process implies that the definition and measurement of distances also vary. Throughout this chapter we use general formula (5-105) although in most of the cases discussed H_k is symmetric and $H_k^T = H_k$.

If H_k is positive definite, then the directions defined by (5-105) are descent directions because then $(\mathbf{s}^k)^T \nabla M(\mathbf{x}^k) = -[\nabla M(\mathbf{x}^k)]^T H_k \nabla M(\mathbf{x}^k) < 0$. Thus, the positive definiteness of the matrices H_k assures, at least theoretically, that the method is stable in the sense of generating directions of descent. As we shall see, precautionary measures have to be taken to ensure this stability in computational practice.

Variable metric techniques differ depending on the scheme used for updating H_k. Different updating schemes are possible because the aforementioned two conditions do not determine H_k uniquely. All these methods may be regarded as second order because H_k is updated in a manner dependent on differences between gradients of the objective function $M(\mathbf{x})$. Also, their computational storage requirement is similar to that of the second-order gradient technique. These methods do, however, involve function and gradient evaluations only, and not second-order differentiation.

We shall proceed to review briefly the arguments leading to the different variable metric methods:

Recall the definition (5-6) of the matrix $\Delta\bar{X}_k$:

$$\Delta\bar{X}_k = [\Delta\mathbf{x}^0, \Delta\mathbf{x}^1, \ldots, \Delta\mathbf{x}^{k-1}]$$

The column vectors $\Delta\mathbf{x}^l$, $l = 0, \ldots, k-1$, are the descent steps defined by (5-4): $\Delta\mathbf{x}^l = \lambda^l\mathbf{s}^l = \mathbf{x}^{l+1} - \mathbf{x}^l$. We now define a matrix Y_k (Sec. 5-5) as $Y_k = [\mathbf{y}^0, \mathbf{y}^1, \ldots, \mathbf{y}^{k-1}]$, where the column vectors $\mathbf{y}^l$, $l = 0, \ldots, k-1$ are given by $\mathbf{y}^l = \nabla M(\mathbf{x}^{l+1}) - \nabla M(\mathbf{x}^l)$.

All the elements of the matrices $\Delta\bar{X}_k$ and Y_k are known at the kth iteration. Hence, they may be used to compute $\mathbf{s}^k$. Furthermore, for the quadratic $M(\mathbf{x}) = a + \mathbf{b}^T\mathbf{x} + \frac{1}{2}\mathbf{x}^TA\mathbf{x}$, we have

$$\mathbf{y}^l = (\mathbf{b}^T + A\mathbf{x}^{l+1}) - (\mathbf{b}^T + A\mathbf{x}^l) = A\Delta\mathbf{x}^l \qquad l = 0, 1, \ldots$$

or equivalently,

$$A^{-1}Y_k = \Delta\bar{X}_k \tag{5-106}$$

If a general nonquadratic function is minimized, the Hessian matrix varies with $\mathbf{x}$. Nevertheless, the variable metric methods tend to make H_k close to the matrix A^{-1}, and therefore H_k is chosen to satisfy

$$H_kY_k = \Delta\bar{X}_k \tag{5-107}$$

Since the descent directions are defined by $\mathbf{s}^k = -H_k^T\nabla M(\mathbf{x}^k)$, we obtain $Y_k^T\mathbf{s}^k = -Y_k^TH_k^T\nabla M(\mathbf{x}^k) = (\Delta\bar{X}_k)^T\nabla M(\mathbf{x}^k)$. If the condition

$$(\Delta\bar{X}_k)^T\nabla M(\mathbf{x}^k) = 0 \tag{5-108}$$

is satisfied, then

$$Y_k^T\mathbf{s}^k = \frac{Y_k^T\Delta\mathbf{x}^k}{\lambda^k} = 0$$

In view of the definition of Y_k and $\mathbf{y}^l$, this implies that for $\lambda^l \neq 0$, $l = 0, \ldots, k-1$

$$\frac{(\Delta\mathbf{x}^l)^TA\Delta\mathbf{x}^k}{\lambda^l} = 0 \qquad l = 0, 1, \ldots, k-1$$

which means that the current step $\Delta\mathbf{x}^k$ is conjugate to all previous steps. It can be shown that condition (5-108) holds, and this also implies that $\nabla M(\mathbf{x}^n) = 0$, since after n steps we obtain a nonsingular matrix $\Delta\bar{X}_n$ (conjugate columns). Similarly, the matrix Y_k has rank n and from (5-107) we get $H_n = A^{-1}$, which means that the directions $\mathbf{s}^k = -H_k^T \nabla M(\mathbf{x}^k)$ approach second-order gradient directions and, therefore, the variable metric method is a quasi-Newton method.

Variants of the Variable Metric Methods

Recursion formulas for matrices H_k satisfying (5-107) can be obtained in a number of ways. One of these has been proposed by Broyden [1967]. He suggested a recursion formula $H_k = H_{k-1} + \alpha_k$. To satisfy (5-107), it is necessary that $H_k \mathbf{y}^{k-1} = \Delta \mathbf{x}^{k-1}$, and hence,

$$\alpha_k \mathbf{y}^{k-1} = \Delta \mathbf{x}^{k-1} - H_{k-1} \mathbf{y}^{k-1}$$

This expression is satisfied if

$$\alpha_k = \Delta \mathbf{x}^{k-1} (\mathbf{q}^{k-1})^T - H_{k-1} \mathbf{y}^{k-1} (\mathbf{z}^{k-1})^T$$

where $\mathbf{q}^{k-1}$ and $\mathbf{z}^{k-1}$ are arbitrary vectors, such that

$$(\mathbf{q}^{k-1})^T \mathbf{y}^{k-1} = (\mathbf{z}^{k-1})^T \mathbf{y}^{k-1} = 1 \tag{5-109}$$

and hence,

$$H_k = H_{k-1} + \Delta \mathbf{x}^{k-1} (\mathbf{q}^{k-1})^T - H_{k-1} \mathbf{y}^{k-1} (\mathbf{z}^{k-1})^T \tag{5-110}$$

Expressions (5-109) and (5-110) lead to the generalized (Davidon) algorithm with

$$\begin{aligned} (\mathbf{q}^{k-1})^T &= \beta^{k-1} (\mathbf{s}^{k-1})^T - \gamma^{k-1} (\mathbf{y}^{k-1})^T H_{k-1} \\ (\mathbf{z}^{k-1})^T &= \delta^{k-1} (\mathbf{y}^{k-1})^T H_{k-1} + \gamma^{k-1} (\Delta \mathbf{x}^{k-1})^T \end{aligned} \tag{5-111}$$

$$\begin{aligned} \beta^{k-1} &= \frac{1 + \gamma^{k-1} (\mathbf{y}^{k-1})^T H_{k-1} \mathbf{y}^{k-1}}{(\mathbf{s}^{k-1})^T \mathbf{y}^{k-1}} \\ \delta^{k-1} &= \frac{1 - \gamma^{k-1} (\Delta \mathbf{x}^{k-1})^T \mathbf{y}^{k-1}}{(\mathbf{y}^{k-1})^T H_{k-1} \mathbf{y}^{k-1}} \end{aligned} \tag{5-112}$$

where γ^{k-1} is an arbitrary scalar, and H_0 is required to be symmetric and nonsingular. In this method H_k, rather than H_k^T, is used in (5-105), but this does not make any difference if H_k is symmetric. Broyden [1967] proves that the aforementioned formulas yield all the matrices H_k positive definite if $\gamma^{k-1} \geq 0$ and H_0 is positive definite.

Broyden [1970] also proves that the generalized algorithm (5-110) through (5-112) has quadratic convergence. Convergence may occur in fewer than n steps, an extreme case being one step when the initial matrix $H_0 = A^{-1}$. Broyden [1970] conjectures, therefore, that it is likely that the precise number of steps needed for convergence depends on H_0. He shows how for quadratic functions the choice of an arbitrary parameter affects convergence and how successive errors depend on H_0. These conclusions are tentatively generalized for nonquadratic functions. The Broyden algorithm (5-110) through (5-112) with $\gamma^{k-1} = 0, k = 0, 1, \ldots$ yields (5-116), Fletcher and Powell's [1963] variant of Davidon's [1959] algorithm. Broyden [1970] proposes and tests a specific algorithm designed to overcome the stability problems exhibited by the Davidon-Fletcher-Powell algorithm.

An argument similar to that yielding (5-110) through (5-112) is provided by Huang [1970]. He shows that search directions satisfying the nonorthogonality condition $[\nabla M(\mathbf{x}^k)]^T \mathbf{s}^k \neq 0$, $k = 0, 1, \ldots$ are descent directions, and that such directions are also mutually conjugate if the following generalized algorithm is used:

$$H_k = H_{k-1} + \rho \frac{\Delta \mathbf{x}^{k-1}(\mathbf{q}^{k-1})^T}{(\mathbf{q}^{k-1})^T \mathbf{y}^{k-1}} - \frac{H_{k-1}\mathbf{y}^{k-1}(\mathbf{z}^{k-1})^T}{(\mathbf{z}^{k-1})^T \mathbf{y}^{k-1}} \tag{5-113}$$

where ρ is an arbitrary constant and

$$\begin{aligned} \mathbf{q}^{k-1} &= C_1 \Delta \mathbf{x}^{k-1} + C_2 H_{k-1}^T \mathbf{y}^{k-1} \\ \mathbf{z}^{k-1} &= K_1 \Delta \mathbf{x}^{k-1} + K_2 H_{k-1}^T \mathbf{y}^{k-1} \end{aligned} \tag{5-114}$$

where C_1, C_2, K_1, and K_2 are scalar coefficients. These scalar coefficients are arbitrary except that K_1 and K_2 must not vanish simultaneously. Different choices of these coefficients can generate infinitely many different specific algorithms. One of these is the Broyden algorithm, which can be obtained from (5-113) and (5-114), assuming that H_k is symmetric and $H_k \mathbf{y}^{k-1} = \Delta \mathbf{x}^{k-1}$.

Huang shows that when a quadratic function is minimized, then for a given $\mathbf{x}^0$ and H_0 the sequences of directions $\mathbf{s}^0, \mathbf{s}^1, \mathbf{s}^2, \ldots$ and of points $\mathbf{x}^0, \mathbf{x}^1, \ldots$ generated by all specific algorithms are the same, independent of the choice of these constants. This is a generalization of the observation made by Myers [1968] concerning the Davidon-Fletcher-Powell algorithm and the conjugate gradient directions generated by (5-70) and (5-71), both of which are specific algorithms that can be generated by the generalized Huang algorithm. Myers has shown that when a quadratic function is minimized and $H_0 = I$, the directions produced by these two specific algorithms are positive scalar multiples of each other, and hence both iteration processes are identical. Furthermore, Myers has shown that projected gradient methods also generate directions identical to those generated by the Davidon-Fletcher-Powell algorithm for a quadratic.

Huang [1970] further examines ten specific algorithms. Numerical tests for these are reported by Huang and Levy [1970]. These algorithms include the Davidon-Fletcher-Powell ($\rho = 1$, $C_1 = 1$, $C_2 = 0$, $K_1 = 0$, and $K_2 = 1$) and others. Also included is the Fletcher-Reeves algorithm defined by (5-70) and (5-71). The Fletcher-Reeves algorithm requires the least computational work, and should be preferred for quadratic functions.

A slightly different development, also motivated by (5-106) and (5-107), was pursued by Pearson [1969]. It leads to a number of algorithms, including

Fletcher and Powell's modification of the Davidon algorithm. These algorithms all involve the computation of directions $\mathbf{s}^k$ by means of (5-105) and a recursion equation similar to Broyden's for the kth matrix H_k. Pearson shows that (5-107) is satisfied if H_k satisfies

$$H_0 = R \quad \text{and} \quad H_k = \Delta\bar{X}_k Y_k^* + R(I - Y_k Y_k^{**}) \qquad (5\text{-}115)$$

where Y_k^* and Y_k^{**} are of the form $(Y_k^T M Y_k)^{-1} Y_k^T M$, $k = 1, 2, \ldots$, where the matrices $\Delta\bar{X}_k$ and Y_k are as before, and R and M are positive definite symmetric matrices with $M = A^{-1}$ or $M = R$. The positive definite matrix A is that associated with the quadratic form (5-9) and the matrix I is the $n \times n$ identity matrix. Pearson [1969] proves Theorem 5-2.

Theorem 5-2 If the descent process using descent directions defined by (5-105) and (5-115) and optimal descent steps is applied to a quadratic function $M(\mathbf{x})$, defined by (5-9) with Hessian matrix A, then it terminates after at most n iterations; and if n iterations are required, then $H_n = A^{-1}$.

The proof is by induction and is based on theory associated with Theorem 5-1.

Equation (5-115) leaves open the choice of R, M, Y_k^*, and Y_k^{**}, and consequently may yield more than one algorithm. If one sets $M = A^{-1}$ in Y_k^* and $M = R$ in Y_k^{**},

$$Y_k^* = (Y_k^T A^{-1} Y_k)^{-1} Y_k^T A^{-1} \qquad \text{and} \qquad Y_k^{**} = (Y_k^T R Y_k)^{-1} Y_k^T R$$

Equation (5-115) then yields the recursion formula

$$H_k = H_{k-1} + \frac{\Delta\mathbf{x}^{k-1}(\Delta\mathbf{x}^{k-1})^T}{(\Delta\mathbf{x}^{k-1})^T\mathbf{y}^{k-1}} - \frac{(H_{k-1}\mathbf{y}^{k-1})(H_{k-1}\mathbf{y}^{k-1})^T}{(\mathbf{y}^{k-1})^T H_{k-1}\mathbf{y}^{k-1}} \qquad (5\text{-}116)$$

Equation (5-116) is the Fletcher and Powell modification of Davidon's algorithm. It yields H_k symmetric and positive definite. The choice $\gamma^{k-1} = 0$ in (5-111) and (5-112) also yields (5-116). Hence, the latter algorithm can be regarded as a special case of the general Broyden recursion formula, and as such, Broyden's proof of the positive definiteness of the matrices H_k also applies to it.

Note that in the projected gradient method, H_k is updated using formula (5-98). This formula is similar to (5-116) used for updating H_k in the variable metric method, but without its second term. Therefore, the projected gradient method is not a quasi-Newton scheme because H_k does not approach A^{-1}. The formula (5-99) suggested by Myers [1968] for resetting this algorithm is motivated by formula (5-116), and is intended as an adjustment to make the algorithm behave in a manner similar to that of a second-order gradient technique. Sorenson [1969] shows that if a quadratic function is minimized, the recursion formula (5-116) is equivalent to [*cf.* (5-98)] $H_k = P_k + Q_k$, where $P_0 = H_0 = I$, $Q_0 = 0$, and

$$\left.\begin{aligned} P_k &= P_{k-1} - \frac{(P_{k-1}\mathbf{y}^{k-1})(P_{k-1}\mathbf{y}^{k-1})^T}{(\mathbf{y}^{k-1})^T P_{k-1}\mathbf{y}^{k-1}} \\ Q_k &= Q_{k-1} + \frac{\Delta\mathbf{x}^{k-1}(\Delta\mathbf{x}^{k-1})^T}{(\mathbf{y}^{k-1})^T\Delta\mathbf{x}^{k-1}} \end{aligned}\right\} \quad \text{for} \quad k = 1, 2, \ldots$$

He also shows that the matrix Q_k has the property that after n steps $Q_n = A^{-1}$, and, furthermore, that for a quadratic function the direction vectors satisfy $\mathbf{s}^k = -P_k \nabla M(\mathbf{x}^k)$. That is, in this case the portion Q_k of H_k that approaches A^{-1} as $k \longrightarrow n$ does not influence $\mathbf{s}^k$. If $P_0 = H_0 = I$, then P_k projects the gradient $\nabla M(\mathbf{x}^k)$ (or any vector) orthogonally to $[\mathbf{y}^0, \mathbf{y}^1, \ldots, \mathbf{y}^{k-1}]$ (Sec. 5-5), and P_n must be a zero matrix, since $[\mathbf{y}^0, \mathbf{y}^1, \ldots, \mathbf{y}^{n-1}]$ are linearly independent. Therefore, of course, $H_n = Q_n = A^{-1}$. If P_0 is any positive definite matrix, then these results hold, in a slightly modified form.

In the general nonquadratic case, H_k is not related to any matrix A^{-1}. Furthermore, the direction $\mathbf{s}^k$ is orthogonal to $\mathbf{y}^{k-1}$, but not also to $\mathbf{y}^{k-2}$, $\mathbf{y}^{k-3}, \ldots, \mathbf{y}^0$ as in the quadratic case. This is so because of the addition to H_k of $\Delta\mathbf{x}^{k-1}(\Delta\mathbf{x}^{k-1})^T/(\mathbf{y}^{k-1})^T\Delta\mathbf{x}^{k-1}$. Sorenson conjectures that this orthogonality of the directions generated by the variable metric method is responsible for the good performance of the method when applied to general nonquadratic functions. In close proximity to the minimum the quadratic approximation of $M(\mathbf{x})$ is more accurate, the influence of the component Q_k of H_k less important, and one is "approaching" the quadratic case and quadratic convergence. It is shown that a modified version of the conjugate gradient method also has this property.

Application of the Variable Metric Methods.

Figure 5-6 depicts the process of minimizing a quadratic function by a variable metric or a conjugate gradient technique. In practice, it is most convenient to use $H_0 = I$. Computational experience indicates that the variable metric methods should be reset in a cyclic fashion: (for example, set $H_k := H_0$ every n iterations). A new cycle should be started when the computation indicates that H_k is no longer positive definite. Tests performed by Pearson [1969], Huang and Levy [1970], and others indicate that the minimizations along the line should be carried out to fairly high precision.

Iteration Steps (Davidon-Fletcher-Powell Method)

1 Initialize, set $k = 0$, evaluate $M(\mathbf{x}^0)$; $\mathbf{x}^0$ is given.
2 Evaluate the gradient $\nabla M(\mathbf{x}^k)$ at the current point.
3 Compute the current approximation H_k, of the inverse A_k^{-1} of the metric

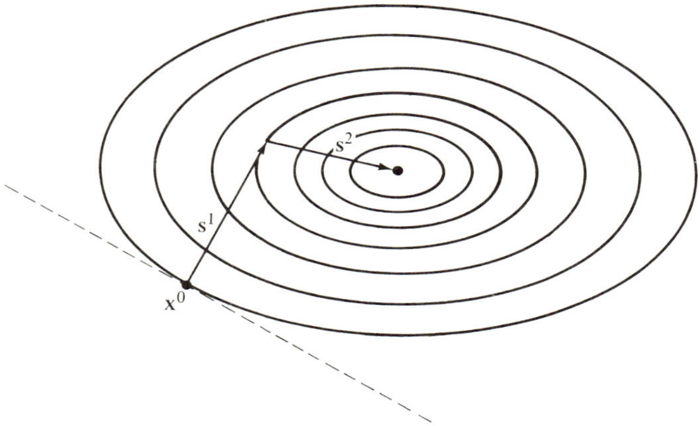

Figure 5-6 Quadratic funtion $M(\mathbf{x}) = a + b^T\mathbf{x} + \frac{1}{2}\mathbf{x}^T A\mathbf{x}$ minimized by the variable metric method or a conjugate gradient method

matrix; $H_0 = I_n$ (the $n \times n$ identity matrix), and

$$H_k = H_{k-1} + \alpha_k \tag{5-117}$$

where

$$\alpha_k = \frac{\Delta\mathbf{x}^{k-1}(\Delta\mathbf{x}^{k-1})^T}{(\Delta\mathbf{x}^{k-1})^T\mathbf{y}^{k-1}} - \frac{(H_{k-1}\mathbf{y}^{k-1})(H_{k-1}\mathbf{y}^{k-1})^T}{(\mathbf{y}^{k-1})^T H_{k-1}\mathbf{y}^{k-1}} \tag{5-118}$$

and where

$$\Delta\mathbf{x}^{k-1} = \mathbf{x}^k - \mathbf{x}^{k-1} \qquad \text{and} \qquad \mathbf{y}^{k-1} = \nabla M(\mathbf{x}^k) - \nabla M(\mathbf{x}^{k-1})$$

Recall that H_k is a symmetric matrix and that only $(1/2)n(n+1)$ distinct terms need to be evaluated and stored.

4 Compute the current descent direction $\mathbf{s}^k$

$$\mathbf{s}^k = -H_k^T\nabla M(\mathbf{x}^k) \tag{5-119}$$

and normalize $\mathbf{s}^k$ so that $\|\mathbf{s}^k\| = 1$.

5 Compute the normalized current derivative of the objective function in the descent direction

$$\delta^k = \frac{|(\mathbf{s}^k)^T\nabla M(\mathbf{x}^k)|}{\|\nabla M(\mathbf{x}^k)\|} \tag{5-120}$$

6 If $\delta^k \leq \tau_1$, $\|\nabla M(\mathbf{x}^k)\| < \tau_2$, and a very accurate value of the solution point is not required, terminate the iteration; otherwise, proceed to step 7 ($\tau_1 > 0$ and $\tau_2 > 0$ are predetermined tolerances).

7 If $(\mathbf{s}^k)^T\nabla M(\mathbf{x}^k) \geq 0$, set $\mathbf{s}^k = -\mathbf{s}^k$ and $H_k = I_n$.

8 Compute the current descent steplength λ^k [Chap. 3 gives methods of optimization along the line $(\mathbf{x}^k + \lambda\mathbf{s}^k)$] that satisfies

$$M(\mathbf{x}^k + \lambda^k\mathbf{s}^k) = \min_{\lambda} M(\mathbf{x}^k + \lambda\mathbf{s}^k) \tag{5-121}$$

9 Compute the current descent step

$$\Delta \mathbf{x}^k = \lambda^k \mathbf{s}^k \tag{5-122}$$

10 Perform the current descent step

$$\mathbf{x}^{k+1} = \mathbf{x}^k + \Delta \mathbf{x}^k \tag{5-123}$$

11 Evaluate the objective function $M(\mathbf{x}^{k+1})$ at the new point.

12 If $M(\mathbf{x}^k) - M(\mathbf{x}^{k+1}) < \epsilon_1$ and $\|\Delta \mathbf{x}^k\| < \epsilon_2$, terminate the iteration; otherwise, proceed to step 13 ($\epsilon_1 > 0$ is a predetermined minimal function improvement and $\epsilon_2 > 0$ a predetermined minimal stepsize).

13 Accept the new point $\mathbf{x}^{k+1}$, and if $k < n$, set $k := k + 1$; otherwise, set $\mathbf{x}^0 = \mathbf{x}^{k+1}$, $k = 0$; return to step 2.

Termination Criteria and Convergence

Convergence theory outlined in Sec. 5-1 applies to this technique. Theorem 5-1 applies if the objective function $M(\mathbf{x})$ is quadratic. Goldfarb [1969] has investigated a more general class of functions. He proved that the variable metric method with optimal steps converges to the global optimum of a strictly convex function whose second partial derivatives are bounded. Some convergence theorems for the Davidon-Fletcher-Powell method applied to nonquadratic functions have been proved by Powell [1969]. Termination criteria in the iteration step are essentially similar to those specified for the second-order gradient technique in Sec. 5-3.

Alternative Variable Metric Techniques

Some of the general theory on which the variable metric methods are based [e.g., (5-110), (5-113) and (5-115)] leaves certain parameters to be specified. Consequently, a number of alternative specific algorithms can be obtained. We present here only those alternatives that have been computationally tested.

The choice of $Y_k^* = Y_k^{**}$ in (5-115) with $M = A^{-1}$ leads to algorithm 2 of Pearson [1969]:

$$\alpha_k = \frac{(\Delta \mathbf{x}^{k-1} - H_{k-1}\mathbf{y}^{k-1})(\Delta \mathbf{x}^{k-1})^T}{(\mathbf{y}^{k-1})^T \Delta \mathbf{x}^{k-1}} \tag{5-124}$$

and if $M = R$, the same approach yields algorithm 3 of Pearson [1969]:

$$\alpha_k = \frac{(\Delta \mathbf{x}^{k-1} - H_{k-1}\mathbf{y}^{k-1})(H_{k-1}^T \mathbf{y}^{k-1})^T}{(\mathbf{y}^{k-1})^T H_{k-1} \mathbf{y}^{k-1}} \tag{5-125}$$

When either of these is used instead of (5-118), the process yields H_k in general unsymmetric and $\mathbf{s}^k = -H_k^T \nabla M(\mathbf{x}^k)$. This is, of course, a disadvantage because twice as much computer storage is then required for H_k. These two processes satisfy Theorem 5-2. Numerical experiments reported by

Pearson [1969] show that formulas (5-124) and (5-125) result in approximately the same computational efficiency as formula (5-118).

A set of ten specific algorithms was derived by Huang [1970] from (5-113) and (5-114) and tested by Huang and Levy [1970]. In addition to (5-118), (5-124), and (5-125), they tested the cases

$$\rho = K_2 = 0, \; K_1 = 1$$

yielding

$$\alpha_k = -\frac{H_{k-1}\mathbf{y}^{k-1}(\Delta\mathbf{x}^{k-1})^T}{(\Delta\mathbf{x}^{k-1})^T\mathbf{y}^{k-1}}$$

and

$$\rho = 0, \; K_1 = 1, \; K_2 = -1$$

yielding

$$\alpha_k = -\frac{H_{k-1}\mathbf{y}^{k-1}(\Delta\mathbf{x}^{k-1} - H_{k-1}^T\mathbf{y}^{k-1})^T}{(\Delta\mathbf{x}^{k-1} - H_{k-1}^T\mathbf{y}^{k-1})^T\mathbf{y}^{k-1}}$$

In all the above cases, H_k is generally not symmetric. Tested were also a number of conjugate gradient techniques such as (5-72) and (5-98) and the rank-one correction matrix (5-128) discussed below. The tests reported by Huang and Levy [1970] indicate that there were no differences between the algorithms when applied to quadratic functions and minor differences when applied to general nonquadratic functions.

A number of algorithms were proposed recently by Greenstadt [1970*a*, *b*] and Goldfarb [1970]. These are based on the updating of the matrix H_k as a variational problem. The fact that algorithms similar to the variable metric technique can be derived on the basis of such considerations is of theoretical interest. The computational advantages of these algorithms are not clear, however. They appear more complicated to code and did not perform any better in computational experiments [Greenstadt, 1970*a*, *b*] than the variable metric technique.

Modification for the Use of Difference Approximations

A modification of the variable metric technique enabling the use of difference approximations of derivatives has been proposed by Stewart [1967]. The modification simply consists of approximating the gradient $\nabla M(\mathbf{x})$ of the objective function by the forward difference formula

$$\nabla M_j \cong \frac{M(\mathbf{x} + \delta_j\mathbf{e}_j) - M(\mathbf{x})}{\delta_j} \qquad j = 1, \ldots, n \tag{5-126}$$

or by the central difference formula

$$\nabla M_j \cong \frac{M(\mathbf{x} + \delta_j\mathbf{e}_j) - M(\mathbf{x} - \delta_j\mathbf{e}_j)}{2\delta_j} \qquad j = 1, \ldots, n \tag{5-127}$$

where ∇M_j is the jth component of $\nabla M(\mathbf{x})$ and $\mathbf{e}_j$ is a vector with all components zero except the jth component, which is equal to unity. Stewart's paper is devoted primarily to the analysis of the stepsize δ_j. A careful choice of the steplength is necessary to balance effects of truncation and cancellation error. The reader is referred to Stewart's paper for further discussion. Numerical experimentation has shown this modification to be an efficient technique for the minimization of functions for which derivatives are not readily available. This would be an alternative to the conjugate directions technique (Sec. 5-4). In fact, a comparison between these two methods reported by Fletcher [1969] shows that the Stewart modification gains superiority as the number of variables increases. Stewart's method works well if the gradient approximation and linear searches are accurate. If either of these is not sufficiently accurate, Stewart's algorithm may converge very slowly or terminate at a nonoptimal point. Cullum [1972] has derived and tested three methods which also do not explicitly use derivatives but do not have this disadvantage.

Variance Methods

An important development related to variable metric methods has been reported by Broyden [1967], Davidon [1968], and Murtagh and Sargent [1970]. They propose algorithms in which the quasi-Newton property is obtained without minimizations along the descent directions. These algorithms may be more economical in terms of function evaluations, and should be further explored and experimented with. Davidon suggested the use of rank-one correction matrices $(H_k - H_{k-1})$; for example:

$$H_k = H_{k-1} + \frac{(\Delta\mathbf{x}^{k-1} - H_{k-1}\mathbf{y}^{k-1})(\Delta\mathbf{x}^{k-1} - H_{k-1}\mathbf{y}^{k-1})^T}{(\Delta\mathbf{x}^{k-1} - H_{k-1}\mathbf{y}^{k-1})^T\mathbf{y}^{k-1}} \tag{5-128}$$

where the matrix H_k is symmetric positive definite and satisfies the relation $H_k\mathbf{y}^{k-1} = \Delta\mathbf{x}^{k-1}$, which motivates the variable metric methods. Formula (5-128) is suggested for updating H_k, which is used in the iteration

$$\mathbf{x}^{k+1} = \mathbf{x}^k - \lambda^k H_k \nabla M(\mathbf{x}^k) \tag{5-129}$$

It has been proven that the method defined by (5-128) and (5-129) terminates within n iterations, and if n iterations have been performed, then $H_n = A^{-1}$ for the quadratic function $M(\mathbf{x}) = a + \mathbf{b}^T x + \frac{1}{2}\mathbf{x}^T A\mathbf{x}$. This is true for an arbitrary choice of λ^k, subject to the condition that λ^k does not lead to a singular H_k or $(\Delta\mathbf{x}^k - H_{k-1}\mathbf{y}^{k-1})^T\mathbf{y}^{k-1} = 0$. If $\lambda^k = 1$ is chosen, for example, H_k may not be positive definite, is which case one may add a different multiple of the correction matrix (positive definite) to H_{k-1}. If the choice of $\lambda^k = 1$ results in $M(\mathbf{x}^{k+1}) > M(\mathbf{x}^k)$, then $\mathbf{x}^{k+1}$ should be rejected as the successor point to $\mathbf{x}^k$.

Wolfe [1967] has observed that if a quadratic function is minimized, we may get $H_n = A^{-1}$ even if (5-129) is not used for the computation of $\mathbf{x}^{k+1}$. This interesting result will now be shown, following the argument given by Powell [1970].

Recall that for a quadratic function we have

$$A^{-1}\mathbf{y}^{l-1} = \Delta\mathbf{x}^{l-1} \qquad l = 1, 2, \ldots, k \tag{5-130}$$

and because the matrix H_l is computed to satisfy

$$H_l\mathbf{y}^{l-1} = \Delta\mathbf{x}^{l-1} \tag{5-131}$$

we obtain for a quadratic

$$A^{-1}\mathbf{y}^{l-1} = H_l\mathbf{y}^{l-1} \tag{5-132}$$

From (5-130) it follows that any vector $\mathbf{v}$ satisfies

$$(\Delta\mathbf{x}^{k-1} - H_{k-1}\mathbf{y}^{k-1})^T\mathbf{v} = (\mathbf{y}^{k-1})^T(A^{-1} - H_{k-1})\mathbf{v} \tag{5-133}$$

It can be shown that if $\mathbf{v}$ is such that

$$A^{-1}\mathbf{v} = H_{k-1}\mathbf{v} \tag{5-134}$$

then from (5-128), (5-133), and (5-134) we obtain

$$A^{-1}\mathbf{v} = H_k\mathbf{v} \tag{5-135}$$

because the product of the second term in (5-128) and $\mathbf{v}$ is zero. Equation (5-132) holds for $l = 1, \ldots, k$, and hence

$$A^{-1}\mathbf{y}^{l-1} = H_k\mathbf{y}^{l-1} \qquad l = 1, \ldots, k \tag{5-136}$$

[compare with (5-106) and (5-107)]. If the vectors $\Delta\mathbf{x}^k$ are linearly independent, then the matrix $Y_n = [\mathbf{y}^0, \mathbf{y}^1, \ldots, \mathbf{y}^{n-1}]$ is nonsingular, and

$$A^{-1}Y_n = H_nY_n \quad \text{implies} \quad H_n = A^{-1} \tag{5-137}$$

This result, which makes the algorithm a quasi-Newton algorithm, can be achieved while having a considerable degree of freedom in computing $\mathbf{x}^{k+1}$ from $\mathbf{x}^k$. Algorithmic details provided by Davidon [1969] deserve further computational experimentation. Recent tests performed by Bard [1970] indicate that the variance methods are better performers than the variable metric methods for minimizing sums of squares and similar problems.

As mentioned, the danger inherent in the variance methods is that H_k may not be positive definite, and therefore directions $\mathbf{s}^k$ which are computed on its basis may not be descent directions. Furthermore, Eq. (5-128) for updating H_{k-1} may render a singular H_k under certain circumstances. The method also may yield an unbounded correction. The variable metric methods, on the other hand, have quadratic convergence only if minimizations along the descent lines are performed, which involves perhaps more computational effort than one would like to invest. A new approach to the problem was suggested by Fletcher [1970]. This approach is motivated by the desire to eliminate the danger of H_k not being positive definite, as well as the need for minimizations along the descent lines, without sacrificing the convergence properties of the variable metric methods. The proposed algorithm generates positive definite H_k, and for quadratic functions the eigenvalues of H_k tend (in some sense) monotonically to those of the inverse of the Hessian of the quadratic function. Results of computational experiments made with the method are quite encouraging. To update H_k, the algorithm uses the variable metric technique and (5-118), as well as

$$\alpha_k = -\frac{\Delta \mathbf{x}^{k-1}(\mathbf{y}^{k-1})^T H_{k-1}}{(\Delta \mathbf{x}^{k-1})^T \mathbf{y}^{k-1}} - \frac{H_{k-1}\mathbf{y}^{k-1}(\Delta \mathbf{x}^{k-1})^T}{(\Delta \mathbf{x}^{k-1})^T \mathbf{y}^{k-1}} + \left(1 + \frac{(\mathbf{y}^{k-1})^T H_{k-1}\mathbf{y}^{k-1}}{(\Delta \mathbf{x}^{k-1})^T \mathbf{y}^{k-1}}\right)\frac{\Delta \mathbf{x}^{k-1}(\Delta \mathbf{x}^{k-1})^T}{(\Delta \mathbf{x}^{k-1})^T \mathbf{y}^{k-1}} \tag{5-138}$$

Discussion

The variable metric techniques have many of the advantages of the second-order gradient techniques without having some of their disadvantages. The advantages include fast convergence near a minimum and finiteness for a quadratic function. Due to the symmetry of H_k, the methods are also comparable in their computer storage requirements. Comparison with other quadratically convergent methods show that the variable metric methods perform better for general nonquadratic functions; thus, quadratic convergence alone does not indicate the full strength of a method. Recall that the generalized conjugate gradient method generates the same directions as the variable metric method for a quadratic. It is likely that the fast convergence of the method when applied to a general nonquadratic function is due to the fact that H_k approximately satisfies $H_k \mathbf{y}^{k-1} = \Delta \mathbf{x}^{k-1}$, which is analogous to $A^{-1}\mathbf{y}^{k-1} = \Delta \mathbf{x}^{k-1}$ for quadratic functions. As has been pointed out, this equation motivates the formulas for updating H_k in the variable metric methods.

The variable metric methods avoid the computation of second-order derivatives of the objective function. Instead, descent directions are generated by means of a simple scheme using first-order derivatives. The methods are basically stable—they generate directions of descent—because the matrices H_k are positive definite (at least in theory). In practice, stability is improved if the algorithm is recycled as recommended. It has also been indicated by a number of authors that the stability of the variable metric methods depends critically on the accuracy of the minimizations along the line. Computational difficulties may be encountered if the Hessian matrix Hess($\mathbf{x}$) is singular close to a solution. This singularity may not be inherent to the problem, but rather caused by poor scaling of the variables. Bard [1968] suggests to overcome this difficulty by recycling (resetting H_k), by proper scaling, and by increased precision of the computation.

Computational experience with unconstrained minimization problems arising from sequential transformations of constrained problems shows that the metric matrices associated with these problems may be ill-conditioned. Pearson [1969] advocates keeping track of the determinant $|H_k|$ to detect trouble. The determinant should not be evaluated without reference to the scaling of the metric matrix. If this determinant approaches zero in the computation, the metric matrix may be ill-conditioned, and the algorithm can be reset or the computation terminated. Pearson [1969] developed some recurrence formulas for these determinants. As can be seen, the same quantities are involved as those used in updating the inverse of the metric matrix. Consequently, not much added computation will be required.

For the unmodified algorithm [(5-117), (5-118)]:

$$|H_k| = |H_{k-1}| \frac{(\Delta x^{k-1})^T \mathbf{y}^k}{(\mathbf{y}^{k-1})^T H_{k-1} \mathbf{y}^{k-1}} \tag{5-139}$$

For the first modification, [(5-117), (5-124)]:

$$|H_k| = |H_{k-1}| \frac{(\Delta \mathbf{x}^{k-1})^T H_{k-1}^{-1} \Delta \mathbf{x}^{k-1}}{(\Delta \mathbf{x}^{k-1})^T \mathbf{y}^{k-1}} \tag{5-140}$$

For the second modification, [(5-117), (5-125)]:

$$|H_k| = |H_{k-1}| \frac{(\Delta \mathbf{x}^{k-1})^T \mathbf{y}^{k-1}}{(\mathbf{y}^{k-1})^T H_{k-1} \mathbf{y}^{k-1}} \tag{5-141}$$

Pearson [1969] has also shown how partial information concerning the Hessian matrix may be used to speed up the computation. This may be of special interest in a number of applications.

5-7 COMPUTING SYSTEMS

Computing systems applying descent techniques are listed in Tables 5-1 to 5-4. Refer also to combination program AESOP (Sec. 4-8). Programs from VIM are listed in the 1971 *Program Library Catalog* of VIM, Users Organization of Control Data Computers.

5-8 NUMERICAL RESULTS

In recent years a number of studies have compared the behavior of different techniques for unconstrained minimization. We briefly review here the main conclusions of some of these computational experiments. Most of the numerical studies of techniques for unconstrained minimization use test problems of two types. The first of these consists of relatively simple problems, similar to that of Wood (Sec. 1-10), which we have used in our illustrations of computations. Problems of this type vary in details such as the number of variables (typically 2 to 20), or possibly points at which the objective function may "misbehave." The second problem type arises when con-

Table 5-1 FIRST-ORDER GRADIENT TECHNIQUES

Title and Reference	*Type, Machine and Storage Locations*	*Experience*	*Remarks*
ALGORITHM 129—MINIFUN Whitley Wasscher	ALGOL 60 program; FORTRAN II version available at Signal Missile Support Agency, White Sands Missile Range, New Mexico	FORTRAN II version has been tested with simple problems	See corrections by Wasscher
ALGORITHM 203—STEEP 1 Wasscher Wallack Varah	ALGOL 60 program	FORTRAN IV version has been tested on IBM 7094 [Wallack]; Burroughs' extended ALGOL version has been tested on B5000 [Varah]	1. See corrections by Varah 2. See ATIVE—for computation of partial derivatives [Haubrich]
ALGORITHM 204—STEEP 2 Wasscher	ALGOL 60 program	None reported	

Table 5-2 CONJUGATE DIRECTIONS TECHNIQUES

Title and Reference	*Type, Machine and Storage Locations*	*Experience*	*Remarks*
VIM E4 NYU BOTM Swenson [1965]	FORTRAN IV program CDC 6600 CHIPPEWA OS 783 words	Has been tested with simple problems Successfully tested on Wood function	
VIM E4 BKY MINSER Beals [1965]	FORTRAN IV program CDC 6600 CHIPPEWA OS 876 words	None reported	
FMND	PL/1 program IBM 360	None reported	Includes modification due to Zangwill [1967]

Table 5-3 CONJUGATE GRADIENT TECHNIQUES

Title and Reference	*Type, Machine and Storage Locations*	*Experience*	*Remarks*
FUMICOG Fletcher and Reeves [1964]	ALGOL 60 program	Has been tested on Pegasus and KDF9 computers	
FMCG, DFMCG System/360 Scientific Subroutine Package (SSP)	FORTRAN IV program CDC 6600 340 words FMCG 722 words DFMCG	Has been thoroughly tested Both successfully tested on Wood problem on CDC 6600	Tests performed on CDC 6600

strained problems are transformed into unconstrained problems, as by means of sequential transformations. Most of the numerical experiments measure the amount of computational work needed for convergence, as defined by the technique. Computational work is typically measured in terms of iteration cycles (sets of n iterations), iterations, function and gradient evaluations, function evaluations, or simply computer time. Reliability of convergence is, of course, also taken into account.

Search and Descent Techniques

Fletcher [1965] compares the performance of search and descent techniques by means of simple test problems. One of the search techniques tested is the modified rotating coordinate method (Sec. 4-4). The descent technique tested is the method of conjugate directions (Sec. 5-4). The main conclusion of the study is that Powell's method of conjugate directions is generally the most powerful and the most efficient. The modified rotating coordinate method, on the other hand, is simple and particularly effective when the objective function in the vicinity of the minimum is not nearly quadratic. As the number n of variables increases (from 2 to 8) the comparison between the two methods becomes less favorable for the former. A third method (search) tested appeared inferior to the other two methods.

A comprehensive study by Box [1966] compares three search methods, three descent methods and two methods for minimizing sums of squares. The search methods tested are rotating coordinates (Sec. 4-3), modified rotating coordinates (Sec. 4-4), and simplex (Sec. 4-5). Descent methods included in the test are Powell's conjugate directions (Sec. 5-4), conjugate gradient (Sec. 5-5), and Davidon-Fletcher-Powell variable metric (Sec. 5-6).

On simple problems, the variable metric technique was consistently the best performer, with the conjugate directions method not far behind. The

Table 5-4 VARIABLE METRIC TECHNIQUES

Title and Reference	*Type, Machine and Storage Locations*	*Experience*	*Remarks*
ALGORITHM 251—Wells	ALGOL 60 program	None reported	See certification by Fletcher
VIM E4 BKY VARMIT Beals	FORTRAN IV program CDC 6600 CHIPPEWA OS 4516 words	Has been thoroughly tested; successfully tested on Wood function	(1) Admits linear constraints, (2) cubic interpolation, based on 2 points and 2 slopes, used for minimization along descent lines; (3) if H_k is not positive definite, its off diagonal elements are set equal to zero and its diagonal elements are set equal to their absolute values; (4) dimensioned for $n \leq 40$
VIM E4 NYU MINI Swenson	FORTRAN IV program CDC 6600 CHIPPEWA OS 889 words	Has been tested with simple problems; successfully tested on Wood function	Has been dimensioned for $n \leq 10$; minor coding changes necessary to increase $n > 10$.
FMFP, DFMFP System/360 Scientific Subroutine Package (SSP)	FORTRAN IV program CDC 6600 444 words FMFP 915 words DFMFP	Has been thoroughly tested; successfully tested on Wood function	Tests performed on CDC 6600.
FMD	PL/1 program IBM 360	None reported	

conjugate gradient and search techniques were found to be less successful. The performance of the search techniques deteriorated in particular when the number of variables increased (from 2 to 20). Box also solved constrained problems that were transformed into unconstrained problems by change of variables.

Another comprehensive study of computer programs has been made by Leon [1966] Eight methods were tested on five simple problems: pattern search (Sec. 4-2) and another similar search method; first-order gradient descent (Sec. 5-2); and PARTAN-accelerated first-order gradient (Sec. 5-2) and another variation thereof. Two other variations of accelerated first-order gradient and a variable metric technique (Sec. 5-6) were also tested. The conclusion reported is that the variable metric technique is the most consistently good performer. The first-order gradient, especially the accelerated version, also worked well in most cases, but was somewhat less reliable. The search methods were not always satisfactory performers.

The general conclusion of these studies, and others, seems to be that the descent techniques are more reliable and efficient performers than the search techniques. This is particularly true for problems with a larger number of variables. The search techniques are simple and efficient for simple problems with relatively few variables.

Various Descent Techniques

Goldfeld, et al. [1968] compared the variable metric technique, the conjugate directions technique, and their accelerated second-order gradient ("quadratic hill climbing") technique. They found the last to be very efficient, and concluded that it should be used whenever possible. The conclusion that a modified second-order gradient technique should be used whenever possible was also reached in various other numerical studies [Fiacco and McCormick, 1968]. When applicable, this method seems to outperform other descent methods in efficiency and reliability. Goldfeld, et al. [1968] found further that the method of conjugate directions had less consistent performance than the other two methods. It should be kept in mind, however, that this method is somewhat more economical in computer storage.

Fiacco and McCormick [1968] compared first- and second-order gradient, conjugate gradient, and variable metric methods. The latter two methods performed about equally well on an unconstrained problem resulting from a sequentially transformed constrained problem. The first-order gradient technique failed to converge in the allotted number of iterations. The second-order gradient technique performed best in terms of number of iterations. If number of function evaluations had been counted, the second-order method would not have been so outstanding, but possibly still the best performer.

Thorough comparisons of several descent methods have been made by Pearson [1969] and by Huang and Levy [1970]. Pearson's tests involved five problems of both the simple and transformed constrained types, and the methods tested were second-order gradient, conjugate gradient and its projected gradient modifications, and variable metric (including three of its variations). It should be noted that the one-dimensional minimizations in all these tests were done by means of the same Fibonacci search. Normal as well as reset modes of algorithms were tested whenever appropriate. In general, for simple problems the conjugate gradient technique or its projected gradient modification outperformed the variable metric techniques. For the transformed constrained problems, on the other hand, the basic (Fletcher-Powell-Davidon) variable metric technique was generally superior. Its two variations also appear competitive, although not sufficiently so to justify the added computer storage requirement due to the lack of symmetry of the matrix H_k. The projected second-order technique did not perform well enough to warrant the added complication. The techniques generally performed better in the reset mode on the somewhat ill-conditioned problems of the second kind (transformed constrained problems). For simple problems of the first type, the normal mode of the algorithms was usually more efficient. Huang and Levy [1970] tested ten conjugate gradient and variable metric methods. Only minor differences between the algorithms were evidenced when applied to nonquadratic problems of the simple type. The behavior of the algorithms with quadratic functions was identical.

Straeter [1969] tested with one simple problem and an optimal control problem reformulated as an unconstrained minimization problem. Methods tested were variable metric (Fletcher-Powell-Davidon), variance (Davidon), conjugate gradient (Fletcher-Reeves), and two modifications of the second-order gradient. It is interesting that the variance technique has shown faster overall convergence; at the same time, however, this technique is critically dependent on the stepsize adjustment, which should be further explored. The modified second-order gradient techniques also exhibited excellent convergence in these tests.

Additional test results are reported in Cragg and Levy [1969] and Tabak [1969].

In conclusion, it seems that the modified second-order gradient (Newton) method should be used whenever analytical derivatives are available, and that steepest descent (even PARTAN-accelerated) should be avoided whenever possible. When modified second-order gradient is not applicable the unmodified variable metric method (Davidon-Fletcher-Powell) is generally the most powerful method. For simple, quadratic, or nearly quadratic problems, however, the conjugate gradient technique (Fletcher-Reeves) performs at least as well as the variable metric technique. Note that the former method requires less computer work and storage than the latter, although the difference for simple problems with a small number of variables is hardly appreciable. Furthermore, for quadratic functions the descent path taken by the conjugate gradient and variable metric methods is identical. Stewart's method appears to consistently outperform other methods that do not require differentiation.

Methods deserving of further exploration include the supermemory gradient, the methods reported by Cullum [1972], and variance techniques, which appear quite promising in terms of function-evaluation economy. As indicated in tests reported by Bard [1970], the variance methods performed better than the variable metric methods in minimizing sums of squares and similar problems. Also, the relative efficiency of the most promising descent techniques should be explored in applications to problems with different cost ratios of gradient to function and Hessian to function evaluation.

Table 5-5 CONJUGATE DIRECTIONS METHOD (Program BOTM)

Number of Function Evaluations	$M(x)$	x_1	x_2	x_3	x_4
93	36.185600	−0.080000	0.273384	−0.080000	0.178121
113	25.355546	−0.503216	0.395633	−0.381286	0.028812
197	7.871582	−1.076158	1.168061	−0.847373	0.724874
294	7.656448	−1.367856	1.881752	−0.304171	0.085705
400	4.958958	−1.169313	1.336883	0.862395	0.714588
499	2.185997	−0.269094	0.072455	1.406969	1.987963
602	0.599048	0.566878	0.287293	1.300133	1.685184
707	1.428105×10^{-2}	0.967681	0.939261	1.042801	1.080083
804	7.680640×10^{-22}	1.000000	1.000000	1.000000	1.000000
824	1.893771×10^{-26}	1.000000	1.000000	1.000000	1.000000

Table 5-6 CONJUGATE GRADIENT METHOD (Program FMCG)

Number of Iterations	$M(x)$	x_1	x_2	x_3	x_4
5	1213.010529	0.587691	−0.378548	0.229160	−0.438303
10	34.986827	0.486076	0.162223	0.364505	0.024141
15	77.300525	0.405511	0.170620	0.301723	0.063669
20	16.263351	0.996862	0.736164	0.836461	0.601020
25	20.608157	0.995723	0.922891	0.951694	0.699803
30	0.835029	1.048746	1.100170	0.949651	0.908802
35	0.465422	1.047096	1.096463	0.952280	0.905501
40	$.768152 \times 10^{-2}$	0.983491	0.984607	0.974682	0.949039
50	2.062719×10^{-6}	1.001369	1.001404	0.999634	0.999123
60	2.220×10^{-14}	1.000000	1.000000	1.000000	1.000000
69*	1.5657×10^{-24}	1.000000	1.000000	1.000000	1.000000

*Total computation involved 138 function or gradient evaluations. Equivalent number of function evaluations if gradients had to be finite difference approximated is 345.

Computer Program Performance

In Tables 5-5 through 5-9 we show the solution process of the Wood test problem (Sec. 1-10). For illustration, several of the programs mentioned in Sec. 5-7 have been used. These tables should not be used for cross comparison, because differences in programming details may interact with differences due to technique. In all cases we used a CDC 6600 and $\mathbf{x}^0 = (-3, -1, -3, -1)^T$; correspondingly, $M(\mathbf{x}^0) = 19{,}192.00$.

Table 5-7 Variable Metric Technique (Program VARMIT)

Number of Function and Gradient Evaluations	$M(x)$	x_1	x_2	x_3	x_4
5	293.460	−0.072468	−0.882175	−0.035394	−0.884484
10	28.6156	0.413965	0.302146	0.106191	0.220110
15	14.5782	0.893294	0.658124	0.533517	0.543084
21	1.84333×10^{-1}	1.08089	1.15421	0.986010	0.969778
31	6.88585×10^{-3}	1.04243	1.08707	0.955879	0.912747
40	3.30280×10^{-12}	1.00000	1.00000	1.00000	0.999999
50	1.06294×10^{-23}	1.00000	1.00000	1.00000	1.00000
54*	1.22436×10^{-25}	1.00000	1.00000	1.00000	1.00000

*Total computation involved 108 function or gradient evaluations. Equivalent number of function evaluations if gradients had to be finite difference approximated is 270.

Table 5-8 Variable Metric Technique (Program MINI)

Number of Cycles	$M(x)$	x_1	x_2	x_3	x_4
5	9.864917	−0.984461	0.931484	−0.989489	0.855584
10	7.874595	−1.031228	1.072893	−0.902690	0.826807
15	7.739916	−1.296334	1.686721	−0.463617	0.212047
20	5.240785	−1.207089	1.459796	0.648191	0.423280
25	1.896485	−0.133257	−0.001857	1.395155	1.940777
30	1.235311×10^{-5}	0.983395	0.968746	1.016079	1.032416
35	2.893473×10^{-17}	1.000000	1.000000	1.000000	1.000000
Final*	4.516377×10^{-21}	1.000000	1.000000	1.000000	1.000000

*Total computation involved 402 function or gradient evaluations. Equivalent number of function evaluations if gradients had to be finite difference approximated is 1005.

Table 5-9 VARIABLE METRIC TECHNIQUE (Program FMFP)

Number of Iterations	$M(x)$	x_1	x_2	x_3	x_4
5	15.739222	0.283971	−0.431158	−0.044207	−0.485854
10	34.780457	2.003307	0.664426	−2.335633	0.610637
15	12.645136	0.531359	0.383609	−0.519955	0.414875
20	7.983407	1.946832	2.797421	0.071755	−0.779927
25	4.520892	1.147623	1.326866	−0.710307	0.507866
30	2.936911	1.362205	1.801406	−0.555218	0.253549
35	2.885340	1.313081	1.720241	−0.547177	0.266803
40	2.697843	1.311281	1.747308	−0.492664	0.233561
50	2.219377	1.421369	1.976403	−0.231066	0.066836
60	1.916536	1.385972	1.924520	−0.116522	−0.003036
80	1.240631	1.418251	2.005981	0.323336	0.030020
100	1.172322	1.307605	1.711064	0.501213	0.221767
125	0.201603	1.162881	1.345251	0.785452	0.624118
150	1.64973×10^{-4}	1.003577	1.007947	0.996851	0.994893
175	2.733876×10^{-25}	1.000000	1.000000	1.000000	1.000000
183*	2.695254×10^{-25}	1.000000	1.000000	1.000000	1.000000

*Total computation involved 366 function or gradient evaluations. Equivalent number of function evaluations if gradients had to be finite difference approximated is 915.

References

ADACHI, N.: "On Variable Metric Algorithms," *Journal of Optimization Theory and Applications*, Vol. 7, 391, 1971.

AKAIKE, H.: "On a Successive Transformation of Probability Distribution and its Application to the Analysis of the Optimum Gradient Method," *Ann. Inst. Statist. Math. Tokyo*, Vol. 11, 1, 1959.

BARD, Y.: "On a Numerical Instability of Davidon-Like Methods," *Math. of Computation*, Vol. 22, 665, 1968.

———: "Comparison of Gradient Methods for the Solution of Nonlinear Parameter Estimation Problems," *SIAM Journal Numer. Anal.*, Vol. 7, 157, 1970.

BEALS, E.: "MINSER," *CDC VIM E4-BKY-MINSER*, 1965.

———: "VARMIT," *CDC VIM E4-BKY-VARMIT.*

BECKMAN, F. S.: "The Solution of Linear Equations by the Conjugate Gradient Method," in A. Ralston and H. S. Wilf (eds.), *Mathematical Methods for Digital Computers*, Vol. 1, John Wiley & Sons, Inc., New York, 1960.

BOX, M. J.: "A Comparison of Several Current Optimization Methods, and the Use of Transformations in Constrained Problems," *Computer Journal*, Vol. 9, 67, 1966.

BROYDEN, C. G.: "Quasi-Newton Methods and Their Application to Function Minimization," *Math. of Computation*, Vol. 21, 368, 1967.

———: "The Convergence of a Class of Double-rank Minimization Algorithms: 1. General Considerations; and 2. The New Algorithm," *J. Inst. Math Applics.*, Vol. 6, 76 and 222, 1970.

CANTRELL, J. W.: "Relation Between the Memory Gradient Method and the Fletcher-Reeves Method," *Journal of Optimization Theory and Applications*, Vol. 4, 67, 1969.

CAUCHY, A. L.: "Methode Generale pour la Resolution des Systemes d'Equations Simultanees," *C. R. Acad. Science*, Paris, Vol. 25, 536, 1847.

CRAGG, E. E. and A. V. LEVY: "Study on a Supermemory Gradient Method for the Minimization of Functions," *Journal of Optimization Theory and Applications*, Vol. 4, 191, 1969.

CULLUM, J.: "Unconstrained Minimization of Functions without Explicit Use of Their Derivatives," submitted to the *IBM Journal of Research and Development*, The Mathematics of Numerical Computation, 1972.

DAVIDON, W. C.: *Variable Metric Method for Minimization*, Report ANL-5990, AEC Research and Development, 1959.

———: "Variance Algorithm for Minimization," *Computer Journal*, Vol. 10, 406, 1968.

———: "Variance Algorithm for Minimization," in R. Fletcher (ed.), *Optimization*, Academic Press Inc., New York, 13, 1969.

FIACCO, A. V. and G. P. MCCORMICK: "Computational Aspects of Unconstrained Minimization Algorithms," *Nonlinear Programming: Sequential Unconstrained Minimization Techniques*, John Wiley & Sons, Inc., New York, Ch. 8, 156, 1968.

FLETCHER, R.: "Function Minimization without Evaluating Derivatives—a Review," *Computer Journal*, Vol. 8, 33, 1965.

———: "A Review of Methods for Unconstrained Optimization, in R. Fletcher (ed.), *Optimization*, Academic Press Inc., New York, 1, 1969.

———: "Certification of Algorithm 251—FUNCTION MINIMIZATION," *Collected Algorithms from CACM.*

———: "A New Approach to Variable Metric Algorithms," *Computer Journal*, Vol. 13, 317, 1970.

——— and M. J. D. POWELL: "A Rapidly Convergent Descent Method for Minimization," *Computer Journal*, Vol. 6, 163, 1963.

——— and C. M. REEVES: "Function Minimization by Conjugate Gradients," *Computer Journal*, Vol. 7, 149, 1964.

FORSYTHE, G. E.: "On the Asymptotic Directions of the S-Dimensional Optimum Gradient Method," *Numerische Mathematik*, Vol. 11, 57, 1968.

——— and T. S. MOTZKIN: "Asymptotic Properties of the Optimum Gradient Method (abstract)," *American Mathematical Society Bulletin*, Vol. 57, 183, 1951.

GOLDFARB, D.: "Sufficient Conditions for the Convergence of a Variable Metric Algorithm," in R. Fletcher (ed.), *Optimization*, Academic Press Inc., New York, 273, 1969.

———: "A Family of Variable Metric Methods Derived by Variational Means," *Math. of Computation*, Vol. 24, 23, 1970.

GOLDFELD, S. M., R. E. QUANDT and H. F. TROTTER: "Maximization by Quadratic Hill Climbing," *Econometrica*, Vol. 34, 541, 1966.

———: *Maximization by Improved Quadratic Hill Climbing and Other Methods*, Economics Research Memorandum No. 95, Princeton University Research Program, 1968.

GOLDSTEIN, A. A. and J. F. PRICE: "An Effective Algorithm for Minimization," *Numerische Mathematik*, Vol. 10, 184, 1967.

GREENSTADT, J.: "On the Relative Efficiencies of Gradient Methods," *Math. of Computation*, Vol. 21, 360, 1967.

———: "Variations on Variable Metric Methods," *Math. of Computation*, Vol. 24, 1, 1970*a*.

———: *A Variable Metric Method Using No Derivatives*, Report No. 320–2991, IBM Scientific Center, New York, 1970*b*.

HAUBRICH, J.G.A.: "Remark on Algorithm 205–ATIVE," *Collected Algorithms from CACM.*

——— "Algorithm 205—ATIVE," *Collected Algorithms from CACM.*

HESTENES, M. and E. STIEFEL: *Method of Conjugate Gradients for Solving Linear Systems*, Report 1659, National Bureau of Standards, 1952.

HUANG, H. Y.: "Unified Approach to Quadratically Convergent Algorithms for Function Minimization," *Journal of Optimization Theory and Applications*, Vol. 5, 405, 1970.

——— and A. V. LEVY: "Numerical Experiments on Quadratically Convergent Algorithms for Function Minimization," *Journal of Optimization Theory and Applications*, Vol. 6, 269, 1970.

KELLEY, H. J., G. E. MYERS and I. L. JOHNSON, JR.: "An Improved Conjugate Direction Minimization Procedure," *AIAA Journal*, Vol. 8, 2091, 1970.

KOWALIK, J. and M. R. OSBORNE: "Descent Methods," *Methods for Unconstrained Optimization Problems*, American Elsevier Publishing Company, New York, Ch. 3, 29, 1968.

LEON, A.: "A Comparison Among Eight Known Optimizing Procedures," in A. Lavi and T. P. Vogl (eds.), *Recent Advances in Optimization Techniques*, John Wiley & Sons, Inc., New York, 23, 1966.

LUENBERGER, D. G.: "The Conjugate Residual Method for Constrained Minimization Problems," *SIAM Journal of Numerical Analysis*, Vol. 7, 390, 1970.

MARQUARDT, D. W.: "An Algorithm for Least Squares Estimation of Nonlinear Parameters," *SIAM Journal*, Vol. 11, 431, 1963.

MATTHEWS, A. and D. DAVIES: "A Comparison of Modified Newton Methods for Unconstrained Optimization," *Computer Journal*, Vol. 14, 293, 1971.

MIELE, A. and J. W. CANTRELL: "Study on a Memory Gradient Method for the Minimization of Functions," *Journal of Optimization Theory and Applications*, Vol. 3, 459, 1969.

Murtagh, B. A. and R. W. H. Sargent: "Computational Experience with Quadratically Convergent Minimization Methods," *Computer Journal*, Vol. 13, 185, 1970.

Myers, G. E.: "Properties of the Conjugate Gradient and Davidon Methods," *Journal of Optimization Theory and Applications*, Vol. 2, 209, 1968.

Pearson, J. D.: "Variable Metric Methods of Minimization," *Computer Journal*, Vol. 12, 171, 1969.

Pierre, D. A.: "Search Techniques and Nonlinear Programming," *Optimization Theory with Applications*, John Wiley & Sons, Inc., New York, Ch. 6, 264, 1969.

Powell, M. J. D. "An Efficient Method for Finding the Minimum of a Function of Several Variables without Calculating Derivatives," *Computer Journal*, Vol. 7, 155, 1964.

———: "On the Convergence of the Variable Metric Algorithm," Rep. No. T. P. 382 A.E.R.E., Harwell, England, 1969.

———: "A Survey of Numerical Methods for Unconstrained Optimization, *SIAM Review*, Vol. 12, 79, 1970.

Shah, B. V., R. J. Buehler, and O. Kempthorne: "Some Algorithms for Minimizing a Function of Several Variables," *SIAM Journal*, Vol. 12, 74, 1964.

Shanno, D. F.: "Parameter Selection for Modified Newton Methods for Function Minimization," *SIAM Journal of Numerical Analysis*, Vol. 7, 366, 1970.

Sorenson, H. W.: "Comparison of Some Conjugate Direction Procedures for Function Minimization," *Journal of The Franklin Institute*, Vol. 288, 421, 1969.

Stewart, G. W., III: "A Modification of Davidon's Minimization Method to Accept Difference Approximations of Derivatives," *Journal of the ACM*, Vol. 14, 72, 1967.

Stiefel, E.: "Über Einige Methoden der Relaxationsrechnung," *Z. Angew. Math. Physik*, Vol. 3, 1, 1952.

———: "Kernel Polynomials in Linear Algebra and Their Numerical Applications," *National Bureau of Standards, Applied Mathematics Series*, Vol. 49, 1958.

Straeter, T. A.: "A Comparison of Gradient Dependent Techniques for the Minimization of an Unconstrained Function of Several Variables," *AIAA Aerospace Computer Systems Conference: Los Angeles, Calif.*, Paper 69–951, or *AIAA Journal*, Vol. 8, 2226, 1969.

"Subroutines fmcg and dfmcg," *IBM-System/360 Scientific Subroutine Package (360A-CM-03X)*, Version II, Section on Mathematics—Nonlinear Equations.

"Subroutines fmfp and dfmfp," *IBM-System/360 Scientific Subroutine Package (360A-CM-03X)*, Version II, Section on Mathematics—Nonlinear Equations.

Swenson, E. V.: *MINI-Minimization Routine, E4-NYU-MINI*, AEC Computing and Applied Mathematics Center, Courant Institute of Mathematical Sciences, New York University, New York, 1955.

———: *BOTM-Minimization Routine*, *E4-NYU-BOTM*, AEC Computing and Applied Mathematics Center, Courant Institute of Mathematical Sciences, New York University, New York, 1965. (Also available as CDC VIM 213.)

TABAK, D.: "Comparative Study of Various Minimization Techniques Used in Mathematical Programming," *IEEE Transactions on Automatic Control*, Vol. AC-14, 572, 1969.

VARAH, J. M.: "Certification of Algorithm 203—STEEP1," *Collected Algorithms from CACM.*

VETTERS, K.: "Quadratische Approximations methoden zur konvexen Optimierung," *ZAMM*, Vol. 50, 479, 1970.

VIM, Users Organization for Control Data Corporation Computers, *Program Library Catalog*, 1971.

WALLACK, P.: "Certification of Algorithm 203—STEEP1," *Collected Algorithms from CACM.*

WASSCHER, E. J.: "Remark on Algorithm 129—MINIFUN," *Collected Algorithms from CACM.*

———: "Algorithm 203—STEEP1," *Collected Algorithms from CACM.*

———: "Algorithm 204—STEEP2," *Collected Algorithms from CACM.*

WELLS, M.: "Algorithm 251—Function Minimization," *Collected Algorithms from CACM.*

WHITLEY, V. W.: "Algorithm 129—MINIFUN," *Collected Algorithms from CACM.*

WOLFE, P.: "Another Variable Metric Method," *IBM Working Paper*, 1967.

———: "Convergence Conditions for Ascent Methods," *SIAM Review*, Vol. 11, 226, 1969 and Vol. 13, 185, 1971.

ZANGWILL, W. I.: "Minimizing a Function without Calculating Derivatives," *Computer Journal*, Vol. 10, 293, 1967.

———: "Unconstrained Problems," *Nonlinear Programming, A Unified Approach*, Prentice-Hall, Inc., Englewood Cliffs, N.J., Ch. 5, 102, 1969.

6 TRANSFORMATION METHODS

A general nonlinear programming problem

$$\min_{\mathbf{x}} \{M(\mathbf{x}) \mid g_j(\mathbf{x}) \geq 0, j = 1, \ldots, m_1; g_j(\mathbf{x}) = 0, j = m_1 + 1, \ldots, m\}$$

can be reduced to a sequence of unconstrained optimization problems. This is achieved by a transformation combining the objective function and constraints to form a new unconstrained function whose minima approximate the solution of the constrained problem. This function will be referred to as a *transformed function*. The unconstrained optimization approach to constrained problems is motivated by practical considerations.

Unconstrained optimization problems are easier to handle; algorithms that can solve these problems have been more extensively studied and are further developed (Chaps. 4 and 5) than algorithms capable of handling constraints explicitly (Chap. 7). It should be pointed out, however, that there are cases in which the transformation methods would not be advantageous since they usually destroy the structure of the original constrained problem. Consider, for instance, a constrained problem that involves nearly linear functions. It seems reasonable to linearize this problem and solve it sequentially using the simplex method for linear programming. Unfortunately, the transformation methods could not take advantage of the structural characteristics of the problem.

6-1 TYPES OF TRANSFORMATIONS

To introduce the concept of transformation let us consider a very simple problem: $\min ax$ subject to $x - 1 \geq 0$, $2 - x \geq 0$, where $a > 0$. The solution is $x^* = 1$. Two types of transformations can be developed: interior point transformation

$$P_I(x, r) = ax + r\left(\frac{1}{x-1} + \frac{1}{2-x}\right) \tag{6-1}$$

and exterior point transformation

$$P_E(x, r) = ax + \frac{1}{r}\{(\min[x-1, 0])^2 + (\min[2-x, 0])^2\} \tag{6-2}$$

where $r > 0$ is a perturbation parameter.

Figure 6-1 illustrates the original problem and transformed functions. The

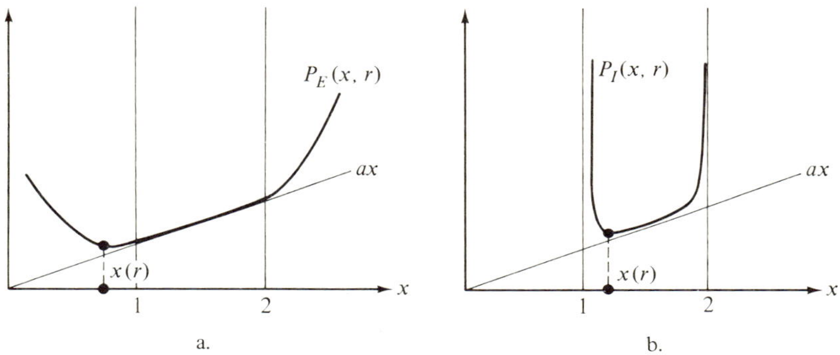

Figure 6-1 Interior and exterior point transformation functions

transformed function $P_I(x, r)$ sets up a barrier against leaving the feasible region. Clearly, all solutions $x(r)$ of min $P_I(x, r)$ are interior points, and we may expect that under suitable conditions $\lim_{r \to 0} x(r) = x^*$. To demonstrate that this is true in the example under consideration, we obtain $x(r)$ for small enough r from the condition that

$$\nabla P_I(x, r) \cong a - \frac{r}{(x-1)^2} = 0 \tag{6-3}$$

$$x(r) \cong 1 + \sqrt{\frac{r}{a}} \tag{6-4}$$

and

$$\lim_{r \to 0} x(r) = 1 \tag{6-5}$$

Generally, the structure of the transformed function in the interior point (barrier) method is:

$$P_I(\mathbf{x}, r) = M(\mathbf{x}) + r\sum_{j=1}^{m} \phi[g_j(\mathbf{x})] \tag{6-6}$$

where $\mathbf{g}(\mathbf{x}) = [g_1(\mathbf{x}), g_2(\mathbf{x}), \ldots, g_m(\mathbf{x})]^T \geq 0$ is a set of inequality constraints, $r > 0$, and $\phi(g_j)$ satisfies the following conditions:

1 $\phi(g_j)$ is continuous, convex, and decreasing in the interval $(0, \infty)$.
2 It becomes infinite on the boundary of the feasible region, $\lim \phi(g_j) = +\infty$ if $g_j(\mathbf{x}) \longrightarrow 0$.

In the second transformation, the exterior point (penalty) method, we have

$$P_E(\mathbf{x}, r) = M(\mathbf{x}) + \frac{1}{r} \sum_{j=1}^{m} \psi[g_j(\mathbf{x})] \tag{6-7}$$

where $\psi(g_j)$ satisfies the following conditions:

1 $\psi(g_j)$ is continuous, convex, and nonincreasing in the interval $(-\infty, 0]$.
2 $\psi(g_j) = 0$ for $g_j \geq 0$ and $\psi(g_j) > 0$ for $g_j < 0$.

Transformation functions $\psi(g_j)$ that handle the equality constraints $g_j(\mathbf{x}) = 0$ are nonincreasing in the interval $(-\infty, 0]$ and nondecreasing in the interval $[0, +\infty)$, and they satisfy $\psi(0) = 0$, $\psi(g_j) > 0$ for $g_j \neq 0$.

A variety of functions ϕ and ψ satisfy the above requirements; some are discussed in this chapter. Without specifying particular cases, the following comments can be made on these two classes of transformations.

Interior Point Methods

Interior point methods require interior points (to the feasible inequality-constrained region) to initialize computations. The necessity of generating such points is common to many optimization methods and presents a computational problem only. If such a point $\mathbf{x}^0$ is not available, it can be obtained by means of methods discussed in Sec. 2-2. Computing difficulties can be expected as $r \longrightarrow 0$ since the ill-conditioning of $P_I(\mathbf{x}, r)$ increases near the boundaries. Note that the interior point method is incapable of handling equality constraints and usually has to be combined with a penalty term to overcome this difficulty.

As an example, consider a problem involving m_1 inequalities $g_j(\mathbf{x}) \geq 0$, $j = 1, \ldots, m_1$, and $m - m_1$ equations $g_j(\mathbf{x}) = 0$, $j = m_1 + 1, \ldots m$. The extended transformation that can handle both types of constraints is:

$$P(\mathbf{x}, r) = M(\mathbf{x}) + r \sum_{j=1}^{m_1} \frac{1}{g_j(\mathbf{x})} + r^{-1/2} \sum_{j=m_1+1}^{m} [g_j(\mathbf{x})]^2 \tag{6-8}$$

Starting with $\mathbf{x}^0 \in R^0 = \{\mathbf{x} \mid g_j(\mathbf{x}) > 0, j = 1, \ldots, m_1\}$, we solve a sequence of problems $\min P(\mathbf{x}, r)$ for $r = r_0, r_1, r_2, \ldots$ with $0 < r_{k+1} < r_k$ and $r_k \longrightarrow 0$. This method is called the *sequential unconstrained minimization technique*. Developed by Fiacco and McCormick [1964*a*, *b*; 1966, 1968], it has been successfully applied to a variety of practical problems [Bracken and McCormick, 1968; Kavlie and Moe, 1969; Moe, 1969] and is described in Secs. 6-2 and 6-3. Other forms of the transformed functions have been suggested by Frisch [1955], Parisot [1960], and Huard [1969].

Exterior Point Methods

In exterior point methods, no restrictions are imposed on starting points; however, these methods do not produce intermediate solutions that are feasible. The exterior point transformation can handle inequalities and equations with equal ease. We have

$$P_E(\mathbf{x}, r) = M(\mathbf{x}) + r^{-1} \sum_{j=1}^{m} [\bar{g}_j(\mathbf{x})]^2 \tag{6-9}$$

where

$$\bar{g}_j(\mathbf{x}) \begin{cases} = \min[g_j(\mathbf{x}), 0] & j = 1, 2, \ldots, m_1 \\ = g_j(\mathbf{x}) & j = m_1 + 1, \ldots, m \end{cases} \tag{6-10}$$

The transformation may have the effect of reducing the order of differentiability at the boundary points of the region $R = \{\mathbf{x} \mid g_j(\mathbf{x}) \geq 0, j = 1, \ldots, m_1\}$. For instance, the transformation $P_E(\mathbf{x}, r)$ given by (6-9) with the quadratic penalty function is once, but not twice, differentiable at the boundary of the feasible region. This property of $P_E(\mathbf{x}, r)$ does restrict the choice of algorithms which may be used to optimize the function $P_E(\mathbf{x}, r)$.

Another disadvantage of exterior point transformations is that the related unconstrained minimization problem becomes progressively ill-conditioned as the penalty coefficient increases. As an example, consider the problem of minimizing a quadratic form subject to linear constraints,

$$\min \mathbf{x}^T B \mathbf{x} \tag{6-11}$$

subject to

$$A\mathbf{x} - \mathbf{b} = 0$$

where the $(n \times n)$ matrix B is positive definite and the $(m \times n)$ matrix A is rectangular with $m < n$. The corresponding exterior point transformation is

$$P(\mathbf{x}, K) = \mathbf{x}^T B \mathbf{x} + (A\mathbf{x} - \mathbf{b})^T K (A\mathbf{x} - \mathbf{b}) \tag{6-12}$$

where K is a diagonal matrix with $K_{ii} = r^{-1}$.

The quadratic term of $P(\mathbf{x}, K)$ is

$$P_q(\mathbf{x}, K) = \mathbf{x}^T (B + A^T K A) \mathbf{x} = \mathbf{x}^T H \mathbf{x} \tag{6-13}$$

Matrix $A^T K A$ has at most rank m and at least $n - m$ zero eigenvalues. Matrix B has all eigenvalues positive and finite. If $K_{ii} \longrightarrow \infty$, the largest eigenvalue of H, $\gamma_{max} \longrightarrow \infty$, but the smallest one γ_{min} is bounded by the largest eigenvalue of B. This follows from Weyl's [1912] theorem, which relates eigenvalues of three matrices M, N, and Q, where $Q = M + N$. The condition number of H, $\gamma_{max}/\gamma_{min}$, increases with increasing penalty coefficient K_{ii}, and it may be difficult to optimize $P(\mathbf{x}, K)$ using methods that are not invariant with respect to the variables scale. Lootsma [1970] considers a general case of the mixed interior-exterior transformation $P(\mathbf{x}, r) = M(\mathbf{x}) + r^p \phi(\mathbf{x}) + r^{-q} \psi(\mathbf{x})$ where $p > 0$, $q > 0$ are integers, and shows that the condition number of the Hessian of $P(\mathbf{x}, r)$ is proportional to r^{-1} for sufficiently small values of r.

In addition to the ill-conditioning problem, the numerical errors in the penalty terms become significant for large penalty coefficients. These two difficulties indicate that we may not be able to satisfy the constraints with any desirable accuracy.

An interesting but nonconclusive application of the penalty method combined with the variable metric method was attempted by Kelley, et al. [1966], who suggest that

$$P_E(\mathbf{x}, K) = M(\mathbf{x}) + \tfrac{1}{2}\mathbf{g}^T K \mathbf{g} \tag{6-14}$$

be minimized in a two-phase method. First, the variable metric method is applied, yielding an approximation to the solution. In the second phase, this approximation is improved by the Newton method using the metric matrix obtained at the last variable metric step. The first phase is terminated when $K\mathbf{g}$ approaches a limiting value (stabilizes). This stopping criterion follows from the observation that the minimum of P_E satisfies

$$\nabla M(\mathbf{x}) + GK\mathbf{g} = 0 \tag{6-15}$$

where $G_{ij} = \partial g_j/\partial x_i$ and the solution of the constrained problem satisfies

$$\begin{aligned} \nabla M(\mathbf{x}) + G\mathbf{u} &= 0 \\ \mathbf{g} &= 0 \end{aligned} \tag{6-16}$$

where $\mathbf{u}$ is a vector of Lagrange multipliers. From (6-15) and (6-16) we get $K\mathbf{g} \longrightarrow \mathbf{u}$, and this explains the first-phase termination criterion.

Another modification of the penalty method has been suggested by Hestenes [1969], who considered the problem of minimizing $M(\mathbf{x})$ subject to $g(\mathbf{x}) = 0$. In this modification, called the multiplier method, the transformation function is

$$P(\mathbf{x}, u, c) = M(\mathbf{x}) + ug(\mathbf{x}) + \frac{1}{2}cg^2(\mathbf{x}) \tag{6-17}$$

where c is a suitably large constant and u is a Lagrange multiplier. The multiplier method is based on the theorem that states that (under quite general conditions) if $\mathbf{x}^*$ is the solution to $\min_{\mathbf{x}} \{M(\mathbf{x}) \mid g(x) = 0\}$, a multiplier u and constant c exist such that $\mathbf{x}^*$ is also a local minimum of the function $P(\mathbf{x}, u, c)$ defined by (6-17).

Hestenes indicates a possible scheme of the method:

1 A suitably large c and initial estimate of u are selected.

2 The problem $\min_{\mathbf{x}} P(\mathbf{x}, u, c)$ is solved.

3 The process is repeated from step 2 with modified u and c.

Various schemes for updating u and c can be given that ensure the convergence of the method.

An initial estimate of u can be obtained from the penalty method. We may begin with the penalty method to get the first estimates of u and $\mathbf{x}^*$ and then switch to the method of multipliers. The method of Hestenes has an important advantage of not involving large coefficients of $g^2(\mathbf{x})$ and has been computationally explored by Miele et al. [1971].

A transformation that does not use the penalty parameter bases on the least-squares structure of the transformed function [Fox and Schmit, 1965; Morrison, 1968] is

$$P(\mathbf{x}, M_k) = [M(\mathbf{x}) - M_k]^2 + \sum_{j=1}^{m} [\bar{g}_j(\mathbf{x})]^2 \tag{6-18}$$

where

$$\bar{g}_j(\mathbf{x})\begin{cases} = \min\,[g_j(\mathbf{x}), 0] \text{ for inequality constraints} \\ = g_j(\mathbf{x}) \text{ for equality constraints} \end{cases}$$

and M_k, $k = 0, 1, 2, \ldots$ is a sequence of constants which under suitable conditions converges to the solution of the original constrained problem. Section 6-4 gives computational details and the main theoretical results for the least-squares transformation.

Recent developments in the theory of the interior and exterior point transformation provide the generalization and validation of this approach and can be found in the papers of Zangwill [1967], Pietrzykowski [1969], Butler and Martin [1962], Beltrami [1967], Fiacco and Jones [1969], and Lootsma [1970]. An extensive survey of the approach can be found in a report by Fiacco [1967].

6-2 INTERIOR POINT TRANSFORMATION AND ITS EXTENSIONS

Fiacco and McCormick [1968] developed the sequential unconstrained minimization techniques (SUMT) based on an earlier suggestion by Carroll [1961]. The method is an interior point technique and it generates a sequence of interior feasible solutions which under suitable assumptions converges to the constrained minimum of the objective function.

In order to obtain a tractable problem of minimizing

$$P(\mathbf{x}, r_k) = M(\mathbf{x}) + r_k \sum_{j=1}^{m} [g_j(\mathbf{x})]^{-1} \tag{6-19}$$

where $g_j(\mathbf{x}) \geq 0$, $j = 1, 2, \ldots, m$ are the constraints of the considered optimization problem, we assume that

1 The interior-feasible region R^0 is not empty, and an interior-feasible point $\mathbf{x}^0$ can be found.
2 The functions $M(\mathbf{x})$ and $g_j(\mathbf{x})$, $j = 1, \ldots, m$ are twice continuously differentiable.
3 The objective function $M(\mathbf{x})$ is bounded below on the feasible region R; that is, $M(\mathbf{x}) \geq L_M > -\infty$, where L_M is a greatest lower bound.

4 The set $S = \{\mathbf{x} \mid M(\mathbf{x}) \leq K, \mathbf{x} \in R\}$ is bounded for every K, which implies that every local minimum is located at a finite point.

Under these conditions the problem

$$\min_{\mathbf{x}} P(\mathbf{x}, r_k) = M(\mathbf{x}) + r_k \sum_{j=1}^{m} [g(\mathbf{x})]^{-1} \tag{6-20}$$

has at least one local solution $\mathbf{x}(r_k)$ which is finite. Stronger results can be obtained if in addition

5 The functions $M(\mathbf{x})$ and $-g_j(\mathbf{x})$ are convex (this implies that $[g_j(\mathbf{x})]^{-1}$ are convex) and at least one of these functions is strictly convex.

If conditions 1 through 5 are satisfied, we have a well-behaved minimization problem with a convex function being optimized on a convex feasible region. For this convex problem Theorem 6-1 has been proved.

Theorem 6-1 If conditions 1 through 5 are satisfied,

$$\lim_{r_k \to 0} \mathbf{x}(r_k) = \mathbf{x}^* \tag{6-21}$$

$$M(\mathbf{x}^*) = \min_{\mathbf{x} \in R^0} M(\mathbf{x}) = L_M \tag{6-22}$$

where $\mathbf{x}(r_k)$ minimizes $P(\mathbf{x}, r_k)$.

In many practical problems some of these conditions do not hold, or it is very difficult to verify if they do. The SUMT method, however, is not limited to the cases where all above conditions are satisfied. If, for instance, the convexity condition does not hold, then the method can be used to find local minima. A number of practical applications of the SUMT method show that the global solution can be identified if suitable starting points are used. Thus, it is important to know something about the problem being solved.

In the process of minimizing (6-19) we use a sequence of positive values of r_k, $k = 1, 2, \ldots$, which are strictly decreasing to zero. Since $\mathbf{x}(r_k) \equiv \mathbf{x}^k$ is a solution of $\min P(\mathbf{x}, r_k)$ we have

$$M(\mathbf{x}^{k+1}) + r_{k+1} \sum_{j=1}^{m} [g_j(\mathbf{x}^{k+1})]^{-1} < M(\mathbf{x}^k) + r_{k+1} \sum_{j=1}^{m} [g_j(\mathbf{x}^k)]^{-1} \tag{6-23}$$

and

$$M(\mathbf{x}^k) + r_k \sum_{j=1}^{m} [g_j(\mathbf{x}^k)]^{-1} < M(\mathbf{x}^{k+1}) + r_k \sum_{j=1}^{m} [g_j(\mathbf{x}^{k+1})]^{-1} \tag{6-24}$$

from which it follows that

$$M(\mathbf{x}^{k+1})(r_{k+1}^{-1} - r_k^{-1}) < M(\mathbf{x}^k)(r_{k+1}^{-1} - r_k^{-1}) \tag{6-25}$$

and

$$M(\mathbf{x}^{k+1}) < M(\mathbf{x}^k) \tag{6-26}$$

Relationship (6-26) shows that the sequence of $M(\mathbf{x}^k)$, $k = 1, 2, \ldots$ approaches the solution from above. From (6-24) and (6-25),

$$\sum_{j=1}^{m} [g_j(\mathbf{x}^{k+1})]^{-1} > \sum_{j=1}^{m} [g_j(\mathbf{x}^k)]^{-1} \tag{6-27}$$

Since $\lim_{k\to\infty} M(\mathbf{x}^k) = L_M$, we have

$$\lim_{k\to\infty} r_k \sum_{j=1}^{m} [g_j(\mathbf{x}^k)]^{-1} = 0 \tag{6-28}$$

Initially the SUMT method did not have the capability of handling equality constraints. It has been extended later by adding a penalty term to the function (6-19). The resulting transformation function is (6-8), repeated here for convenience:

$$P(\mathbf{x}, r_k) = M(\mathbf{x}) + r_k \sum_{j=1}^{m_1} [g_j(\mathbf{x})]^{-1} + r_k^{-1/2} \sum_{j=m_1+1}^{m} [g_j(\mathbf{x})]^2 \tag{6-29}$$

where $g_j(\mathbf{x}) \geq 0$, $j = 1, 2, \ldots, m_1$, and $g_j(\mathbf{x}) = 0$, $j = m_1 + 1, \ldots, m$. Similar conditions to these for the problem without equalities make possible the convergence proof for the problem of minimizing the transformed function (6-29).

Iteration Steps

Given is an initial interior point $\mathbf{x}^0 \in R^0 = \{\mathbf{x} \mid g_j(\mathbf{x}) > 0, j = 1, 2, \ldots, m_1\}$. Set $k = 1$ and $r_1 > 0$.

1 Starting from $\mathbf{x}^{k-1}$, minimize the function

$$P(\mathbf{x}, r_k) = M(\mathbf{x}) + r_k \sum_{j=1}^{m_1} [g_j(\mathbf{x})]^{-1} + r_k^{-1/2} \sum_{j=m_1+1}^{m} [g_j(\mathbf{x})]^2 \tag{6-30}$$

where $r_k > 0$ is a perturbation parameter. This results in $\mathbf{x}(r_k) \equiv \mathbf{x}^k \in R^0$. Methods of Chaps. 4 and 5 are applicable to this problem.

2 Set

$$r_{k+1} = r_k/c \tag{6.31}$$

where $c > 1$ and $k := k + 1$.

3 Restart from step 1.

Remarks For constrained problems with inequalities only, Fiacco and McCormick [1964*b*] suggest that a good parameter r_1 should minimize $\| \nabla P(\mathbf{x}^0, r) \|$. This gives

$$\frac{d}{dr}\left[\nabla M(\mathbf{x}^0) + r\nabla \sum_{j=1}^{m} [g_j(\mathbf{x}^0)]^{-1}\right]^2 = 0 \tag{6-32}$$

$$r_1 = -\frac{[\nabla M(\mathbf{x}^0)]^T \nabla p(\mathbf{x}^0)}{[\nabla p(\mathbf{x}^0)]^2} \tag{6-33}$$

where

$$p(\mathbf{x}) = \sum_{j=1}^{m} [g_j(\mathbf{x})]^{-1} \tag{6-34}$$

Formula (6-33) gives good values of r_1 if $\mathbf{x}^0$ is not too close to the boundaries. Otherwise, it may generate a value of r_1 that is too small. If r_1 given by (6-33) is not positive, we can move along $-\nabla M(\mathbf{x}^0)$ until r_1 is positive or an unconstrained minimum is found.

Another formula that estimates r_1 for convex problems is:

$$r_1 = \left\{\frac{\nabla M(\mathbf{x}^0)^T H^{-1} \nabla M(\mathbf{x}^0)}{[\nabla \Sigma g_j^{-1}(\mathbf{x}^0)]^T H^{-1} [\nabla \Sigma g_j^{-1}(\mathbf{x}^0)]}\right\}^{1/2} \tag{6-35}$$

where the elements of H are

$$h_{il} = \frac{\partial^2 \Sigma [g_j(\mathbf{x}^0)]^{-1}}{\partial x_i \partial x_l} \qquad i, l = 1, 2, \ldots, n \tag{6-36}$$

The values of r_k are reduced at each stage by a factor of 10 to 50; that is, $c = r_k/r_{k+1} = 10$ to 50. An empirical approach to the problem of finding a suitable value of r_1 has been suggested by Box, et al. [1969].

Termination Criteria and Convergence

The function $P(\mathbf{x}, r_k)$ is minimized for a sequence of decreasing values of r_k, and $\mathbf{x}(r_k)$ is expected to converge to the solution of the constrained problem. A number of criteria can be applied to terminate the minimization process; these include

1 $||\mathbf{x}(r_{k-1}) - \mathbf{x}(r_k)|| < \epsilon_1$ (6-37)

2 $\max_i |x_i(r_{k-1}) - x_i(r_k)| < \epsilon_2$ (6-38)

3 $M[\mathbf{x}(r_{k-1})] - M[\mathbf{x}(r_k)] < \epsilon_3$ (6-39)

4 $\dfrac{M[\mathbf{x}(r_{k-1})] - M[\mathbf{x}(r_k)]}{|M[\mathbf{x}(r_k)]|} < \epsilon_4$ (6-40)

where ϵ_i, $i = 1, 2, 3, 4$ are suitably selected positive numbers. They are problem dependent and bear some relation to c (relationship 6-31).

In the sequential minimization process we may decide not to solve every unconstrained problem completely, but stop after a finite number of steps and begin the next stage with a reduced r_k. Computational experience shows that very little is gained at the end of the optimization process of $P(\mathbf{x}, r_k)$. It is not known, however, how far the minimization process should be continued before r_k is reduced.

A very convenient termination criterion can be applied to the convex problems with inequality constraints. It is shown that

$$M[\mathbf{x}(r_k)] - r_k \sum_{j=1}^{m} \{g_j[\mathbf{x}(r_k)]\}^{-1} \leq L_M \leq M[\mathbf{x}(r_k)] \tag{6-41}$$

where L_M is the value of $M(\mathbf{x})$ at the solution point. From (6-41) we get

$$r_k \sum_{j=1}^{m} \{g_j[\mathbf{x}(r_k)]\}^{-1} < \eta \tag{6-42}$$

as a possible termination criterion with a suitably small chosen η.

A more refined criterion than that of (6-37) through (6-40) is based on an extrapolation formula [Fiacco and McCormick, 1964*b*, 1966]. It is assumed that the sequence $M[\mathbf{x}(r_k)]$ is a continuous function of r. Practical experience and some theoretical results indicate that the values of $M_{\min}(r)$ can be approximated by a function linear in $r^{1/2}$ as r approaches zero. If we assume that

$$M_{\min}(r) \cong \alpha + \beta r^{1/2}$$

and apply this approximation to r_{k-1} and r_k, we obtain

$$M[\mathbf{x}(r_{k-1})] = \alpha + \beta r_{k-1}^{1/2} \tag{6-43}$$

$$M[\mathbf{x}(r_k)] = \alpha + \beta\left(\frac{r_{k-1}}{c}\right)^{1/2} \tag{6-44}$$

$$M_{\min}(0) \cong \alpha = \frac{c^{1/2}M[\mathbf{x}(r_k)] - M[\mathbf{x}(r_{k-1})]}{c^{1/2} - 1} \tag{6-45}$$

$M_{\min}(0)$ can be used as a good approximation to the best available $M(\mathbf{x})$. The computational process is terminated when the values of $M[\mathbf{x}(r_k)]$ and extrapolated $M_{\min}(0)$ are sufficiently close.

Modifications

The basic transformation formula given by (6-30) includes all inequality functions in the second term. However, this can be modified to include only constraints that are likely to be violated. We may start the optimization procedure without the barrier function term in (6-30), and if during the optimization process any $g_j(\mathbf{x}) > 0$ constraints are violated, we return to the last interior-feasible point and build up the second term of (6-30). It also is possible to drop those constraints from which the current $\mathbf{x}^k$ moves away. This modification complicates coding, but a substantial computational effort can be saved if only few inequalities are actively constraining the optimization problem.

It is also possible to avoid the perturbation parameter in the SUMT method by using the transformation [Fiacco and McCormick, 1965]

$$T(\mathbf{x}, \mathbf{x}^{k-1}) = [M(\mathbf{x}^{k-1}) - M(\mathbf{x})]^{-1} + \sum_{j=1}^{m} [g_j(\mathbf{x})]^{-1} \tag{6-46}$$

The method handles only problems with inequality constraints. Possible advantages of this formulation are that the objective function must decrease monotonically from the starting point $\mathbf{x}^0$, and once this point is selected the entire process is fully determined. On the other hand, we lose control over the sequential minimization process because the perturbation parameter is not present in T. Other transformations with similar properties have been described by Huard [1967].

Another possible modification is to convert the original problem min $M(\mathbf{x})$

subject to $g_j(\mathbf{x}) \geq 0, j = 1, 2, \ldots, m$ into the equivalent one: min $M(\mathbf{x})$ subject to $g_j(\mathbf{x}) - s_j = 0$, $s_j \geq 0$, $j = 1, 2, \ldots, m$. This formulation is equivalent to the sequential minimization of

$$P(\mathbf{x}, r_k) = M(\mathbf{x}) + r_k \sum_{j=1}^{m} s_j^{-1} + r_k^{-1/2} \sum_{j=1}^{m} [g_j(\mathbf{x}) - s_j]^2 \tag{6-47}$$

The formulation (6-47) is referred to as the *slackened unconstrained minimization technique* (SLUMT) [Fiacco and McCormick, 1967] and is a nontrivial variant of SUMT since the initializing phase to find a feasible starting point is not needed. The method produces unfeasible solutions in the sequential minimization process.

A special modification suited to solve problems with many linear constraints is described in Sec. 6-3.

Gisvold [1970] has proposed and experimented with a modified SUMT transformation capable of handling nonlinear discrete programming problems.

Special Acceleration Techniques

Considerations similar to those of the preceding section suggest the possibility of using information gathered at every two solutions $\mathbf{x}(r_{k-1})$ and $\mathbf{x}(r_k)$ to find an extrapolated vector $\mathbf{x}_E(r_{k+1})$. This extrapolation can serve as a good starting point to minimize $P(\mathbf{x}, r_{k+1})$.

Assuming that

$$\mathbf{x}(r) = \mathbf{a} + \mathbf{b}r^{1/2} \tag{6-48}$$

and fitting this function to the two known solutions $\mathbf{x}(r_{k-1})$, $\mathbf{x}(r_k)$, we obtain

$$\mathbf{a} = [c^{1/2}\mathbf{x}(r_k) - \mathbf{x}(r_{k-1})](c^{1/2} - 1)^{-1} \tag{6-49}$$

$$\mathbf{b} = \frac{\mathbf{x}(r_{k-1}) - \mathbf{a}}{r_{k-1}^{1/2}} \tag{6-50}$$

From these relationships, the extrapolated $\mathbf{x}_E(r_{k+1})$ can be obtained

$$\mathbf{x}_E(r_{k+1}) = \mathbf{a} + \mathbf{b}r_{k+1}^{1/2} \tag{6-51}$$

If $\mathbf{x}_E(r_{k+1})$ is interior-feasible, it can be used as a starting point to minimize $P(\mathbf{x}, r_{k+1})$. Otherwise, it has to be abandoned or used only to find a feasible starting point along the vector $\mathbf{s} = \mathbf{x}(r_k) - \mathbf{x}_E(r_{k+1})$. The function $P(\mathbf{x}, r_{k+1})$ can also be minimized along this vector, and the solution will be a feasible and good starting point for minimizing $P(\mathbf{x}, r_{k+1})$.

The efficiency of the SUMT method can be considerably improved by a special treatment of the function $P(\mathbf{x}, r)$ in the one-dimensional minimization problem $\min_\lambda P(\mathbf{x} + \lambda\mathbf{s}, r)$ subject to $(\mathbf{x} + \lambda\mathbf{s}) \in R^0$, interior points with respect to inequality constraints. As described in Chap. 3 there are two approaches to this minimization, the golden section and low-order polynomial fit. The golden section method requires typically 5 to 12 function evaluations, and those evaluations are time-consuming since $P(\mathbf{x}, r)$ incorporates the objective and constraint functions of the optimization problem. On the other hand, due to the rapid increase of $P(\mathbf{x}, r)$ in the vicinity of the boundary (Fig. 6-2), the low-order approximations are not suitable.

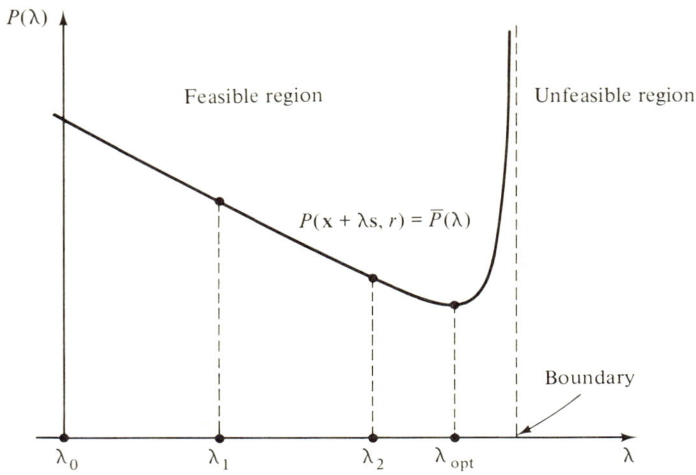

Figure 6-2 Typical form of $P(\lambda)$ close to a boundary

Moe [1969] and Lund [1970] suggested a method where the objective function and constraints [not $P(\mathbf{x}, r)$] are approximated by linear and quadratic polynomials in λ. The function P is then expressed in terms of these approximations and its minimum is searched for. This approach is particularly efficient if the objective and constraining functions are computationally expensive. Numerical experience with a number of practical problems indicates that the computational effort involved in this method amounts to about one-third the effort required if the golden section method is applied to the same problem.

The approximation method works in the following manner:

1 Values of the functions $M(\mathbf{x})$ and $g_j(\mathbf{x})$ are computed at $\mathbf{x}$ and $\mathbf{x} + \lambda_1\mathbf{s}$, where $\mathbf{x}$ is the last solution, $\mathbf{s}$ is the search direction (see step 1 of the iteration), and $\lambda_1 = \Delta\lambda$ is a suitable increment.

2 Linear approximations to $M(\mathbf{x})$ and $g_j(\mathbf{x})$ along the vector $\mathbf{s}$ are computed, using values at $\mathbf{x}$ and at $\mathbf{x} + \lambda_1\mathbf{s}$:

$$M_1(\lambda) = a + b\lambda \tag{6-52}$$

$$g_{1j}(\lambda) = a_j + b_j\lambda \qquad j = 1, 2, \ldots, m \tag{6-53}$$

Then approximations to P and $\nabla_\lambda P$ are computed:

$$P_1(\lambda, r) = a + b\lambda + r \sum_{j=1}^{m} (a_j + b_j\lambda)^{-1} \tag{6-54}$$

$$\nabla_\lambda P_1(\lambda, r) = b - r \sum_{j=1}^{m} b_j(a_j + b_j\lambda)^{-2} \tag{6-55}$$

3 A next step λ_2 is taken to the left or right of λ_1 (if $\mathbf{s}$ is a descent direction, then $\lambda_1 > 0$, $\lambda_2 > 0$) depending on the value of $\nabla_\lambda P_1(\lambda_1)$.

4 The functions $M(\lambda)$ and $g_j(\lambda)$ are approximated by parabolas that fit exactly at $\lambda = 0$, λ_1, and λ_2, using values at $\mathbf{x}$, $\mathbf{x} + \lambda_1\mathbf{s}$ and $\mathbf{x} + \lambda_2\mathbf{s}$:

$$M_2(\lambda) = \hat{a} + \hat{b}\lambda + \hat{c}\lambda^2 \tag{6-56}$$

$$g_{2j}(\lambda) = \hat{a}_j + \hat{b}_j\lambda + \hat{c}_j\lambda^2, \qquad j = 1, 2, \ldots, m \tag{6-57}$$

then one computes

$$P_2(\lambda, r) = \hat{a} + \hat{b}\lambda + \hat{c}\lambda^2 + r \sum_{j=1}^{m} [\hat{a}_j + \hat{b}_j\lambda + \hat{c}_j\lambda^2]^{-1} \tag{6-58}$$

$$\nabla_\lambda P_2(\lambda, r) = \hat{b} + 2\hat{c}\lambda - r \sum_{j=1}^{m} (\hat{b}_j + 2\hat{c}_j\lambda)(\hat{a}_j + \hat{b}_j\lambda + \hat{c}_j\lambda^2)^{-2} \tag{6-59}$$

5 If $\min_\lambda P(\lambda, r)$ has been bracketed by 0 and λ_1, 0 and λ_2, or λ_1 and λ_2, then it can be located more precisely—if necessary, by sequential bisection. [The minimum can be recognized by condition $\nabla_\lambda P(\lambda, r) = 0$.] Otherwise, additional trial values of λ are necessary to find a bracketing interval.

The minimizing λ obtained from the first approximation of P_2 is usually quite accurate and can be accepted as a minimum of $P(\mathbf{x}, r)$ along $\mathbf{s}$. This reduces computing work to three evaluations of $M(\mathbf{x})$ and $g_j(\mathbf{x})$, plus additional work involved in finding polynomial coefficients and values of $\nabla_\lambda P$.

A similar method of accelerating the linear search in the SUMT method has been explored by Fletcher and McCann [1969].

Hybrid SUMT-Powell Method

Powell [1969] attempted to improve the penalty and barrier methods that tend to involve large numbers $\{r^{-1}$ and $[g_j(\mathbf{x})]^{-1}\}$. This attempt is designed to solve the minimization problem with the equality constraints $g_j(\mathbf{x}) = 0$, $j = 1, \ldots, m$, but can be extended to include inequalities. The method resembles the penalty method, but more parameters are introduced to the transformed unconstrained function, namely $\mathbf{l} = (l_1, \ldots, l_m)^T$ and $\mathbf{s} = (s_1, \ldots, s_m)^T$. The P function is

$$P(\mathbf{x}, \mathbf{l}, \mathbf{s}) = M(\mathbf{x}) + \sum_{j=1}^{m} l_j[g_j(\mathbf{x}) + s_j]^2 \tag{6-60}$$

The method is based on Theorem 6-2.

Theorem 6-2 If the value of $\mathbf{x}$ that minimizes (6-60) is $\hat{\mathbf{x}}(\mathbf{l}, \mathbf{s})$, the same value solves the constrained problem

$$\min M(\mathbf{x}) \tag{6-61}$$

subject to

$$g_j(\mathbf{x}) = g_j[\hat{\mathbf{x}}(\mathbf{l}, \mathbf{s})] \qquad j = 1, \ldots, m \tag{6-62}$$

The proof is immediate, since if the theorem does not hold and $\bar{\mathbf{x}}(\mathbf{l}, \mathbf{s})$ minimizes $M(\mathbf{x})$ subject to (6-62), then $P(\bar{\mathbf{x}}, \mathbf{l}, \mathbf{s}) < P(\hat{\mathbf{x}}, \mathbf{l}, \mathbf{s})$, which is a contradiction.

The main consequence of Theorem 6-2 is that the constrained problem is solved if one obtains values of $(\mathbf{l}, \mathbf{s})$ such that

$$g_j[\hat{\mathbf{x}}(\mathbf{l}, \mathbf{s})] = 0 \qquad j = 1, 2, \ldots, m \tag{6-63}$$

This can be done by an iterative adjustment process of the $2m$ parameters $\mathbf{l}$ and $\mathbf{s}$. An essential property of this method is that $\mathbf{l}$ can be kept under control. The system (6-63) is solved by adjusting $\mathbf{s}$ while $\mathbf{l}$ is fixed. Powell shows that the adjustment procedure is simply

$$s_j^{k+1} = s_j^k + g_j(\hat{\mathbf{x}}) \qquad j = 1, 2, \ldots, m \tag{6-64}$$

where k is an iteration number and $\hat{\mathbf{x}}$ minimizes $P(\mathbf{x}, \mathbf{l}, \mathbf{s})$ in the previous step. This relation can be used if l_j are sufficiently large numbers. Thus, the method offers more flexibility than the penalty method, uses moderate values of l_j, and has very good convergence properties. The basic elements of the Powell algorithm can be described as follows:

1 Find an unconstrained minimum of $P(\mathbf{x}, \mathbf{l}^k, \mathbf{s}^k)$ given by (6-60) and call it $\hat{\mathbf{x}}^k$. The initial choice is $\mathbf{l}^0 = (1 \ldots, 1)^T$, $\mathbf{s}^0 = (0, \ldots, 0)^T$

2 If

$$|g_j(\hat{\mathbf{x}}^k)| \le \epsilon \qquad i = 1, \ldots, m \tag{6-65}$$

where $\epsilon > 0$ is a sufficiently small number, the computation is terminated.

3 If the rate of convergence is sufficiently rapid, compute

$$\mathbf{s}^{k+1} = \mathbf{s}^k + \mathbf{g}(\hat{\mathbf{x}}^k) \tag{6-66}$$

and return to step 1 with $k := k + 1$; otherwise, increase $\mathbf{l}^k$ (e.g., $\mathbf{l}^{k+1} = 10\mathbf{l}^k$) and return to step 1 with $k := k + 1$.

Osborne and Ryan [1970*b*] have successfully combined the SUMT and Powell methods. They minimize the following objective function:

$$T(\mathbf{x}, \mathbf{l}^k, \mathbf{s}^k, r_K) = M(\mathbf{x}) + r_K \sum_\alpha \phi_\alpha[g_\alpha(\mathbf{x})] + \sum_\beta l_\beta^k[g_\beta(\mathbf{x}) + s_\beta^k]^2 \tag{6-67}$$

where k and K are iteration numbers for the Powell and SUMT methods, respectively, and

$$\phi_\alpha[g_\alpha(\mathbf{x})] = \log\{G_\alpha - \log[g_\alpha(\mathbf{x})]\} \tag{6-68}$$

with

$$G_\alpha \ge 1 + \max_{\mathbf{x} \in R^0} \log[g_\alpha(\mathbf{x})] \tag{6-69}$$

In a reasonably well-scaled problem, we can set $G_\alpha = 10$.

At each kth SUMT iteration, the inequalities included in the SUMT term are examined. Constraints recognized as nonbinding $[g_\alpha(\mathbf{x}^K) \longrightarrow \text{const} > 0$ as $r_K \longrightarrow 0]$ are dropped from the computations. Any recognized as binding $[g_\alpha(\mathbf{x}^K) \longrightarrow 0$ as $r_K \longrightarrow 0]$ are transferred to the Powell term of P and treated as equalities. (Constraints transferred in error to the Powell formulation can be transferred back.) These decisions are based on the information provided by the Lagrange multiplier estimates

$$u_\alpha = -r \frac{\partial \phi_\alpha[g_\alpha(\mathbf{x})]}{\partial g_\alpha}$$

The hybrid SUMT-Powell algorithm can be stated briefly as follows:

1. Set $K = k = 0, \quad \mathbf{s}^k = (0, \ldots, 0)^T \quad \mathbf{l}^k = (1, 1, \ldots, 1)^T$.
2. Using an unconstrained optimization method, find the minimum of $T(\mathbf{x}, \mathbf{l}^k, \mathbf{s}^k, r_K)$ defined by (6-67).
3. If there are equality constraints or recognized binding inequalities, go to step 4; otherwise, go to step 5.
4. Apply Powell's algorithm to the problem

$$\min\{M(\mathbf{x}) + r_K \sum_{\infty} \phi_\alpha[g_\alpha(\mathbf{x})]\} \tag{6-70}$$

   subject to $\quad g_\beta(\mathbf{x}) = 0 \quad$ (equations and binding inequalities).
   Here we minimize (6-67) by using the method of Powell. If there is no convergence, return to step 2 with $k := k + 1$; otherwise, proceed to step 5.
5. If there are no equality constraints, the computation is terminated. Otherwise, the inequalities in the SUMT term are examined and according to certain rules they are dropped, retained, or transferred to the Powell term of T. Return to step 2 with $K := K + 1$ unless K exceeds a predetermined number.

The method has a number of advantages over the usual SUMT technique.

1. Nonbinding constraints are recognized at the earliest possible stage of the optimization process and dropped from the computation. This decreases the constraint evaluation effort.
2. Binding constraints are transferred to the third term of $T(\mathbf{x}, \mathbf{l}^k, \mathbf{s}^k, r_K)$ and handled as equalities by the efficient Powell method. Usually four to five SUMT iterations are sufficient to classify the constraints as binding or nonbinding.
3. We avoid the difficulties of minimizing the badly scaled objective functions, which commonly occur due to singularity of $P(\mathbf{x}, r)$ when $r \longrightarrow 0$, in the SUMT method.
4. Osborne and Ryan [1970*a*] show also that the transformation (6-68) produces accurate solutions in fewer SUMT iterations than are required by the penalty functions $[g_\alpha(\mathbf{x})]^{-1}$ or $-\log[g_\alpha(\mathbf{x})]$.

5 The minimization of $T(\mathbf{x}, \mathbf{l}^k, \mathbf{s}^k, r_K)$ in step 2 of the algorithm is performed by the variable metric method of Davidon (Sec. 5-6). A special matrix update procedure is used to compute good estimates of the new Hessian inverses, which enables the Davidon method to converge with ultimate second-order convergence.

Computational experience with the hybrid SUMT-Powell method indicates that the method is much more efficient (in terms of computer time and number of function evaluations) than the usual SUMT method.

Discussion

The SUMT method has been extensively studied and successfully applied to a number of practical problems. It is applicable to nonconvex problems and can handle nonlinear constraints. The method gives feasible intermediate solutions and a final solution with arbitrary accuracy if a sufficiently small perturbation parameter is used. It has been shown that the condition number (ratio of largest to smallest eigenvalue) of the Hessian matrix of $P(\mathbf{x}, r) = M(\mathbf{x}) + r \sum_{j=1}^{m} [g_j(\mathbf{x})]^{-1}$ may tend to infinity for $r \longrightarrow 0$ [Fletcher and McCann, 1969; Murray, 1967]. Therefore, we may have numerical difficulties in optimizing $P(\mathbf{x}, r)$. Other disadvantages are:

1 Any special structure of constraints (e.g., linearity) is destroyed.
2 The rapid change of $P(\mathbf{x}, r)$ in the vicinity of the boundary complicates the one-dimensional optimization problem which has to be handled by special techniques.
3 Close to boundaries the terms $[g_j(\mathbf{x})]^{-1}$ are very large, and at the final phase of minimization we get $r_k \longrightarrow 0$, which causes the function $P(\mathbf{x}, r)$ to be very sensitive to variable changes and introduces round-off errors.

In general, the SUMT method is considered suitable for solving constrained optimization problems with few highly nonlinear constraints. If, however, the objective function is minimized subject to many linear constraints, it would be preferable to use a method that handles the linear constraints separately (Sec. 6-3).

Suggested Research Topics

1 What is the best formulation of the barrier function $\phi[g_j(\mathbf{x})]$?
2 What is an appropriate minimization strategy of $P(\mathbf{x}, r)$? Complete minimization for every r_k or several steps followed by a reduction of r_k?
3 What is a proper choice of the initial value of r_1 and the optimal way in which this perturbation parameter is reduced? This question has been

partially answered on the basis of the theoretical and experimental results [Fiacco and McCormick, 1964*b*; Kowalik and Osborne, 1968].

6-3 MODIFIED INTERIOR POINT TRANSFORMATION

Due to the structure of the transformation function $P(\mathbf{x}, r)$ the SUMT method is not well suited to solve problems with many linear constraints. Therefore, it has been suggested that the linear constraints should not be included in the transformed objective function $P(\mathbf{x}, r)$. One possible way of handling problems involving many linear constraints is through the combination of the SUMT with a feasible direction method such as those of Zoutendijk and Rosen (Chap. 7). This strategy is called the modified interior point transformation.

Consider the problem

$$\min M(\mathbf{x}) \tag{6-71}$$

subject to

$$g_j(\mathbf{x}) \geq 0 \qquad j = 1, \ldots, m$$
$$l_j(\mathbf{x}) \geq 0 \qquad j = m + 1, \ldots, p$$

where $l_j(\mathbf{x}) = \mathbf{a}_j^T\mathbf{x} - b_j$. Zoutendijk [1966] suggested forming an equivalent problem

$$\min P(\mathbf{x}, r) = M(\mathbf{x}) + r \sum_{j=1}^{m} [g_j(\mathbf{x})]^{-1} \tag{6-72}$$

subject to

$$l_j(\mathbf{x}) \geq 0 \qquad j = m + 1, \ldots, p$$

which can be conveniently solved by the method of feasible directions based on the use of linear programming.

Iteration Steps

1 Given is an interior-feasible point $\mathbf{x}^0$, and r_K. Set $k = 0$.

2 Compute

$$\nabla P(\mathbf{x}^k, r_K) = \nabla P_{k,K} = \nabla M(\mathbf{x}^k) - r_K \sum_{j=1}^{m} \nabla g_j(\mathbf{x}^k)/[g_j(\mathbf{x}^k)]^2 \tag{6-73}$$

3 Solve the linear programming problem that yields the optimal feasible direction $\mathbf{s}^k$,

$$\min(\nabla P_{k,K})^T\mathbf{s}^k \tag{6-74}$$

subject to

$$\mathbf{a}_j^T\mathbf{s}^k \geq 0$$

where j takes only those values of indices $m + 1, \ldots, p$ for which $\mathbf{a}_j^T\mathbf{x}^k - b_j = 0$ (active constraints). Constraints include also a normalization condition, such as $-1 \leq s_i \leq 1$, $i = 1, 2, \ldots, n$ and n is the number of variables (the appendix gives techniques for this problem).

4 Solve the constrained problem in the variable λ:

$$\min_{\lambda}\{P(\mathbf{x}^k + \lambda\mathbf{s}^k) \mid l_j(\mathbf{x}^k + \lambda\mathbf{s}^k) \geq 0\} \tag{6-75}$$

where j refers to those nonactive constraints $l_j(\mathbf{x})$ at $\mathbf{x}^k$ which satisfy $\mathbf{a}_j^T\mathbf{s}^k < 0$. Call the solution $\lambda_{\min}$. Solution methods for this problem are discussed in Chap. 3.

5 Determine $\mathbf{x}^{k+1} = \mathbf{x}^k + \lambda_{\min}\mathbf{s}^k$. Set $k := k + 1$ and repeat from step 2.

Zoutendijk expects that it may be worthwhile to make only one or few steps in the linearized subproblem with fixed r_K. If we decide to take only one step, then $k = K$.

Termination Criteria and Convergence

Termination criteria are essentially the same as for the SUMT method. The convergence of this procedure has not been proven.

Modifications and Acceleration Techniques

An improved version of Zoutendijk's feasible direction method is described in Chap. 7.

Discussion

The method combines the advantages of the SUMT and feasible direction procedures. It preserves the structure of the linear constraints and implements the fast and efficient simplex method in the solution of subproblems. However, some of the disadvantages of SUMT are also preserved, for example, difficulties with the minimization process in the vicinity of the boundary constraints, round-off error problems, and the necessity of an interior-feasible starting point.

The method looks attractive for problems with a nonlinear objective function, and many linear and a few nonlinear constraints.

6-4 LEAST-SQUARES TRANSFORMATION

The least-squares transformation has the form:

$$T(\mathbf{x}, M_k) = [M(\mathbf{x}) - M_k]^2 + \sum_{j=1}^{m} [g_j(\mathbf{x})]^2 \tag{6-76}$$

where the corresponding constrained problem is $\min_{\mathbf{x}}\{M(\mathbf{x}) \mid g_j(\mathbf{x}) = 0, j = 1, \ldots, m\}$, and M_k, $k = 1, 2, \ldots$ is a sequence of constant parameters. Let us denote the solution to min $T(\mathbf{x}, M_k)$ by $\mathbf{x}^k$ and the solution of the constrained problem by $\mathbf{x}^*$. Let $T(\mathbf{x}^k)$ denote $T(\mathbf{x}^k, M_k)$.

We first observe that if $M_k < M(\mathbf{x}^*)$, then min $T(\mathbf{x}, M_k) = T(\mathbf{x}^k) > 0$. This gives the basis for the procedure suggested by Morrison [1968] where $T(\mathbf{x}, M_k)$ is sequentially minimized with increasing values M_k, $i = 1, 2, \ldots$. These converge under suitable smoothness conditions to the solution; that is,

$$\lim_{k\to\infty} \mathbf{x}^k = \mathbf{x}^* \quad \text{and} \quad \lim_{k\to\infty} M_k = M(\mathbf{x}^*).$$

Assuming that the solution $\mathbf{x}^*$ exists and that the minimum $\mathbf{x}^k$ is well defined, Morrison has proved that:

1 If $M_k \leq M(\mathbf{x}^*)$, then $M(\mathbf{x}^k) \leq M(\mathbf{x}^*)$ and $g(\mathbf{x}^k) \neq 0$; that is, the $\mathbf{x}^k$ that minimizes $T(\mathbf{x}, M_k)$ corresponds to a solution of the relaxed problem (with unsatisfied equality constraints) and it gives a smaller value of $M(\mathbf{x})$ than $M(\mathbf{x}^*)$. Therefore, to solve the constrained problem $\min_{\mathbf{x}}\{M(\mathbf{x}) \mid g_j(\mathbf{x}) = 0, j = 1, \ldots, m\}$, it is sufficient to find a solution to $\min_{\mathbf{x}} T(\mathbf{x}, M_k)$ that satisfies $\mathbf{g}(\mathbf{x}) = 0$ with an acceptable accuracy. Clearly, we have to keep $M_k \leq M(\mathbf{x}^*)$ since we expect that a whole range of values $M_k > M(\mathbf{x}^*)$ exists that would give $\min_{\mathbf{x}} T(\mathbf{x}, M_k) = 0$.

2 If $M_k < M_{k+1}$, then $M(\mathbf{x}^k) \leq M(\mathbf{x}^{k+1})$.

Morrison has also shown that if we define a sequence of increasing values

$$M_{k+1} = M_k + [T(\mathbf{x}^k)]^{1/2} \tag{6-77}$$

then $M_{k+1} \leq M(\mathbf{x}^*)$ is guaranteed if $M_k \leq M(\mathbf{x}^*)$. The consequence of (6-77) and the last two inequalities is that starting from $M_1 < M(\mathbf{x}^*)$, we can generate a sequence $M_1, M_2, \ldots$ which under suitable conditions converges from below to $M(\mathbf{x}^*)$. The main steps of an algorithm which yields such values are:

1 Given are $\mathbf{x}^0$ and $M_1 < M(\mathbf{x}^*)$; let $k = 1$.

2 Starting from $\mathbf{x}^{k-1}$, minimize

$$T(\mathbf{x}, M_k) = [M(\mathbf{x}) - M_k]^2 + \sum_{j=1}^{m} [g_j(\mathbf{x})]^2$$

Call this minimum $\mathbf{x}^k$.

3 Compute $M_{k+1} = M_k + [T(\mathbf{x}^k)]^{1/2}$.

4 If $T(\mathbf{x}^k) < \epsilon$, terminate the process; otherwise, set $k := k + 1$ and return to step 2 (ϵ is a predetermined tolerance).

Unfortunately, this procedure is increasingly inefficient near a solution. A different method for generating a sequence $\{M_k\}$ has been found to give much better results [Kowalik, Osborne, and Ryan, 1969]. This accelerated method uses:

$$M_{k+1}^{ACC} = M_k + \frac{T(\mathbf{x}^k)}{\left\{T(\mathbf{x}^k) - \sum_{j=1}^{m} [g_j(\mathbf{x}^k)]^2\right\}^{1/2}} \tag{6-78}$$

Clearly, the sequence given by (6-78) promises a faster convergence since

$$M_{k+1}^{ACC} \geq M_{k+1} \tag{6-79}$$

If, in addition, $M_{k+1}^{ACC} \leq M(\mathbf{x}^*)$ is guaranteed when $M_k \leq M(\mathbf{x}^*)$, then the M_{k+1}^{ACC} can be used to improve the optimization process. A theoretical investigation of this concept has shown that the corrections M_{k+1}^{ACC} should satisfy the condition $M_{k+1}^{ACC} \leq M(\mathbf{x}^*)$ if $M_k \leq M(\mathbf{x}^*)$ at least in a small neighborhood of $\mathbf{x}^*$, providing that the optimization problem satisfies certain regularity requirements. Most of the practical problems fall into this category.

However, it is possible that we occasionally may get

$$M_{k+1}^{ACC} > M(\mathbf{x}^*) \tag{6-80}$$

and therefore we may use the M_k given by (6-77) to obtain a sequence of M approximations approaching $M(\mathbf{x}^*)$ from below. Clearly, an algorithm that uses (6-78) as frequently as possible, and (6-77) if M_{k+1}^{ACC} is excessive, is faster than the one based on (6-77).

To start the computational process, an initial value $M_1 < M(\mathbf{x}^*)$ is needed. This number can be either available or generated in the initial phase of the algorithm designed to find a lower bound for $M(\mathbf{x}^*)$. Consider the connected set $S = \{\mathbf{x} \mid g_j(\mathbf{x}) = 0, j = 1, \ldots, m\}$ containing $\mathbf{x}^*$ and $\mathbf{x}^{\max}$, which maximizes $M(\mathbf{x})$, for $\mathbf{x} \in S$. We note that for $M(\mathbf{x}^*) \leq M \leq M(\mathbf{x}^{\max})$

$$\min T(\mathbf{x}, M) = 0 \tag{6-81}$$

Thus, starting with $\mathbf{x}^0 \in S$ and M satisfying $M < M(\mathbf{x}^{\max})$, we minimize $T(\mathbf{x}, M)$, and if $\min_{\mathbf{x}} T(\mathbf{x}, M)$ is sufficiently small, we try again with a smaller value of M. This process is continued until $\min T(\mathbf{x}, M) > 0$, and the first M_k giving this result is accepted as a lower bound for $M(\mathbf{x}^*)$. The second phase starts from M_k and generates a sequence of increasing values $\{M_k\}$.

The inequality constraints $\bar{g}_j(\mathbf{x}) \geq 0$ can be handled by a formal conversion $g_j(\mathbf{x}) = \min\,[0, \bar{g}_j(\mathbf{x})]$.

Iteration Steps

Initial Phase

1 Choose $\mathbf{x}^0 \in S$ defined above and set $M_0 = M(\mathbf{x}^0)$, $k = 1$.

2 Find $\min T(\mathbf{x}, M_k) = [M(\mathbf{x}) - M_k]^2 + \sum_{j=1}^{m} [g_j(\mathbf{x})]^2$ and call the solution $\mathbf{x}^k$. The value of M_k is $M_k = M_{k-1} - \Delta M$, where ΔM is given initially. Techniques for this problem are discussed in Chap. 5 and the appendix.

3 If $[T(\mathbf{x}^k)]^{1/2} < \epsilon$ (test on zero), then $\Delta M := 2\Delta M$, $k := k + 1$, and return to step 2 of this phase; otherwise, $BU = M_{k-1}$ and proceed to the iteration phase. [The value BU is a current upper bound for $M(\mathbf{x}^*)$.]

Iteration Phase

1 Set $BL = M_k$. [The value BL is a current lower bound for $M(\mathbf{x}^*)$.]

2 Compute $\bar{M}_{k+1} = M_k + [T(\mathbf{x}^k)]^{1/2}$ and

$$M_{k+1}^{ACC} = M_k + \frac{T(\mathbf{x}^k)}{\left\{T(\mathbf{x}^k) - \sum_{j=1}^{m} [g_j(\mathbf{x}^k)]^2\right\}^{1/2}}$$

3 Set $k := k + 1$, $BL = \bar{M}_k$. If $M_k^{ACC} < BU$, then $M_k = M_k^{ACC}$; otherwise, $M_k = \bar{M}_k$.

4 Minimize $T(\mathbf{x}, M_k) = [M(\mathbf{x}) - M_k]^2 + \sum_{j=1}^m [g_j(\mathbf{x})]^2$ and call the solution $\mathbf{x}^k$. Techniques for this problem are discussed in Chap. 5 and the appendix.

5 Test for convergence: If $[T(\mathbf{x}^k)]^{1/2} \geq \epsilon$, return to step 2 of this phase; otherwise, $BU = M_k$. If $BU - BL < \epsilon$, terminate computations; otherwise, $M_k = BL$ and return to step 4 of this phase.

Remarks: To start computations it is required that $\mathbf{x}^0 \in S$, which establishes an upper bound for $M(\mathbf{x}^*)$, ensuring that M_{k+1}^{ACC} is not accepted if $M_{k+1}^{ACC} > M(\mathbf{x}^{\max})$. If such a value were accepted, the algorithm would produce a sequence of M diverging to $+\infty$. Note also that when M_k is returned to step 2 of the iteration phase, always $M_k \leq M(\mathbf{x}^*)$.

Termination Criteria and Convergence

In addition to the convergence test in step 5, the algorithm should have a provision for terminating if the improvement of M_k becomes too small. It may happen that BU overestimates $M(\mathbf{x}^*)$ by a sizable amount, and the test $BU - BL < \epsilon$ may not work.

The minimum of $F(\mathbf{x}, M_k)$ has to be computed to high accuracy at each stage, since the condition $\min_{\mathbf{x}} T(\mathbf{x}, M_k) \geq \epsilon$ indicates whether $M_k \geq M(\mathbf{x}^*)$ or $M_k < M(\mathbf{x}^*)$.

A validation of the method has been provided by Theorem 6-3.

Theorem 6-3 Assume that for $\mathbf{x}$ in a closed bounded region ψ:

1 $M(\mathbf{x}), g_j(\mathbf{x}), \; j = 1, \ldots, m$ are continuous.

2 A solution $\mathbf{x}^*$ to problem $\min_{\mathbf{x}}\{M(\mathbf{x}) \mid g_j(\mathbf{x}) = 0, \; j = 1, \ldots, m\}$ exists.

3 A minimum of $T(\mathbf{x}, M_k)$ can be found by unconstrained minimization techniques and that $T(\mathbf{x}^k) = \min_{\mathbf{x}} T(\mathbf{x}, M_k) > 0$ and $M(\mathbf{x}^k) < M(\mathbf{x}^*)$ imply each other.

Then the algorithm converges and the limit points of the set $\mathbf{x}^k$, $k = 1, 2, \ldots$ are solutions to the problem $\min_{\mathbf{x}}\{M(\mathbf{x}) \mid g_j(\mathbf{x}) = 0, \quad j = 1, \ldots, m\}$.

Discussion

The accelerated method has been tested on several problems and performed much better than the method that used $M_{k+1} = M_k + [T(\mathbf{x}^k)]^{1/2}$ throughout the entire process. It can be expected that the acceleration helps at least at the final steps of the process, when the nonaccelerated version loses its efficiency.

In the same numerical study, the accelerated least-squares algorithm has given results at least comparable with those obtained using the SUMT transformation. This observation may not be conclusive, due to the limited experience with the least-squares method, but certain advantages over SUMT are apparent:

1 The least-squares method does not manipulate the very large numbers that cause the difficulties discussed in Sec. 6-1.

2 It does not need a feasible point with respect to the inequality constraints.

On the other hand, the function $T(\mathbf{x}, M_k)$ is not twice differentiable (if it involves transformed inequalities) and this limits the scope of the applicable unconstrained optimization methods.

6-5 COMPUTING SYSTEMS

Computing systems applying sequential transformations are listed in Tables 6-1 and 6-2.

6-6 NUMERICAL RESULTS

Our numerical experiments have mainly been restricted to testing the SUMT transformation used in conjunction with the second-order gradient (Newton) method. The program, called RAC-SUMT, has been developed by McCormick, et al. [1965]. It uses the golden section search method to locate the minimum along the search vectors:

$$\mathbf{s}^k = -[\text{Hess}(\mathbf{x}^k)]^{-1}\nabla M(\mathbf{x}^k) \tag{6-82}$$

if $\text{Hess}(\mathbf{x}^k)$ is positive definite; otherwise, $\mathbf{s}^k = -\nabla M(\mathbf{x}^k)$. There is an option to use an extrapolation based on the last two or three solutions—$\mathbf{x}(r_k)$ and $\mathbf{x}(r_{k-1})$; or $\mathbf{x}(r_k)$, $\mathbf{x}(r_{k-1})$, and $\mathbf{x}(r_{k-2})$ (Sec. 6-2). Tables 6-3 to 6-8 summarize the results of our tests for the Rosen-Suzuki problem described in Sec. 1-10.

To illustrate the impact of an initial choice of r_1, we have tried the following values: $r_1 \cong 152.3$ given by (6-35), $r_1 \cong 62.59$ given by (6-33) and the arbitrary choice $r_1 = 1$. Cumulative evaluation of the computing effort given in Table 6-8 indicates that of these three values the last one ($r = 1$) has been the most appropriate. The case reported in Table 6-3 has been computed twice: with and without extrapolation. As expected, the first computation was more efficient; this is indicated in the first two rows of Table 6-8. Further improvements have been achieved by increasing the value

Table 6-1 SUMT METHOD

Title and Reference	Type, Machine and Storage	Experience	Remarks
RAC—SUMT, [McCormick, et al., 1965], SHARE Number SDA 3189 sequential unconstrained minimization technique	FORTRAN IV CDC 6600, main program uses approximately 19300 words plus user supplied subroutines	Various applications indicate success (Sec. 6-6)	1. Program coded by authors of the method 2. Allows for 100 variables and 200 constraints 3. Can be adapted to any machine with FORTRAN compiler

Table 6-2 MIXED PENALTY-BARRIER FUNCTION METHOD

Title and Reference	Type, Machine and Storage	Experience	Remarks
ALGOL 60 procedure for constrained minimization via mixed penalty-barrier function [Lootsma, 1970]; unconstrained optimization by the Davidon variable metric method	ALGOL 60	Program has been checked by the author on the Electrologica X8 computer in the Mathematical Centre, Amsterdam.	1. Author F. A. Lootsma 2. The procedure is designed to solve $\min\{M(\mathbf{x})\mid g_i(\mathbf{x})\geq 0,\quad i=1,\ldots,m,\ h_j(\mathbf{x})=0,\ j=1,\ldots,p\}$ via penalty-barrier function $P(\mathbf{x},r)=M(\mathbf{x})-r\sum_{I_1}\ln g_i(\mathbf{x})+r^{-1}[l(\mathbf{x})+e(\mathbf{x})]$ where $l(\mathbf{x})=\sum_{I_2}\{\min[0,g_i(\mathbf{x})]\}^2$ $e(\mathbf{x})=\sum_j h_j^2(\mathbf{x})$ and $I_1=\{i\mid g_i(\mathbf{x}^0)>0,\quad 1\leq i\leq m\}$ $I_2=\{i\mid g_i(\mathbf{x}^0)\leq 0,\quad 1\leq i\leq m\}$ where $\mathbf{x}^0$ is a starting point.

Table 6-3 SUMT TRANSFORMATION AND NEWTON METHOD; [Rosen-Suzuki], $\mathbf{x}^0 = (0, 0, 0, 0)^T$, $r_k/r_{k+1} = 4$, extrapolation generated but not used.

r	$M(\mathbf{x})$	$P(\mathbf{x}, r)$	x_1	x_2	x_3	x_4
152.3	−14.008	53.108	−0.2060	0.5034	0.7357	0.2147
9.520	−35.930	25.884	−0.0446	0.7960	1.6653	−0.5224
Extrapolation (3 solutions)	−44.337	−43.498	0.0208	0.9512	2.0566	−0.9313
0.1487	−43.051	−42.021	−0.0098	0.9638	1.9607	−0.9455
Extrapolation (3 solutions)	−44.019	−44.057	−0.0060	1.0038	1.9966	−1.0164
1.453×10^{-4}	−43.971	−43.942	−0.0005	0.9995	1.9986	−0.9989
Extrapolation (3 solutions)	−44.000	−44.000	0.0000	1.0000	1.9999	−0.9999

Table 6-4 SUMT TRANSFORMATION AND NEWTON METHOD; [Rosen-Suzuki], $\mathbf{x}^0 = (0, 0, 0, 0)^T$, $r_k/r_{k+1} = 4$, 3-point extrapolation.

r	$M(\mathbf{x})$	$P(\mathbf{x}, r)$	x_1	x_2	x_3	x_4
62.59	−23.532	10.269	−0.1372	0.6125	1.1306	−0.0522
3.912	−38.881	−32.861	−0.0268	0.8524	1.7922	−0.6747
Extrapolation	−44.163	−44.066	0.0043	0.9645	2.0267	−0.9807
.0153	−43.700	−43.388	−0.0044	0.9911	1.9863	−0.9855
Extrapolation	−43.999	−44.008	−0.0004	1.0042	1.9989	−1.0007
2.388×10^{-4}	−43.963	−43.925	−0.0006	0.9995	1.9981	−0.9984
Extrapolation	−44.000	−44.000	0.0001	1.0007	1.9998	−1.0001

of $c = r_k/r_{k+1}$, from 4 to 16 and 50. The authors of RAC-SUMT recommend use of $c = 16$ (and $c = 4$ in nonconvex problems), but in our example faster reduction of r was better (Tables 6-5 to 6-7).

To illustrate the performance of the SUMT-RAC in the case where all the constraints are inactive at the minimum, we have solved the Wood problem with the linear constraints $-10 \leq \mathbf{x} \leq 10$. The minimum shown in Table 6-9 was obtained after 284 function and constraint evaluations, and 34 function gradient and 26 function second partial derivative evaluations. Table 6-10 summarizes the solution of the Colville optimization problem. As indicated earlier, the SUMT method is not suitable for problems with linear constraints, but we generated these results for the purpose of comparison

Table 6-5 SUMT TRANSFORMATION AND NEWTON METHOD; [Rosen-Suzuki], $\mathbf{x}^0 = (0, 0, 0, 0)^T$, $r_k/r_{k+1} = 4$, 3-point extrapolation.

r	$M(\mathbf{x})$	$P(\mathbf{x}, r)$	x_1	x_2	x_3	x_4
1	−41.468	−38.624	−0.0156	0.9125	1.8992	−0.8249
0.25	−42.762	−41.406	−0.0112	0.9527	1.9496	−0.9255
Extrapolation (2-point)	−44.026	−44.188	−0.0067	0.9928	2.0001	−1.0161
0.0156	−43.697	−43.381	−0.0044	0.9910	1.9862	−0.9854
Extrapolation	−43.999	−44.008	−0.0005	1.0044	1.9988	−1.0007
2.441×10^{-4}	−43.962	−43.924	−0.0006	0.9994	1.9981	−0.9983
Extrapolation	−44.(9)...	−44.000	0.0000	1.0007	1.9998	−1.0001

Table 6-6 SUMT TRANSFORMATION AND NEWTON METHOD; [Rosen-Suzuki], $\mathbf{x}^0 = (0, 0, 0, 0)^T$, $r_k/r_{k+1} = 16$, 3-point extrapolation.

r	$M(\mathbf{x})$	$P(\mathbf{x}, r)$	x_1	x_2	x_3	x_4
1	−41.468	−38.624	−0.0156	0.9125	1.8992	−0.8349
0.0625	−43.390	−42.738	−0.0075	0.9779	1.9737	−0.9678
Extrapolation (2-point)	−44.017	−44.110	−0.0048	0.9996	1.9986	−1.0120
3.906×10^{-3}	−43.849	−43.694	−0.0024	0.9964	1.9928	−0.9931
Extrapolation	−44.000	−44.006	−0.0004	1.0028	1.9992	−1.0008
2.441×10^{-4}	−43.962	−43.924	−0.0006	0.9994	1.9981	−0.9983
Extrapolation	−43.(9)...	−44.000	0.0000	1.0003	1.9999	−0.(9)...

Table 6-7 SUMT TRANSFORMATION AND NEWTON METHOD; [Rosen-Suzuki], $\mathbf{x}^0 = (0, 0, 0, 0)^T$, $r_k/r_{k+1} = 50$, 3-point extrapolation.

r	$M(\mathbf{x})$	$P(\mathbf{x}, r)$	x_1	x_2	x_3	x_4
1	−41.468	−38.624	−0.0156	0.9125	1.8992	−0.8349
0.02	−43.657	−43.298	−0.0050	0.9894	1.9845	−0.9832
Extrapolation (2-point)	−44.010	−44.068	−0.0032	1.0021	1.9985	−1.0077
4×10^{-4}	−43.952	−43.903	−0.0008	0.9991	1.9976	−0.9978
Extrapolation	−43.999	−43.001	0.0000	1.0007	1.9998	−1.0001

Table 6-8 CUMULATIVE EVALUATION

	Number of Evaluations		
Table	*Function and Constraints*	*Function Gradients Added to Constraints Gradients*	*Second Partial Derivatives*
Table 6–3 Without extrapolation	370	143	36
With extrapolation	351	134	25
Table 6–4	323	124	23
Table 6–5	241	88	18
Table 6–6	204	61	15
Table 6–7	188	43	14

Table 6-9 SUMT TRANSFORMATION AND NEWTON METHOD; [Wood] with linear constraints, $\mathbf{x}^0 = (-3, -1, -3, -1)^T$, $r_k/r_{k+1} = 100$, 3-point extrapolation.

r	$M(\mathbf{x})$	$P(\mathbf{x}, r)$	x_1	x_2	x_3	x_4
76.504×10^4	41.059	61.208×10^4	0.6492×10^{-3}	0.1212×10^{-1}	0.6523×10^{-3}	0.1219×10^{-1}
76.504×10^2	17.900	61.510×10^2	0.5363	0.3602	0.5296	0.3611
76.504	5.438×10^{-3}	61.816	0.9938	0.9884	0.9939	0.9886
Extrapolation	4.307×10^{-1}	-30.436×10^{-1}	1.0492	1.0649	1.0501	1.0650
0.76504	5.441×10^{-7}	0.6182	0.9999	1.0000	0.9999	1.0000
Extrapolation	2.368×10^{-5}	-34.362×10^{-3}	1.0002	1.0006	1.0001	1.0005

Table 6-10 SUMT TRANSFORMATION AND NEWTON METHOD; [Colville], $\mathbf{x}^0 = (0, 0, 0, 0, 1)^T$, $r_k/r_{k+1} = 50$, 3-point extrapolation.

r	$M(\mathbf{x})$	$P(\mathbf{x})$	x_1	x_2	x_3	x_4	x_5
1	−26.26	69.19	0.1763	0.2575	0.2889	0.5698	0.4271
0.02	−31.24	−29.70	0.2684	0.3209	0.3743	0.4676	0.2677
4×10^{-4}	−32.19	−32.03	0.2953	0.3311	0.3962	0.4352	0.2328
Extrapolation (2-point)	−32.34	−32.41	0.2997	0.3327	0.3997	0.4299	0.2271
8×10^{-6}	−32.33	−32.30	0.2993	0.3331	0.3994	0.4293	0.2254
Extrapolation (3-point)	−32.3486	−32.3486	0.3000	0.3334	0.4000	0.4283	0.2241

with the other methods of Chap. 7. The minimum point was obtained at the expense of 310 function and constraint evaluations, and 25 function gradient and 16 second partial derivative evaluations.

Tables 6-11 and 6-12 illustrate the performance of the least-squares method with the Morrison parameters and with the acceleration.

Table 6-11 LEAST-SQUARES METHOD WITH THE MORRISON PARAMETERS SEQUENCE; [Rosen-Suzuki], $\mathbf{x}^0 = (0, 0, 0, 0)^T$.

k	$M(\mathbf{x}^k)$	x_1	x_2	x_3	x_4
0	−44.810	0.05034	1.0236	2.0555	−0.9555
1	−44.460	0.03172	1.0147	2.0337	−0.9663
2	−44.273	0.01943	1.0090	2.0203	−0.9772
3	−44.160	0.01171	1.0054	2.0122	−0.9855
4	−44.094	0.00699	1.0033	2.0073	−0.9911
5	−44.056	0.00417	1.0019	2.0043	−0.9946
10	−44.004	0.00030	1.0001	2.0003	−0.9996
15	−44.000	0.00017	1.0001	1.9999	−1.0001
20	−44.000	0.00002	1.0002	1.9999	−1.0001
21	−44.000	0.00000	1.0001	1.9999	−1.0000

Table 6-12 LEAST-SQUARES METHOD WITH THE ACCELERATED PARAMETERS SEQUENCE; [Rosen-Suzuki], $\mathbf{x}^0 = (0, 0, 0, 0)^T$.

k	$M(\mathbf{x}^k)$	x_1	x_2	x_3	x_4
0	−44.810	.05034	1.0236	2.0555	−0.9555
1	−44.029	.00200	1.0010	2.0023	−0.9971
2	−44.000	.00010	1.0000	1.9999	−1.0001
3	−44.010	.00110	1.0005	2.0011	−0.9985
4	−44.009	.00063	1.0003	2.0007	−0.9991
5	−44.005	.00039	1.0002	2.0004	−0.9995
8	−44.001	.00003	0.9999	2.0002	−0.9998
10	−44.000	.00000	1.0001	1.9999	−1.0000

References

BELTRAMI, E. J.: "A Computational Approach to Necessary Conditions in Mathematical Programming," *ICC Bulletin*, Vol. 6, 265, 1967.

BOX, M. J., D. DAVIES and W. H. SWANN: "Non-Linear Optimization Techniques," *ICI Ltd. Monograph No. 5*, Oliver and Boyd, Ltd., Edinburgh, p. 51, 1969.

BRACKEN, J. and G. P. MCCORMICK: *Selected Applications and Nonlinear Programming*, John Wiley & Sons, Inc., New York, 1968.

BUTLER, T. and A. V. MARTIN: "On a Method for Minimizing Functionals," *Journal of Math. and Phys.*, Vol. 41, 291, 1962.

CARROLL, C. W.: "The Created Response Surface Technique for Optimizing Nonlinear Restrained Systems," *Operations Research*, Vol. 9, 169, 1961.

FIACCO, A. V.: *Historical Survey of Sequential Unconstrained Methods for Solving Constrained Minimization Problems*, Technical Paper, RAC–TP–267, Research Analysis Corporation, McLean, Va., 1967.

——— and A. P. JONES: "Generalized Penalty Methods in Topological Spaces," *SIAM Journal of Appl. Math.*, Vol. 17, 996, 1969.

——— and G. P. MCCORMICK: "The Sequential Unconstrained Minimization Technique for Nonlinear Programming, A Primal-Dual Method," *Management Science*, Vol. 10, 360, 1964*a*.

———: "Computational Algorithm for the Sequential Unconstrained Minimization Technique for Nonlinear Programming," *Management Science*, Vol. 10, 601, 1964*b*.

———: "SUMT Without Parameters," *Systems Research Memorandum 121*, Northwestern University, Evanston, Ill., 1965.

———: "Extensions of SUMT for Nonlinear Programming: Equality Constraints and Extrapolation," *Management Science*, Vol. 12, 816, 1966.

———: "The Slacked Unconstrained Minimization Technique for Convex Programming," *SIAM Journal of Appl. Math.*, Vol. 15, 505, 1967.

———: *Nonlinear Programming—Sequential Unconstrained Minimization Techniques*, John Wiley & Sons, Inc., New York, 1968.

FLETCHER, R. and A. P. MCCANN: "Acceleration Techniques for Nonlinear Programming," in R. Fletcher (ed.), *Optimization*, Academic Press Inc., New York, Ch. 13, 203, 1969.

FOX, R. L. and L. A. SCHMIT, JR.: "Advances in the Integrated Approach to Structural Synthesis," *Journal of Spacecraft and Rockets*, Vol. 3, 858, 1966.

FRISCH, K. R.: *The Logarithmic Potential Method of Convex Programming*, Oslo: Memo of University Institute of Economics, 1955.

GISVOLD, K. M.: *A Method for Non-linear Mixed-integer Programming and its Application to Design Problems*, SKB II/M18, Norges Tekniske Hogskole, Trondheim, 1970.

HESTENES, M. R.: "Multiplier and Gradient Methods," *Journal of Optimization Theory and Applications*, Vol. 4, 303, 1969.

HUARD, P.: "Resolution of Mathematical Programming with Nonlinear Constraints by the Method of Centers," in J. Abadie (ed.), *Nonlinear Programming*, North Holland Publishing Co., Ch. 8, 207, 1967.

———: "Programmation Mathematique Convexe," in R. Fletcher (ed.), *Optimization*, Academic Press Inc., New York, Ch. 5, 49, 1969.

KAVLIE, D. and J. MOE: "Application of Nonlinear Programming to Optimum Grillage Design with Non-Convex Sets of Variables," *International Journal for Numerical Methods in Engineering*, Vol. 1, 351, 1969.

KELLEY, H. J., W. F. DENHAM, I. L. JOHNSON and P. O. WHEATLEY: "Accelerated Gradient Method for Parameter Optimization with Nonlinear Constraints," *Journal of Astronautical Sciences*, Vol. 13, 166, 1966.

KOWALIK, J. and M. R. OSBORNE: *Methods for Unconstrained Optimization Problems*, American Elsevier Publishing Company, New York, 1968.

KOWALIK, J., M. R. OSBORNE and D. M. RYAN: "A New Method for Constrained Optimization Problems," *Operations Research*, Vol. 17, 973, 1969.

LOOTSMA, F. A.: *Boundary Properties of Penalty Functions for Constrained Minimization*, Doctor Tech. Sc. Thesis, Eindhoven, Netherlands, 1970.

LUND, S.: *Tanker Frame Optimization by Means of SUMT Transformation and Behavior Models*, SKB II/M17, Norges Tekniske Hogskole, Trondheim, 1970.

MCCORMICK, G. P., W. C. MYLANDER and A. V. FIACCO: *Computer Program Implementing the Sequential Unconstrained*

Minimization Technique for Nonlinear Programming, Technical Paper RAC–TP–151, Research Analysis Corporation, McLean, Va., 1965.

MIELE, A., E. E. CRAGG, R. R. IYER and A. V. LEVY: "Use of the Augmented Penalty Function in Mathematical Programming Problems," Parts 1 and 2, *Journal of Optimization Theory and Applications*, Vol. 8, 115 and 131, 1971.

MOE, J.: "Design of Shipstructures by Means of Nonlinear Programming Techniques," *Conference Proceedings No. 36 of the AGARD Symposium on Structural Optimization*, Istanbul, Turkey, 1969.

MORRISON, D. D.: "Optimization by Least Squares," *SIAM Journal Numer. Anal.*, Vol. 5, 83, 1968.

MURRAY, W.: "Ill-Conditioning in Barrier and Penalty Functions Arising in Constrained Nonlinear Programming," Princeton University: in *Proceedings of the Sixth International Symposium on Mathematical Programming*, 1967.

OSBORNE, M. R. and D. M. RYAN: "On Penalty Function Methods for Nonlinear Programming Problems," *Journal of Math. Analysis and Applications*, Vol. 31, 559, 1970*a*.

———: *An Algorithm for Nonlinear Programming*, Technical Report No. 35, Computer Centre, The Australian National University, Canberra, 1970*b*.

PARISOT, G. R.: "Les Programmes Logarithmiques Application aux Calculs des Programmes Convexes," *UNESCO, NS, ICIP*, D. 1, 5. IBM, France, 1960.

PIETRZYKOWSKI, T.: "An Exact Potential Method for Constrained Maxima," *SIAM Journal Numer. Anal.*, Vol. 6, 299, 1969.

POWELL, M.J.D.: "A Method for Nonlinear Constraints in Minimization Problems," in R. Fletcher (ed.), *Optimization*, Academic Press Inc., New York, Ch. 19, 283, 1969.

WEYL, H.: "Das Asymptotische Verteilungsgesetz der Eigenwerte Linearer Partieller Differentialgleichungen," *Math. Ann.* (Leipzig) Vol. 71, 1912.

ZANGWILL, W. I.: "Non-Linear Programming Via Penalty Functions," *Management Science*, Vol. 13, 344, 1967.

ZOUTENDIJK, G.: "Nonlinear Programming: A Numerical Survey," *SIAM Journal, Theory and Applic. Cont.* Vol. 4, 194, 1966.

7 METHODS FOR CONSTRAINED PROBLEMS

A general nonlinear programming problem is:

$$\min_{\mathbf{x}} M(\mathbf{x}) \tag{7-1}$$

subject to

$$g_j(\mathbf{x}) \geq 0 \qquad j = 1, \ldots, m_1 \qquad\qquad g_j(\mathbf{x}) = 0 \qquad j = m_1 + 1, \ldots, m$$

If no inequality constraints appear in the formulation, we have the "classical" optimization problem, which can be solved using the Lagrangian multiplier method. The method involves the Lagrangian function

$$L(\mathbf{x}, \mathbf{v}) = M(\mathbf{x}) - \sum_{j=1}^{m} v_j \, g_j(\mathbf{x}) \tag{7-2}$$

where $\mathbf{v} = (v_1, \ldots, v_m)^T$ are the Lagrangian multipliers.

Consider now a problem with the differentiable functions $M(\mathbf{x})$, $g_j(\mathbf{x})$, and where all constraints are equations. The necessary (but not sufficient) conditions give at the minimum:

$$\begin{aligned} \frac{\partial L}{\partial x_i} &= \frac{\partial M(\mathbf{x})}{\partial x_i} - \sum_{j=1}^{m} v_j \frac{\partial g_j(\mathbf{x})}{\partial x_i} = 0 \qquad & i = 1, \ldots, n \\ \frac{\partial L}{\partial v_j} &= g_j(\mathbf{x}) = 0 & j = 1, \ldots, m \end{aligned} \tag{7-3}$$

This constitutes a set of $m + n$ equations in the m Lagrangian multipliers v_j and n variables x_i. Thus, the classical optimization problem can be reduced to the problem of solving a set of nonlinear equations. In the presence of inequalities, we have the additional requirements that

$$g_j(\mathbf{x}) \geq 0, \; v_j g_j(\mathbf{x}) = 0 \qquad j = 1, \ldots, m_1 \tag{7-4}$$

The solutions to (7-3) and (7-4), must be searched to find min $M(\mathbf{x})$. Special combinatorial algorithms have been developed that reduce the problem to one with equalities only by searching for active inequalities. Still, the combinatorial nature of this approach and difficulties involved in solving nonlinear sets of equations limit practical use of the Lagrangian method to problems with only few constraints.

Certain constrained optimization problems have received special attention; for example,

1 Linear and quadratic optimization problems that may be solved using the simplex or simplex-like algorithms (appendix).
2 All integer or mixed integer continuous problems, formulated in the same way as the linear problems; in addition, all or some of the problem variables must take on only integer (or discrete) values [Balinski, 1965; Trauth and Woosley, 1969].
3 Problems of a special structure, such as separable programs or decomposable problems [Sec. 2-2; Schwartz, 1969].

Special solution methods exist for all the above problems. The applicability of these methods is restricted to the particular problem formulations.

7-1 TYPES OF METHODS

In this chapter we confine our attention to the general nonlinear programming problem with continuous variables and do not discuss any special problems (except problems with linear constraints) for which the computational procedures are based on the particular form of the problem. We describe the following four categories of generally applicable methods, all of which attack the original (untransformed) problem directly:

1 *Simple heuristic search* method, which is similar to the simplex method described in Chap. 4 (the complex method in Sec. 7-2).
2 *Constraint approximation* methods where the objective function and constraints are linearized and the approximate linear problem is solved by using linear programming techniques. The point obtained as solution is used to construct a new linear approximation. The two types of methods in this group are:
   a. The cutting plane method (Sec. 7-3) where each linearization adds extra constraints that cut off a part of the infeasible region.
   b. The approximate programming method (Sec. 7-4) where we relinearize all the functions in every iteration and limit solution of the linearized problem to a small vicinity of the current iteration point.
3 *Methods of feasible directions*, which operate inside the feasible region and generate a sequence of feasible points $\mathbf{x}^1, \mathbf{x}^2, \ldots \mathbf{x}^k, \mathbf{x}^{k+1}, \ldots$ such that for all k

$$M(\mathbf{x}^{k+1}) \leq M(\mathbf{x}^k) \tag{7-5}$$

where
$$\mathbf{x}^{k+1} = \mathbf{x}^k + \lambda^k \mathbf{s}^k \tag{7-6}$$
and $\lambda^k > 0$. In methods of feasible directions, the move from $\mathbf{x}^k$ to $\mathbf{x}^{k+1}$ is accomplished in two stages: first, a direction-finding problem is solved (the vector $\mathbf{s}^k$ is computed) and second, the steplength λ^k is found.

Definition A direction vector $\mathbf{s}^k$ is *feasible* at
$$\mathbf{x} \in R, [R = \{\mathbf{x} \mid g_j(\mathbf{x}) \geq 0, \text{ all } j\}]$$
if $\alpha > 0$ exists such that $(\mathbf{x} + \beta \mathbf{s}^k) \in R$ for all $0 < \beta \leq \alpha$.

This means that we do not immediately violate any constraints when making a sufficiently small step along this direction. Clearly, any direction $\mathbf{s}^k$ is feasible when $\mathbf{x}^k$ is an interior-feasible point.

When interior steps are performed, methods of feasible directions reduce to the steepest descent or related methods, and all the comments and suggestions concerning these methods apply. If, however, $\mathbf{x}^k$ is on the boundary of R, then the vectors $\mathbf{s}^k$ may be directed to the outside of the feasible region and no step can be made in these directions without immediately violating some constraints. A special feature of the methods of feasible directions is that they can also perform boundary steps.

In addition to the feasibility property, we want the direction $\mathbf{s}^k$ to be usable—that is, a descent direction yielding a reduced objective function value in the vicinity of $\mathbf{x}^k$.

Definition A feasible direction $\mathbf{s}^k$ is *usable* if $\bar{\lambda} > 0$ exists such that for all λ satisfying $0 < \lambda \leq \bar{\lambda}$ we have
$$M(\mathbf{x}^k + \lambda \mathbf{s}^k) \leq M(\mathbf{x}^k) \tag{7-7}$$
or if $M(\mathbf{x})$ is differentiable,
$$\left.\frac{dM(\mathbf{x}^k + \lambda \mathbf{s}^k)}{d\lambda}\right|_{\lambda=0} = (\mathbf{s}^k)^T \nabla M(\mathbf{x}^k) \leq 0 \tag{7-8}$$
Any feasible vector $\mathbf{s}^k$ satisfying relation (7-7) is called *feasible-usable* and can serve our purposes.

4 *The reduced gradient* method, which extends the simplex method to the case where the objective function and constraints are nonlinear.

7-2 COMPLEX METHOD

The complex method, first introduced by Box [1965], is similar to the simplex method (Sec. 4-5) and can handle constrained problems. On the basis of practical experience, certain improvements have been suggested by Guin [1968] that increase the chances of reaching the optimum. They have been included in the algorithm presented in this section. The method searches for the minimum value of the objective function $M(\mathbf{x})$ in a feasible region defined by the lower and upper bounds (explicit constraints) $b_i \leq x_i \leq u_i$, $i = 1, 2, \ldots, n$, and constraint functions $g_j(\mathbf{x}) \geq 0, j = 1, 2, \ldots, m$ (implicit constraints). The method cannot handle nonlinear equalities.

Iteration Steps

1 This method uses a set of $\sigma \geq n + 1$ points of which one point satisfies all constraints. The remaining $\sigma - 1$ points are obtained by the use of pseudorandom numbers r_i in the relation $x_i = b_i + r_i(u_i - b_i)$ where r_i are uniformly distributed over the interval [0, 1]. These points satisfy the lower-upper bound constraints.

If some implicit constraints are violated, then the trial point is moved halfway toward the centroid of the remaining, already accepted, points.

The centroid $\bar{\mathbf{x}}^0$ is given by

$$\bar{\mathbf{x}}^0 = \frac{1}{s} \sum_{k=1}^{s} \mathbf{x}^k \tag{7-9}$$

where $\mathbf{x}^1, \mathbf{x}^2, \ldots, \mathbf{x}^s$ are available feasible points. Ultimately, a feasible point can be found and, by repeating this procedure, $\sigma - 1$ points in the feasible region can be generated.

2 The objective function $M(\mathbf{x})$ is evaluated at each vertex, and the vertex $\mathbf{x}^h$ at which the function $M(\mathbf{x})$ assumes the largest value is reflected by computing

$$\mathbf{x}^r = (1 + \alpha)\mathbf{x}^0 - \alpha\mathbf{x}^h, \alpha \geq 1 \tag{7-10}$$

where $\mathbf{x}^0$ is the centroid of the remaining vertices

$$\mathbf{x}^0 = \frac{1}{\sigma - 1} \sum_{\substack{k=1 \\ k \neq h}}^{\sigma} \mathbf{x}^k \tag{7-11}$$

Box found that $\alpha = 1.3$ was a good choice of the overreflection coefficient, but this choice is not critical. The use of the overreflection coefficient $\alpha > 1$ compensates for the tendency of the complex to shrink, which is caused by the moves toward the centroid.

3 If $M(\mathbf{x}^r) < M(\mathbf{x}^h)$ and $\mathbf{x}^r$ is feasible, the point $\mathbf{x}^r$ replaces $\mathbf{x}^h$ and we return to step 2. If $M(\mathbf{x}^r) \geq M(\mathbf{x}^h)$, the overreflection coefficient is reduced, $\alpha := \alpha/2$, and the new $\mathbf{x}^r$ is computed and tried. This is repeated (if necessary) until $\alpha \leq \tau$, where $\tau = 10^{-5}$ is a satisfactory value. If the relation $M(\mathbf{x}^r) < M(\mathbf{x}^h)$ does not hold even for that small value of α, then the projected point $\mathbf{x}^r$ is replaced by the original value of $\mathbf{x}^h$ and the second worst vertex is reflected instead. This process keeps the complex moving toward the minimum unless the centroid is very close to it. If a reflected vertex is not feasible, then this point is moved halfway toward the centroid until it becomes feasible. The method will progress so long as the complex has not collapsed into its centroid.

Remark If the search is performed in a nonconvex region, it may happen that the centroid of all the feasible points is not feasible itself. In this case, all the points of the simplex are discarded except $\mathbf{x}^l$ at which the objective function assumes the lowest value. The new complex is generated by applying

$$x_i^k = x_i^l + r_i(x_i^0 - x_i^l) \qquad i = 1 \ldots, n \tag{7-12}$$

where $\mathbf{x}^0$ is the old infeasible centroid. This construction is restricted to a more favorable subspace and it will help the search to be continued. An occasional infeasibility is treated in the same way as when the complex is moved.

Termination Criteria and Convergence

The process is terminated when the complex shrinks to an acceptable small size or if:

$$\left\{\frac{1}{\sigma}\sum_{k=1}^{\sigma}[M(\mathbf{x}^0) - M(\mathbf{x}^k)]^2\right\}^{1/2} < \epsilon \tag{7-13}$$

where $\epsilon > 0$ is a predetermined small number. It is important for convergence that the complex retain its full dimensionality. Box successfully applied $\sigma = 2n$ as the number of vertices of the complex.

Discussion

The complex method is to some extent scale independent, since an initial complex is roughly scaled to the order of the problem variables. No use is made of the first derivatives of $M(\mathbf{x})$ or $g_j(\mathbf{x})$; hence, there is relatively little work per step. The computer code is very simple and requires only a moderate amount of storage.

The method becomes inefficient rapidly as the number of variables increases. Another disadvantage is that it is incapable of solving problems with equality constraints. A method that is a direct modification of the simplex method and can handle equality and inequality constraints is described in Sec. 4-5.

7-3 CUTTING PLANE METHOD

The cutting plane method [Cheney and Goldstein, 1959; Kelley, 1960] is based on the useful property of convex problems that the linearized constraints always lie entirely outside the feasible region $R = \{\mathbf{x} \mid g_j(\mathbf{x}) \geq 0,$

all $j\}$. We therefore can approximate the region R by linearized envelopes produced by the first terms of the Taylor's series expansions of the constraint functions and solve this approximate problem by the LP simplex method when the objective function is linear. If the solution is not sufficiently accurate, the binding constraints have to be better represented by an additional number of approximating hyperplanes. Typically, the cutting plane method proceeds as follows:

1 The constraints are linearized in the neighborhood of a starting point.
2 The resulting linear problem (we assume that the objective function is linear) is solved by the LP simplex method.
3 The solution is substituted in the original nonlinear constraints, and the most violated constraint is detected.
4 This constraint is linearized about the optimum point of the previous linear problem, and the resulting linear inequality is added to the linearized problem.
5 We return to step 2 and keep repeating steps 2, 3, and 4 until all nonlinear constraints are satisfied to a desired degree of accuracy.

It is sufficient to consider a linear objective function since an arbitrary objective function $M(\mathbf{x})$ can be handled by introducing a new objective function x_{n+1} and a new constraint $x_{n+1} - M(\mathbf{x}) \geq 0$. Therefore, no generality is lost if the problem under consideration is assumed to have a linear objective function $\mathbf{a}^T\mathbf{x}$ and a set of nonlinear constraints $g_j(\mathbf{x}) \geq 0$, $j = 1, 2, \ldots m$.

Iteration Steps

1 Initialize, set $k = 0$; $\mathbf{x}^0$ is given (not necessarily feasible).
2 Solve the linear subproblem (an approximation of the given problem):

$$\min \mathbf{a}^T\mathbf{x} \tag{7-14}$$

subject to

$$g_j(\mathbf{x}^k) + [\nabla g_j(\mathbf{x}^k)]^T(\mathbf{x} - \mathbf{x}^k) \geq 0 \qquad j = 1, 2, \ldots, m \tag{7-15}$$

(A technique for this linear problem is discussed in the appendix.) A possible infinite solution could be prevented by means of appropriate upper-bounding the variables of the problem: $|x_i| \leq u_i$, where u_i, $i = 1, 2 \ldots, n$ are given constants whose values depend on the problem being solved. They should be large enough not to exclude the optimum solution of the original problem. Call the resulting point $\mathbf{x}^{k+1}$.

3 If $\mathbf{x}^{k+1}$ satisfies all the original constraints—that is, $g_j(\mathbf{x}^{k+1}) \geq -\tau$, all j, where τ is a predetermined small positive number, then the computation is terminated since $\mathbf{x}^{k+1}$ is the solution of the nonlinear problem. This follows from the fact that the set $R = \{\mathbf{x} \mid g_j(\mathbf{x}) \geq 0, \text{ all } j\}$ is contained in the set defined by linearized constraints in step 1 if the problem is convex.

Otherwise, we select the most violated constraint $g_J(\mathbf{x}^{k+1}) = \min_j g_j(\mathbf{x}^{k+1})$, relinearize it with respect to $\mathbf{x}^{k+1}$,

$$g_J(\mathbf{x}^{k+1}) + [\nabla g_J(\mathbf{x}^{k+1})]^T(\mathbf{x} - \mathbf{x}^{k+1}) \geq 0 \tag{7-16}$$

and add it to the linear subproblem.

4 The new linear subproblem is solved, and the resulting point is tested again in step 3. The process is then repeated if necessary with $k := k + 1$.

Remark We note that the linear subproblem continuously expands. However, only one linear constraint at a time is added to the previous set of constraints. This enables us to use the dual simplex method, which starts iterations from the last solution (Appendix).

Convergence

Every new linearization cuts off the old optimum but not any portion of the convex set R (Fig. 7-1). The sequence of approximate solutions $\mathbf{x}^k$ are, in general, all infeasible and give nondecreasing values of $M(\mathbf{x}^k) \leq M(\mathbf{x}^*)$ where $\mathbf{x}^*$ solves the original problem. Convergence has been proved by Kelley [1960]; see also Collatz and Wetterling [1966].

Modification and Acceleration Techniques

In case that all constraints are linear, the cutting plane method can be accelerated by a device suggested by Wolfe [1961]. He also showed that by this augmented procedure the quadratic programming problem (satisfying certain mild conditions) can be solved in a finite number of steps. Zoutendijk [1966] suggested several modifications that could accelerate the method. We can, for instance, check the nonlinear constraints cyclically (in every iteration) and accept for linearization the first one that gives $g_j(\mathbf{x}^k) < -\tau_k$, where τ_k is gradually reduced. He argues that this would introduce more iterations but with less computational effort per step. He also posed a question (the answer to which should be explored): Is it better to solve each subproblem completely or perform only few simplex iterations for a fixed k?

Reinschmidt, et al. [1966] considered the possibility of dropping inactive constraints to reduce the size of the linear subproblems. They have not found any simple rule, however, for determining the constraints that could be safely eliminated from the subproblem. Another potential improvement of the method would be to start the process with linear constraints only and add other constraints gradually when they are violated (Fig. 7-1).

Discussion

The cutting plane method is efficient in solving nearly linear convex problems. Considerable computational advantage is taken of the fact that the

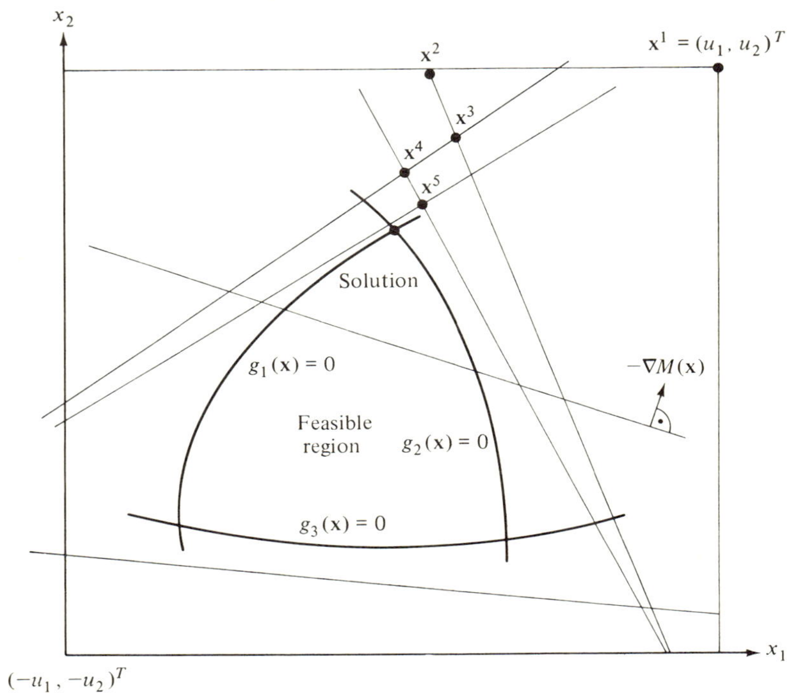

Figure 7-1 Cutting plane method

linear subproblem changes very little from step to step. The dual simplex method enables the efficient solution of problems generated by the addition of constraints to a linear problem that has already been solved. Assuming that the simplex linear programming routine is available, the method is simple to code and has an easy termination criterion.

An obvious disadvantage of the method is that its applicability is restricted to convex problems. Another undesirable feature is that the method can be plagued by round-off errors if the optimum solution does not coincide with a vertex of the feasible region. In this case, the successive cutting planes become more and more parallel, and the expanding computational process deteriorates due to ill-conditioning. Finally, the fact that the intermediate solutions are infeasible may be undesirable in certain applications.

Besides mildly nonlinear convex problems, it has been suggested that this technique be applied to nonlinear integer programming problems. A linear integer programming problem is then solved in each iteration.

A variant of the cutting plane method, which is applicable under less restrictive assumptions than those required by the original cutting plane method, has been proposed by Veinott [1967]. Another method closely related to the cutting plane method has been proposed by Hartley and Hocking [1963].

7-4 METHOD OF APPROXIMATE PROGRAMMING (MAP)

The Griffith and Stewart [1961] method of approximate programming (MAP) is based on the use of linear programming. As in the cutting plane method, the objective function and constraints are linearized by taking the first terms of the Taylor expansion about the current point $\mathbf{x}^k$. However, in the MAP procedure we relinearize all the nonlinear relations at each iteration and no part of the preceeding linear subproblem is retained. The original nonlinear problem thus is locally approximated by linear terms, permitting the solution of nonconvex programming problems without any modifications. To assure that the approximation is adequate, the region of permissible solutions must be kept small and consequently the method is characterized by many iterations with short steps. The method proceeds roughly as follows:

1 The objective function and constraints are linearized in the neighborhood of an arbitrary starting point. We impose additional constraints of the form $|x_i - x_i^0| \leq \delta_i^0$, which limit variations of the variables x_i, $i = 1, \ldots, n$ (Fig. 7-2).

2 The linearized problem is solved and the process is repeated from step 1 with all the nonlinear functions relinearized about the solution of the previous linearized problem. This process is repeated with gradually decreasing bounds δ_i^k until no significant improvement occurs and all the constraints are satisfied within desirable accuracy.

Iteration Steps

1 Initialize, set $k = 0$; $\mathbf{x}^0$ is given.

2 Solve the linear subproblem min $[\nabla M(\mathbf{x}^k)]^T\mathbf{x}$, subject to

$$g_j(\mathbf{x}^k) + [\nabla g_j(\mathbf{x}^k)]^T(\mathbf{x} - \mathbf{x}^k) \geq 0, \qquad j = 1, 2, \ldots, m;$$

and
$$|x_i - x_i^k| < \delta_i^k, \text{ all } i,$$

where $\delta_i^k > 0$ are small predetermined numbers limiting the steps. The resulting point is called $\mathbf{x}^{k+1}$. (A technique for this linear problem is discussed in the appendix.)

3 Set $k := k + 1$ and repeat from step 2 with gradually decreasing δ_i^k until the two conditions are satisfied: (*a*) $|M(\mathbf{x}^k) - M(\mathbf{x}^{k+1})| < \tau_1$; and (*b*) $g_j(\mathbf{x}^{k+1}) \geq -\tau_2$, where $\tau_1 > 0$, $\tau_2 > 0$ are preset tolerances.

Remarks Successive solutions may occasionally start to oscillate between the vertices of the feasible region. This difficulty is usually eliminated by the gradual reduction

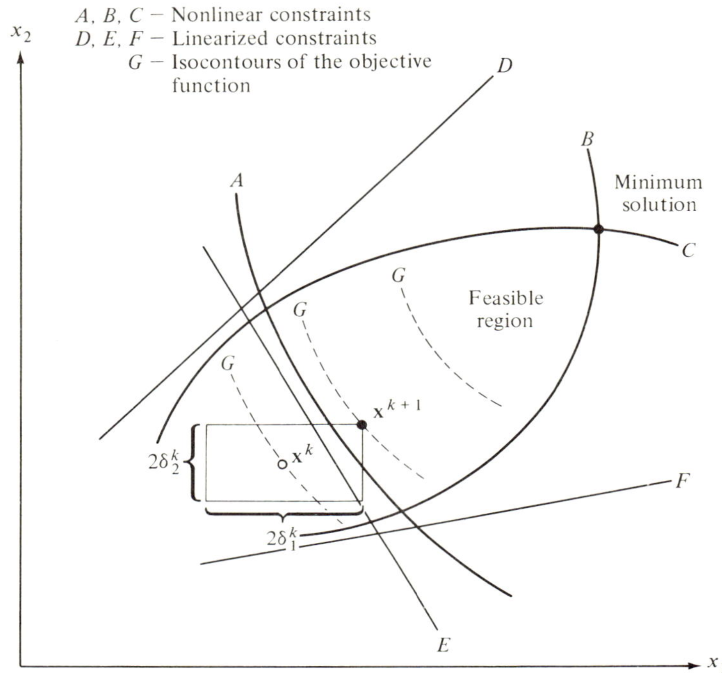

Figure 7-2 Method of approximate programming

of the step sizes δ_i^k. The equality constraints $g_j(\mathbf{x}) = 0, j = m + 1, \ldots, p$ are handled in the same way as inequalities. Sometimes, however, the imposition of large penalty on the artificial slack variable in the linear programming is necessary to assure that the equality relation holds.

The starting point $\mathbf{x}^0$ must be feasible for nonconvex problems. Otherwise, the feasible region of the linearized subproblem may be empty.

Convergence

No proof of convergence has been published. It is likely, however, according to Zoutendijk [1966], that the proof can be obtained for the convex problem by formalizing the choice of δ_i^k.

Discussion

The method of approximate programming is applicable to nonconvex problems and produces feasible or nearly feasible intermediate solutions with good accuracy. As in the cutting plane method, the simplex linear programming routine is used in each iteration. The approximate programming method has been applied successfully to a wide spectrum of nonlinear problems (convex and nonconvex) with continuous and differentiable functions [Pope, 1969].

Experience must be employed in the choice of the stepsize bounds δ_i^k. There is not sufficient evidence to indicate the best choice of the initial step sizes and how to reduce these values. If a poor starting point and too small step sizes δ_i^k are used, then many short steps and slow convergence may occur. Taking too small δ_i^k, we also risk that the current linear programming problem may have no feasible solutions (Fig. 7-2).

The approximate programming method differs from the cutting plane method in that there is no link between subsequent linear programming subproblems. Each time the simplex algorithm is restarted, the existing inverse matrix (computed in the previous step) is disregarded. The method is inefficient for quadratic programs.

An extension of the method using local quadratic programming approximations has been developed by Wilson [1963] and discussed by Beale [1967]. Another algorithm for minimization of a function subject to linear constraints where a quadratic approximation is made to the objective function was suggested by Fletcher [1970]. In this method the quadratic approximation is updated in each iteration and a strategy is given for expanding or contracting the region in which the approximation is adequate.

7-5 METHODS OF FEASIBLE DIRECTIONS

We say that $g_j(\mathbf{x})$ is *an active constraint* with respect to $\mathbf{x}^k$ if $g_j(\mathbf{x}^k) = 0$ and denote a set of all the active constraints by J_A. We restrict the discussion to problems with inequality constraints only.

The direction $\mathbf{s}^k$ is feasible if this vector satisfies

$$(\mathbf{s}^k)^T \nabla g_j(\mathbf{x}^k) - C_j \sigma \geq 0 \qquad j \in J_A \tag{7-17}$$

where $\sigma > 0$ is a slack variable and

$$C_j \begin{cases} > 0 & \text{for the nonlinear constraints} \\ = 0 & \text{for the linear constraints} \end{cases}$$

Let us assume that $M(\mathbf{x})$ is differentiable and the feasible vector $\mathbf{s}^k$ is also usable; that is,

$$(\mathbf{s}^k)^T \nabla M(\mathbf{x}^k) \leq 0 \tag{7-18}$$

Figure 7-3 shows two feasible usable sectors for the nonlinear constraints $g_1(\mathbf{x}) \geq 0$ and $g_2(\mathbf{x}) \geq 0$. Note that the sector corresponding to $\mathbf{x}^k$ is feasible only if $\sigma > 0$, $C_j > 0$.

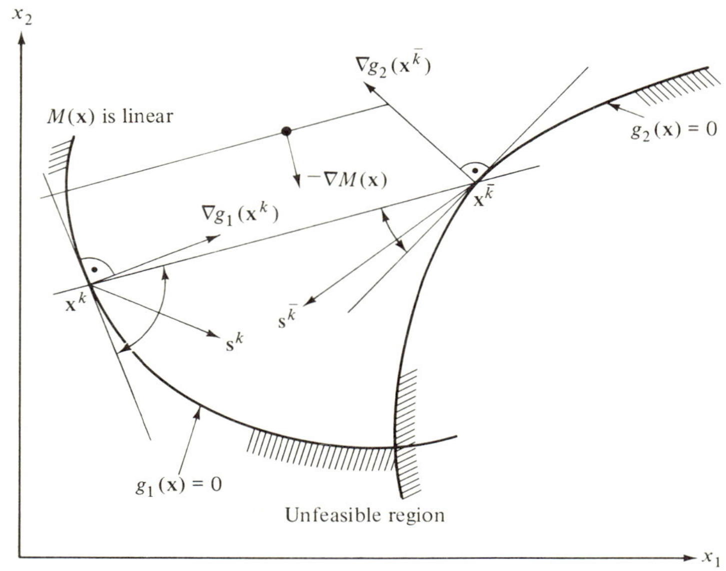

Figure 7-3 Usable feasible sectors

To take into account that $C_j = 0$ for linear and $C_j > 0$ for nonlinear constraints, we define our optimization problem as follows:

$$\min M(\mathbf{x}) \tag{7-19}$$

subject to

$$g_j(\mathbf{x}) \geq 0 \qquad j \in J_N \tag{7-20}$$

$$a_j^T\mathbf{x} \geq b_j \qquad j \in J_L \tag{7-21}$$

where J_N, J_L are sets of indices of the nonlinear and linear constraints. Let us also denote by J_{AN} and J_{AL} the subsets of indices of the active nonlinear and linear constraints. With this notation, the direction-finding problem can now be formulated in the following manner: given $\mathbf{x}^k$, find $\mathbf{s}^k$ to satisfy

$$\max \sigma, \text{ such that } \sigma > 0 \text{ and}$$

$$(\mathbf{s}^k)^T\nabla g_j(\mathbf{x}^k) - C_j\sigma \geq 0 \qquad j \in J_{\mathrm{AN}} \tag{7-22}$$

$$(\mathbf{s}^k)^T\mathbf{a}_j \geq 0 \qquad j \in J_{\mathrm{AL}} \tag{7-23}$$

$$(\mathbf{s}^k)^T\nabla M(\mathbf{x}^k) + \sigma \leq 0 \tag{7-24}$$

and $\mathbf{s}^k$ is normalized by one of the following:

$$(\mathbf{s}^k)^T\mathbf{s}^k = 1 \tag{7-25}$$

$$-1 \leq s_i^k \leq 1 \qquad i = 1, \ldots, n \tag{7-26}$$

$$[\nabla M(\mathbf{x}^k)]^T\mathbf{s}^k \leq 1 \tag{7-27}$$

These requirements bound the length of $\mathbf{s}^k$.

Any solution of the above requirements with $\sigma > 0$ gives a feasible-usable $\mathbf{s}^k$. If we select all $C_j = 1$, then we can interpret our auxiliary optimization subproblem (7-22) through (7-27) as an attempt to find a direction $\mathbf{s}^k$ in which the constraint functions $g_j(\mathbf{x})$ increase about the same amount as the objective function decreases in some immediate vicinity of $\mathbf{x}^k$. It is desirable that this decrease be maximal.

In the case when only linear constraints are active, the auxiliary optimization problem reduces to max σ, such that $\sigma > 0$ and

$$(\mathbf{s}^k)^T\mathbf{a}_j \geq 0 \qquad j \in J_{\text{AL}} \tag{7-28}$$

$$(\mathbf{s}^k)^T\nabla M(\mathbf{x}^k) + \sigma \leq 0 \tag{7-29}$$

and $\mathbf{s}^k$ normalization is as in (7-25) through (7-27).

Both auxiliary problems are linear optimization problems provided that a linear $\mathbf{s}$ normalization is selected. Furthermore, if this condition is $-1 \leq s_i^k \leq 1$ (equivalent to $0 \leq \hat{s}_i^k \leq 2$ where $\hat{s}_i^k = s_i^k + 1$), both auxiliary problems are linear programming problems with upper bounded variables. They can be solved by an efficient, special-purpose simplex method that treats the upper constraints in a special manner, not as constraints. If the auxiliary problem leads to $\sigma_{\max} > 0$, then $M(\mathbf{x})$ can be improved within the feasible region. If, however, $\sigma_{\max} = 0$, then the Kuhn-Tucker optimality conditions (Sec. 1-6) are satisfied at $\mathbf{x}^k$, which is accepted as the optimal solution.

Iteration Steps

1 Initialize, set $k = 0$, evaluate $M(\mathbf{x}^0)$ and $g_j(\mathbf{x}^0), j = 1, \ldots, m$; $\mathbf{x}^0$ is given (feasible).

2 If $g_j(\mathbf{x}^k) > 0$, $j = 1, 2, \ldots, n$ ($\mathbf{x}^k$ is an interior feasible point), set the current direction

$$\mathbf{s}^k = -\nabla M(\mathbf{x}^k) \tag{7-30}$$

normalize $\mathbf{s}^k$ so that $\|\mathbf{s}^k\| = 1$ and proceed to step 5; otherwise, proceed to step 3.

3 Solve the direction-finding problem max σ, such that $\sigma > 0$ and

$$(\mathbf{s}^k)^T\nabla g_j(\mathbf{x}^k) - C_j\sigma \geq 0 \qquad j \in J_{\text{AN}} \tag{7-31}$$

$$(\mathbf{s}^k)^T\mathbf{a}_j \geq 0 \qquad j \in J_{\text{AL}} \tag{7-32}$$

$$(\mathbf{s}^k)^T\nabla M(\mathbf{x}^k) + \sigma \leq 0 \tag{7-33}$$

and one of the following

$$(\mathbf{s}^k)^T\mathbf{s}^k = 1 \tag{7-34}$$

$$-1 \leq s_i^k \leq 1 \qquad i = 1, \ldots, n \tag{7-35}$$

$$[\nabla M(\mathbf{x}^k)]^T\mathbf{s}^k \leq 1 \tag{7-36}$$

where C_j, $j \in J_{AN}$ are positive constraint weighting coefficients and J_{AN}, J_{AL} are the sets of indices of the nonlinear and linear active constraints, respectively. If (7-35) or (7-36) is chosen, step 3 is a linear programming problem; methods for its solution are discussed in the appendix.

4 If σ determined in step 3 satisfies $\sigma < \tau$, terminate the computation and accept $\mathbf{x}^k$ as the solution point; otherwise, proceed to step 5 (τ is a predetermined tolerance).

5 Compute the steplength λ^k [Chap. 3 gives methods for optimization along the line $(\mathbf{x}^k + \lambda\mathbf{s}^k)$].
   a. Optimal step alternative: find λ^k that minimizes $\{M(\mathbf{x}^k + \lambda\mathbf{s}^k) \mid g_j(\mathbf{x}^k + \lambda\mathbf{s}^k) \geq 0, \quad j = 1, \ldots m\}$.
   b. Nonoptimal step alternative: find λ^k that satisfies $M(\mathbf{x}^k + \lambda^k\mathbf{s}^k) \leq M(\mathbf{x}^k)$ and $g_j(\mathbf{x}^k + \lambda^k\mathbf{s}^k) \geq 0, \quad j = 1, \ldots, m$.

6 Perform the step and set $\mathbf{x}^{k+1} = \mathbf{x}^k + \lambda^k\mathbf{s}^k$.

7 Evaluate the objective function $M(\mathbf{x})$ at the new point.

8 If $M(\mathbf{x}^k) - M(\mathbf{x}^{k+1}) < \epsilon_1$ and $\|\mathbf{x}^{k+1} - \mathbf{x}^k\| < \epsilon_2$, terminate the iterations; otherwise, proceed to step 9 ($\epsilon_1 > 0$ is a predetermined minimal function improvement, and $\epsilon_2 > 0$ is a predetermined minimal stepsize).

9 Set $k := k + 1$ and repeat from step 2.

Convergence and Acceleration Techniques

Special precautions are necessary to guarantee and speed up the convergence of the feasible direction methods. Careful investigation of these algorithms shows that such a procedure as the one above may be very slow or even nonconvergent due to the so-called "jamming phenomenon." Jamming occurs when the algorithm generates a sequence of $\mathbf{x}^k$ converging to a point that is not a solution. Instead of moving toward the solution, all points $\mathbf{x}^k$ jam into a corner of the feasible region from which the method cannot extricate them. This was first observed by Zoutendijk [1960], and examples of procedures that produce the jamming (when used to solve certain numerical problems) have been given by Wolfe [1966] and Zangwill [1967].

Another common feature of all gradient methods is that sometimes a large number of very short steps is taken in strongly alternating directions. This is caused by a rapid change of the gradient vector in the feasible region. Small steps may also occur when the algorithm progresses along the boundaries. To prevent these inefficiencies and to secure convergence we should try to stabilize the search directions and keep iterative solutions away from the boundaries by including in the set of active constraints those that are likely to be approached.

Let us introduce a positive parameter $\epsilon > 0$ and for $\mathbf{x}^k \in R$ (feasible region) the following sets of indices:

$$J_{AN}(\mathbf{x}^k,\epsilon) = \{j \mid \epsilon \geq g_j(\mathbf{x}^k) \geq 0\} \tag{7-37}$$

$$J_{AL}(\mathbf{x}^k,\epsilon) = \{j \mid \epsilon \geq \mathbf{a}_j^T\mathbf{x}^k - b_j \geq 0\} \tag{7-38}$$

and

$$J(\mathbf{x}^k,\epsilon) = J_{AN}(\mathbf{x}^k,\epsilon) + J_{AL}(\mathbf{x}^k,\epsilon) \tag{7-39}$$

These include as a particular case the sets of strictly active constraints: $J_{AN} = J_{AN}(\mathbf{x}^k, 0)$ and $J_{AL} = J_{AL}(\mathbf{x}^k, 0)$. Since we want to avoid slow creeping along boundaries and jamming, we propose to solve the direction-finding subproblems (7-22) through (7-27), and (7-28) and (7-29), where $J_{AN}(\mathbf{x}^k,\epsilon)$, $J_{AL}(\mathbf{x}^k, \epsilon)$ replace J_{AN} and J_{AL}, respectively. Parameter $\epsilon > 0$ should gradually be reduced when $\mathbf{x}^k$ approaches the optimal solution which is indicated by small values of σ, the solution of the direction-finding problems.

In another modification of the technique the constraints encountered twice during the optimization process are kept in the set of active constraints for a certain number of iterations. The following strategy has been successful in practice:

1 If at the current step of the iterative process the approximate solution $\mathbf{x}^k$ is on the linear boundary which has been met twice before, then the condition

$$(\mathbf{s}^k)^T\mathbf{a}_j \geq 0 \tag{7-40}$$

is added to direction-finding subproblems in all subsequent iterations. If, however, the variable σ has not been reduced by a significant amount from the previous iteration, only active constraints are entered.

2 If, at the current iteration the point $\mathbf{x}^k$ is on the nonlinear boundary approached previously, then we require

$$(\mathbf{s}^k)^T\nabla g_j(\mathbf{x}^k) \geq 0 \tag{7-41}$$

in all direction-finding problems following the first one in which

$$(\mathbf{s}^k)^T\nabla g_j(\mathbf{x}^k) - C_j\sigma \geq 0 \tag{7-42}$$

is required. Requirement (7-41) is updated when this constraint becomes active again.

3 In both cases 1 and 2 the "antijamming" inequalities are deleted if the current point is feasible or if σ is less than some predetermined number, which can be gradually reduced.

Inside the feasible region the convergence can be accelerated by introducing the principle of conjugate directions as an additional requirement in the direction-finding problem. This requirement is

$$[\nabla M(\mathbf{x}^{l+1}) - \nabla M(\mathbf{x}^l)]^T\mathbf{s}^q = 0 \tag{7-43}$$

where $l = r, r + 1, \ldots, q - 1$, and $\mathbf{x}^r$ is the last point located on the boundary of the feasible region. All the subsequent points $\mathbf{x}^{r+1}, \mathbf{x}^{r+2}, \ldots, \mathbf{x}^q$ are interior-feasible. The condition given by (7-43) produces conjugate search directions when a quadratic is minimized and guarantees convergence in a finite number of iterations. In a more general problem it is not possible to achieve finiteness, but it may be expected that the application of this principle improves the convergence properties of the method.

In the interior of the feasible region we can also use the Fletcher and Reeves formula (5-71), producing conjugate gradient directions if $M(\mathbf{x})$ is quadratic. Convergence of the feasible directions method has been proved by Zoutendijk, the author of the method. More general conditions which are necessary and sufficient for convergence of nonlinear programming algorithms have been developed by Topkis and Veinott [1967] and recently generalized by Zangwill [1969]. It can be verified that the Zoutendijk method satisfies these convergence conditions.

Discussion

Zoutendijk's [1960] feasible directions method offers an efficient way of reducing the nonlinear programming problem to a sequence of linear programming problems (if a linear normalization is used). Furthermore, it is finite for quadratic programming problems and in more general cases can handle nonconvex problems. From the practical point of view, it offers the additional advantage of providing feasible intermediate approximations to the solution. The method has been implemented to solve practical problems of considerable size and proved to be successful. The following are disadvantages of the method:

1 A rather complicated computer program, which has to include antijamming precautions.
2 Inability to handle the equality constraints without special devices (Sec. 2-2).

There are also several open questions that possibly can be answered on the basis of additional computational experience. These include: (1) appropriate choice of parameters $C_j > 0$ and antijamming precautions (both problems are probably heavily formulation dependent); (2) the type of $\mathbf{s}$ normalization that leads to the most efficiently solvable subproblems; and (3) whether in the steplength determination subproblem we should try to find λ^k minimizing $M(\mathbf{x}^k + \lambda\mathbf{s}^k)$ or just satisfy $M(\mathbf{x}^k + \lambda^k\mathbf{s}^k) < M(\mathbf{x}^k)$ to achieve the best overall efficiency.

7-6 MODIFIED FEASIBLE DIRECTIONS METHOD

The modified feasible directions method was proposed by Zoutendijk [1966]. It combines the linearization technique of the cutting plane method with the feasible directions concept developed by Zoutendijk [1960]. The method generates at every iteration step three types of points:

1 Interior-feasible points $\mathbf{x}^k$, such that $M(\mathbf{x}^{k+1}) < M(\mathbf{x}^k)$, and the sequence of which converges to the solution.

2 Infeasible points $\mathbf{y}^k$ with nondecreasing values of $M(\mathbf{y}^k)$, giving a lower bound for a minimum solution; that is, $M(\mathbf{y}^{k+1}) \geq M(\mathbf{y}^k)$ (valid only for convex problems).
3 Boundary points $\mathbf{z}^k$ giving at each step an upper bound for a minimum.

Two separate cases will be considered:

1 Minimization of a linear objective function $M(\mathbf{x})$ subject to $g_j(\mathbf{x}) \geq 0$, where all $g_j(\mathbf{x})$ are concave and linear constraints $l_j(\mathbf{x}) \geq 0$ (convex problem).
2 The same as (1), but with at least one constraint function $g_j(\mathbf{x})$ that is not concave (nonconvex problem).

Any problem with a nonlinear objective function $M(\mathbf{x})$ can be converted to (1) or (2) by adding the constraint $-M(\mathbf{x}) + x_{n+1} \geq 0$ to the problem and minimizing x_{n+1}, where n is a number of the original variables x_i, and x_{n+1} is a new variable. To start the computation a feasible initial point is needed.

Iteration Steps

Case 1

1 Start with a feasible point, which is interior, with respect to nonlinear constraints: $\mathbf{x}^0 \in R = \{\mathbf{x} \mid g_j(\mathbf{x}) > 0, \quad j = 1, \ldots, m$ and $l_j(\mathbf{x}) \geq 0, j = m + 1, \ldots, p\}$; set $k = 0$.
2 Solve the linear subproblem L_k:

$$\min M(\mathbf{x}) \text{ or } \mathrm{x}_{n+1} \text{ if } M(\mathbf{x}) \text{ is nonlinear} \tag{7-44}$$

subject to

$$l_j(\mathbf{x}) \geq 0, \qquad j = m + 1, \ldots, p \quad \text{and} \quad |x_i| \leq u_i, \qquad i = 1, \ldots, n,$$

where u_i are chosen sufficiently large. Refer to the solution of subproblem L_k as $\mathbf{y}^k$. (A technique for this linear problem is discussed in the appendix.)

3 Determine the boundary point $\mathbf{z}^k$ by computing $\mathbf{z}^k = \mathbf{x}^k + \lambda^k(\mathbf{y}^k - \mathbf{x}^k)$, where λ^k is the solution of

$$\max\{\lambda \mid \mathbf{x}^k + \lambda(\mathbf{y}^k - \mathbf{x}^k) \text{ is feasible}\}. \tag{7-45}$$

Since $\mathbf{z}^k$ is the boundary point, it follows that for one or more j, $g_j(\mathbf{x}^k) = 0$, $j = j_p, j_q, j_r, \ldots$.

4 Linearize the constraints $g_j(\mathbf{x}) \geq 0$, $j = j_p, j_q, j_r, \ldots$, with respect to $\mathbf{z}^k$ and obtain the tangent planes

$$[\nabla g_j(\mathbf{z}^k)]^T\mathbf{x} \geq [\nabla g_j(\mathbf{z}^k)]^T\mathbf{z}^k \tag{7-46}$$

Add these relations to the constraints of the linear subproblem L_k and call the new subproblem L_{k+1}.

5 Calculate

$$\mathbf{x}^{k+1} = \beta\mathbf{x}^k + (1 - \beta)\mathbf{z}^k \tag{7-47}$$

where $0 < \beta < 1$ (say, $\beta = 1/2$).

6 Evaluate $M(\mathbf{x})$ at the new point $\mathbf{x}^{k+1}$. If $M(\mathbf{x}^k) - M(\mathbf{x}^{k+1}) < \epsilon_1$ and $\|\mathbf{x}^{k+1} - \mathbf{x}^k\| < \epsilon_2$, where $\epsilon_1 > 0$, $\epsilon_2 > 0$, terminate the iteration; otherwise, proceed to step 7.
7 Set $k := k + 1$ and repeat from step 2.

Remark Figure 7-4 illustrates the first two steps of the modified feasible directions method used to solve a simple problem with $M(\mathbf{x})$ linear, and two constraints.

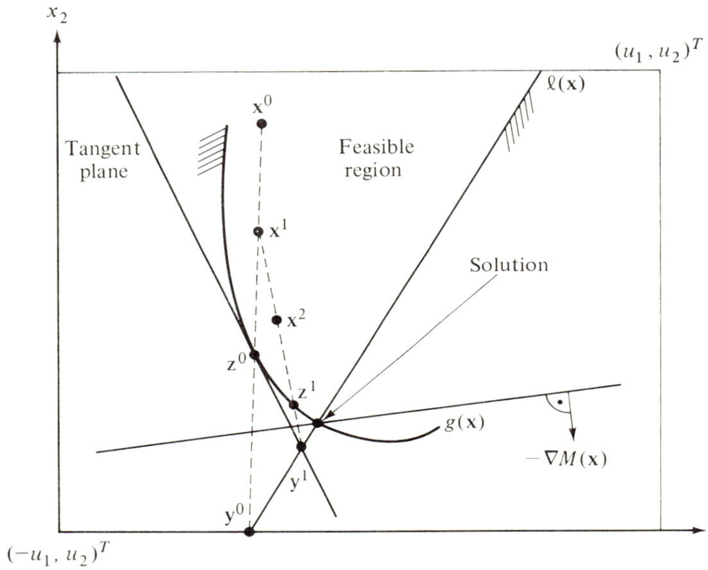

Figure 7-4 Modified feasible directions method

Case 2. In a nonconvex problem, cutting off parts of the feasible region by the sequential linearization should be avoided. If for some j, $g_j(\mathbf{y}^k) > 0$, then any active linearization of $g_j(\mathbf{y}^k)$ in the linear subproblem should be dropped out in the next iteration. Zoutendijk [1966] suggests that this test be performed from time to time, not necessarily in every step.

Termination Criterion and Convergence

For convex problems we get upper and lower bounds of the solution [$M(\mathbf{z}^k)$ and $M(\mathbf{y}^k)$, respectively]. It is therefore natural to terminate computations if

$$M(\mathbf{z}^k) - M(\mathbf{y}^k) < \tau \tag{7-48}$$

where τ is some predetermined tolerance. In nonconvex problems we may stop when

$$M(\mathbf{x}^k) - M(\mathbf{x}^{k+1}) \leq \epsilon_1 \qquad \text{and} \qquad \|\mathbf{x}^{k+1} - \mathbf{x}^k\| \leq \epsilon_2 \tag{7-49}$$

where $\epsilon_1 > 0$, $\epsilon_2 > 0$. The convergence proof for convex problems has been given by Zoutendijk [1960, 1966].

Modifications and Acceleration Techniques

Zoutendijk [1966] suggested that for a problem with $M(\mathbf{x})$ nonlinear and all $g_j(\mathbf{x})$ concave, it may be better to search for a minimum of $M(\mathbf{x})$ along the line $(\mathbf{x}^k - \mathbf{y}^k)$; that is,

$$\hat{\mathbf{x}}^{k+1} = \mathbf{x}^k + \lambda^k(\mathbf{x}^k - \mathbf{y}^k) \tag{7-50}$$

and λ^k is the solution of the problem of optimization along a line (see Chap. 3 for techniques)

$$\min_{\lambda} M[\mathbf{x}^k + \lambda(\mathbf{x}^k - \mathbf{y}^k)] \tag{7-51}$$

If feasible, the point $\hat{\mathbf{x}}^{k+1}$ will be accepted as $\mathbf{x}^{k+1}$ and step 5 is not executed. The linear function to be minimized in L_k will be $[\nabla M(\mathbf{x}^k)]^T\mathbf{x}$. To speed up the convergence we may use the principle of conjugate directions by incorporating in L_k temporarily the linear equation

$$[\nabla M(\mathbf{x}^{k+1}) - \nabla M(\mathbf{x}^k)]^T(\mathbf{x} - \mathbf{x}^{k+1}) = 0 \tag{7-52}$$

where $\mathbf{x}^{k+1}$ has been found as an interior minimum of $M(\mathbf{x})$ on the line $\mathbf{x}^k - \mathbf{y}^k$.

Other possible modifications include: (1) dropping nonactive linearizations from the linear subproblem; and (2) varying the parameter β if step 5 is used. A similar procedure has been given by Kleibohm [1966].

Discussion

Some of the advantages of the method are similar to those of the feasible directions method; it can handle nonconvex problems and uses a fast linear programming routine to solve its central problem. The modification is more troublesome than the feasible directions method if used to solve nonconvex or highly nonlinear problems. This would suggest that the method may be recommended for problems with linear or near-linear constraints.

The method has a very useful property of generating bounds of the optimal solution, which are applied in terminating computations. Its computational performance is not known to the authors of this book, but it is possible to foresee some computational problems similar to those encountered with the cutting plane method. Bad conditioning of linear problems may arise due to near-dependency of constraints, which apply close to the solution. The method is characterized by a considerable amount of work per step and a rather complicated computer program. It shares with the feasible directions method the disadvantage of being unable to handle nonlinear equations directly (without any special tricks; Sec. 2-2).

7-7 GRADIENT PROJECTION METHOD

The gradient projection method of Rosen [1960, 1961], in contrast to Zoutendijk's method of feasible directions, does not require solving auxiliary linear optimization problems. It uses projections of the objective function gradient into the manifold of the constraints that are currently active. The method works with direction vectors $\mathbf{s}$ that are feasible and usable—that is, vectors satisfying

$$\mathbf{s}^T \nabla M(\mathbf{x}) < 0 \tag{7-53}$$

$$\mathbf{s}^T \nabla g_j(\mathbf{x}) = 0 \qquad j = j_1, j_2, \ldots, j_q \tag{7-54}$$

The latter is required for all active constraints. We assume here that all the constraints are linear $g_j(\mathbf{x}) = \mathbf{a}_j^T \mathbf{x} - b_j \geq 0$, and that indices of the active constraints are $j = j_1, j_2, \ldots, j_q$. It is convenient to introduce the matrix of constraint gradients

$$N_q = [\nabla g_{j_1}, \nabla g_{j_2}, \ldots, \nabla g_{j_q}] \tag{7-55}$$

so that the feasibility condition can be concisely stated by

$$N_q^T \mathbf{s} = 0 \tag{7-56}$$

where $j_1, j_2, \ldots, j_q \in J_c$ (the set of active constraints). We seek the direction vector $\mathbf{s}$ satisfying

$$\min_{\mathbf{s}} \mathbf{s}^T \nabla M(\mathbf{x}) \tag{7-57}$$

and

$$\mathbf{s}^T \mathbf{s} - 1 = 0 \tag{7-58}$$

to assure a finite solution.

The Lagrangian function (Sec. 1-7) associated with Prob. (7-56) through (7-58) whose stationary point is equivalent to the solution of the original problem with equality constraints is

$$L(\mathbf{s}, \mathbf{\Lambda}, \omega) = \mathbf{s}^T \nabla M(\mathbf{x}) + \mathbf{\Lambda}^T N_q^T \mathbf{s} + \omega(\mathbf{s}^T \mathbf{s} - 1) \tag{7-59}$$

where $\mathbf{\Lambda} = [\Lambda_1, \ldots, \Lambda_q]^T$ and ω are the Lagrangian multipliers. Optimality conditions give

$$\frac{\partial L}{\partial \mathbf{s}} = \nabla M(\mathbf{x}) + N_q \mathbf{\Lambda} + 2\omega \mathbf{s} = 0 \tag{7-60}$$

$$\frac{\partial L}{\partial \mathbf{\Lambda}} = N_q^T \mathbf{s} = 0 \tag{7-61}$$

$$\frac{\partial L}{\partial \omega} = \mathbf{s}^T \mathbf{s} - 1 = 0 \tag{7-62}$$

From (7-60) and (7-61) we get

$$\mathbf{s} = -\frac{1}{2\omega}[\nabla M(\mathbf{x}) + N_q\mathbf{\Lambda}] \tag{7-63}$$

$$N_q^T\mathbf{s} = -\frac{1}{2\omega}[N_q^T\nabla M(\mathbf{x}) + N_q^TN_q\mathbf{\Lambda}] = 0 \tag{7-64}$$

and since $\omega \neq 0$,

$$N_q^T\nabla M(\mathbf{x}) + N_q^TN_q\mathbf{\Lambda} = 0 \tag{7-65}$$

$$\mathbf{\Lambda} = -(N_q^TN_q)^{-1}N_q^T\nabla M(\mathbf{x}) \tag{7-66}$$

Returning to the expression for $\mathbf{s}$ we have (disregarding the scaling constant ω):

$$\mathbf{s} = -[I - N_q(N_q^TN_q)^{-1}N_q^T]\nabla M(\mathbf{x}) \tag{7-67}$$

The matrix

$$P_q = I - N_q(N_q^TN_q)^{-1}N_q^T \tag{7-68}$$

is called a *projection matrix:* it projects every vector into the intersection of the $q \leq n$ hyperplanes (manifold of linear active constraints). It is assumed that all columns of N_q are linearly independent, from which it follows that $(N_q^TN_q)$ is nonsingular and can be inverted. The normalized directions $\mathbf{s}^k$ can be found from

$$\mathbf{s}^k = \frac{-P_q\nabla M(\mathbf{x}^k)}{\|P_q\nabla M(\mathbf{x}^k)\|} \tag{7-69}$$

If $\mathbf{s}^k \neq 0$, then it is possible to find $\mathbf{x}^{k+1} = \mathbf{x}^k + \lambda^k\mathbf{s}^k$ such that $\mathbf{x}^{k+1}$ is feasible and $M(\mathbf{x}^{k+1}) < M(\mathbf{x}^k)$. If, however, $\mathbf{s}^k = 0$, then from (7-67) we have

$$-\nabla M(\mathbf{x}^k) = N_q\mathbf{\Lambda} \tag{7-70}$$

where

$$\mathbf{\Lambda} = -(N_q^TN_q)^{-1}N_q^T\nabla M(\mathbf{x}^k) \tag{7-71}$$

That is, the negative gradient of the objective function can be expressed as a linear combination of the gradients of the active constraints. If all components of $\mathbf{\Lambda}$ are nonnegative, then the first necessary condition of Kuhn-Tucker for $\mathbf{x}^k$ to be the minimum is satisfied and the computation is terminated. If this condition does not hold, the computation is continued after the projection matrix is modified by deleting from the matrix N the column corresponding to the most negative component of $\mathbf{\Lambda}$. By releasing the active constraint that corresponds to the negative component of $\mathbf{\Lambda}$, we can obtain a lower value of $M(\mathbf{x})$. It may also occur that $\mathbf{x}^{k+1}$, which gives the manifold optimal value of $M(\mathbf{x})$, is located at a new constraint (hyperplane). We then form a new manifold by adding this constraint to the set of active constraints. A considerable computational problem is therefore introduced by the periodical updating of the projection matrix. This problem is discussed later in this section.

Iteration Steps

1 Initialize, set $k = 0$; $\mathbf{x}^0$ is given (interior-feasible or boundary).

2 If $g_j(\mathbf{x}^k) > 0$, $j = 1, \ldots, m$; set

$$\mathbf{s}^k = -\nabla M(\mathbf{x}^k) \tag{7-72}$$

and proceed to step 4; otherwise, proceed to step 3.

3 Compute the current direction $\mathbf{s}^k$

$$\mathbf{s}^k = \frac{-P_q \nabla M(\mathbf{x}^k)}{\|P_q \nabla M(\mathbf{x}^k)\|} \tag{7-73}$$

where

$$P_q = I - N_q(N_q^T N_q)^{-1} N_q^T \tag{7-74}$$

and N_q includes the gradients of all currently active constraints.

4 If $\mathbf{s}^k \neq 0$, compute $\lambda_{\min}^k$ satisfying

$$M(\mathbf{x}^k + \lambda_{\min}^k \mathbf{s}^k) = \min_{\lambda} M(\mathbf{x}^k + \lambda \mathbf{s}^k) \tag{7-75}$$

$$0 < \lambda_{\min}^k < \hat{\lambda} \tag{7-76}$$

where $\hat{\lambda}$ is the largest step that may be taken from $\mathbf{x}^k$ along $\mathbf{s}^k$ without violating any constraint. This value can be computed from

$$\mathbf{a}_j^T(\mathbf{x}^k + \hat{\lambda}\mathbf{s}^k) - b_j = 0 \tag{7-77}$$

for all j such that $\mathbf{a}_j^T \mathbf{s}^k < 0$ and $j \notin J_c$ (nonactive constraints). We have to accept the smallest $\hat{\lambda}$ from the set of all these values; that is,

$$\hat{\lambda} = \min_{j \notin J_c} \left[\frac{\mathbf{a}_j^T \mathbf{x}^k - b_j}{-\mathbf{a}_j^T \mathbf{s}^k} \right] \tag{7-78}$$

Clearly, since $\mathbf{a}_j^T \mathbf{x}^k - b_j > 0$ for $j \notin J_c$ and $-\mathbf{a}_j^T \mathbf{s}^k > 0$, then $\hat{\lambda} > 0$. The new approximation to the solution is given by

$$\mathbf{x}^{k+1} = \mathbf{x}^k + \lambda_{\min}^k \mathbf{s}^k \tag{7-79}$$

Two cases should now be considered:

a. If $\lambda_{\min}^k = \hat{\lambda}$, some new constraints (one or more) become active $[g_j(\mathbf{x}^{k+1}) = 0]$ and should be added to the matrix N_q. Go to step 5.

b. If $\lambda_{\min}^k < \hat{\lambda}$, the matrix N_q (and P_q) remains unaltered. Set $k := k + 1$ and repeat from step 2.

5 If $\mathbf{s}^k = 0$, we check the components of the vector $\mathbf{\Lambda}(\mathbf{x})$,

$$\mathbf{\Lambda}(\mathbf{x}^k) = -(N_q^T N_q)^{-1} N_q^T \nabla M(\mathbf{x}^k) \tag{7-80}$$

Two cases are possible:

a. All components of $\mathbf{\Lambda}$ are nonnegative, which indicates that a minimum has been found and the computation is terminated.

b. If some components of $\mathbf{\Lambda}$ are negative, then the column of ∇g_j corresponding to the smallest component is deleted from N_q. Set $k := k + 1$ and return to step 2.

Remarks

The method can easily handle linear equality constraints. Suppose that we have $g_j(\mathbf{x}) = \mathbf{a}_j^T\mathbf{x} - \mathbf{b}_j \geq 0$, $j = 1, \ldots, m_1$ and $g_j(\mathbf{x}) = \mathbf{a}_j^T\mathbf{x} - \mathbf{b}_j = 0$, $j = m_1 + 1, \ldots, m$. We reduce the n-dimensional space E^n of the original problem to the manifold determined by the intersection of the $m - m_1$ hyperplanes $g_j(\mathbf{x}) = 0$, $j = m_1 + 1, \ldots, m$. That means that in addition to satisfying m_1 inequality constraints, all the feasible points must lie in this manifold. With this restriction, the problem with equality and inequality constraints can be treated as one having only inequality constraints. Computationally, this can be accomplished by initially forming the matrix $(N_q^TN_q)^{-1}$ where $N_q = [\nabla g_{m_1+1}, \ldots, \nabla g_m]$ and keeping the vectors $\nabla g_{m_1+1}, \ldots, \nabla g_m$ in N_q throughout the whole computing process. This guarantees that all vectors $\mathbf{s}^k$ lie in the manifold determined by the intersection of the $m - m_1$ hyperplanes.

The convergence to the global solution of a convex problem has been proven by Rosen [1960, 1961] provided the objective function and constraints satisfy certain minimal conditions.

Computational Aspects

The periodical updating of the projection matrix is a considerable computational problem. Fortunately, the subsequent matrices N differ usually by one column ∇g_j, which is either dropped from the set of active constraints or is added to it. It is possible to avoid the complete recomputation of $(N_q^TN_q)^{-1}$ from its definition, which takes $0(q^3)$ multiplications, and use a more efficient recursive procedure which generates the new inverse in only $0(q^2)$ multiplications. Such a method is based on the partitioned form of an inverse.

Suppose that the inverse $(N_q^TN_q)^{-1}$ is known and that ∇g_{j_q} is to be deleted from $N_q = [\nabla g_{j_1}, \nabla g_{j_2}, \ldots, \nabla g_{j_q}]$ and the new inverse $(N_{q-1}^TN_{q-1})^{-1}$ is sought where $N_{q-1} = [\nabla g_{j_1}, \nabla g_{j_2}, \ldots, \nabla g_{j_{q-1}}]$. Let

$$N_q^TN_q = \begin{bmatrix} A & \mathbf{z} \\ \mathbf{z}^T & a \end{bmatrix} \tag{7-81}$$

where

$$A = N_{q-1}^TN_{q-1} \tag{7-82}$$

The desired inverse of A can be computed from the available submatrices of

$$(N_q^T N_q)^{-1} = \begin{bmatrix} B & \mathbf{u} \\ \mathbf{u}^T & b \end{bmatrix} \tag{7-83}$$

The relationship

$$\begin{bmatrix} B & \mathbf{u} \\ \mathbf{u}^T & b \end{bmatrix} \begin{bmatrix} A & \mathbf{z} \\ \mathbf{z}^T & a \end{bmatrix} = I \tag{7-84}$$

gives

$$BA + \mathbf{u}\mathbf{z}^T = I \tag{7-85}$$

$$B\mathbf{z} + a\mathbf{u} = 0 \tag{7-86}$$

$$\mathbf{u}^T A + b\mathbf{z}^T = 0 \tag{7-87}$$

$$\mathbf{u}^T\mathbf{z} + ab = 1 \tag{7-88}$$

From (7-85) and (7-87) we get

$$A^{-1} = B + \mathbf{u}\mathbf{z}^T A^{-1} \tag{7-89}$$

and

$$A^{-1} = B - b^{-1}\mathbf{u}\mathbf{u}^T \tag{7-90}$$

This procedure can be generalized in the case when ∇g_{jl} other than ∇g_{jq} is dropped from N. It is sufficient to interchange the lth and qth row and column of $(N_q^T N_q)^{-1}$ before relationship (7-90) is applied.

In a similar way, we can obtain a procedure for computing $(N_q N_q)^{-1}$ when a column is added to N_{q-1}. We assume the inverse $(N_{q-1}^T N_{q-1})^{-1}$ and ∇g_{jq} are known and it is desired to compute $(N_q^T N_q)^{-1}$.

We have

$$(N_q^T N_q) = (N_{q-1}, \nabla g_{jq})^T (N_{q-1}, \nabla g_{jq}) = \begin{bmatrix} A & \mathbf{z} \\ \mathbf{z}^T & a \end{bmatrix}$$

where

$$A = N_{q-1}^T N_{q-1} \tag{7-91}$$

$$\mathbf{z} = N_{q-1}^T \nabla g_{jq} \tag{7-92}$$

$$a = \nabla g_{jq}^T \nabla g_{jq} \tag{7-93}$$

From (7-85) and (7-87) we get

$$B = A^{-1} - \mathbf{u}\mathbf{z}^T A^{-1} = A^{-1} + bA^{-1}\mathbf{z}\mathbf{z}^T A^{-1} = A^{-1} + b\mathbf{w}\mathbf{w}^T \tag{7-94}$$

where

$$\mathbf{w} = A^{-1}\mathbf{z} \tag{7-95}$$

$$\mathbf{u} = -bA^{-1}\mathbf{z} = -b\mathbf{w} \tag{7-96}$$

The scalar b can be found from (7-88)

$$b = a^{-1}(1 - \mathbf{u}^T\mathbf{z}) = a^{-1}(1 + b\mathbf{z}^T A^{-1}\mathbf{z})$$

and

$$\begin{aligned} b &= (a - \mathbf{z}^T A^{-1}\mathbf{z})^{-1} = \{\nabla g_{jq}^T [I - N_{q-1}(N_{q-1}^T N_{q-1})^{-1} N_{q-1}^T] \nabla g_{jq}\}^{-1} \\ &= (\nabla g_{jq}^T P_{q-1} \nabla g_{jq})^{-1} = |P_{q-1} \nabla g_{jq}|^{-2} \end{aligned} \tag{7-97}$$

The last equality holds because

$$P_{q-1}\nabla g_{jq} = P_{q-1}(P_{q-1}\nabla g_{jq}) \tag{7-98}$$

which is an obvious property of the projection matrix. The computational procedure can be summarized as follows:

1 Compute the auxiliary vector

$$\mathbf{w} = (N_{q-1}^T N_{q-1})^{-1} N_{q-1}^T \nabla g_{jq} \tag{7-99}$$

and the scalar

$$c = | P_{q-1} \nabla g_{j_q} |^2 \tag{7-100}$$

2 The segments of the matrix

$$(N_q^T N_q)^{-1} = \begin{bmatrix} B & \mathbf{u} \\ \mathbf{u}^T & b \end{bmatrix} \tag{7-101}$$

are given by the relations

$$\begin{aligned} b &= c^{-1} \\ \mathbf{u} &= -c^{-1}\mathbf{w} \\ B &= (N_{q-1}^T N_{q-1})^{-1} + c^{-1}\mathbf{w}\mathbf{w}^T \end{aligned} \tag{7-102}$$

This procedure can also be used recursively to obtain the initial inversion of $(N_q^T N_q)^{-1}$ and P_q from the set of active constraints. An additional advantage of using this recursion is its ability to select the largest set of linearly independent active constraints from the set of all active constraints. It is clear from (7-102) that $(N_q^T N_q)^{-1}$ cannot be obtained if $P_{q-1}\nabla g_{j_q} = 0$; ∇g_{j_q} is linearly dependent on the set of the vectors $\nabla g_{j_1}, \nabla g_{j_2}, \ldots, \nabla g_{j_{q-1}}$ and should be ignored if this equation is satisfied.

Unfortunately, the matrix $(N_q^T N_q)$ is frequently ill-conditioned with respect to the inverse problem, and if $(N_q^T N_q)^{-1}$ is computed without special precautions, it may be seriously influenced by round-off errors. This is a well-known difficulty that also appears in the linear least-squares problem. It is desirable to compute P_q without inverting numerically the matrix $(N_q^T N_q)$. The matrix N_q is decomposed in the following manner, $N_q = QR$ where Q is an orthogonal matrix, and $R = \begin{pmatrix} \hat{R} \\ 0 \end{pmatrix}$, where $\hat{R}(q \times q)$ is upper triangular and 0 is an $(n - q) \times q$ zero matrix. Thus, we have

$$N_q^T N_q = R^T Q^T Q R = R^T R = \hat{R}^T \hat{R} \tag{7-103}$$

which is the Choleski decomposition of $N_k^T N_k$. Now it is easy to compute P_q since $R\hat{R}^{-1} = \begin{pmatrix} I_q \\ 0 \end{pmatrix}$ where I_q is the $(q \times q)$ unit matrix

and
$$P_q = I - QR(\hat{R}^T \hat{R})^{-1} R^T Q^T = Q(I - J_q)Q^T \tag{7-104}$$

where
$$J_q = \begin{pmatrix} I_q & 0 \\ 0 & 0 \end{pmatrix} \tag{7-105}$$

There are a number of ways of achieving the decomposition QR; a very efficient one is based on the Householder transformation [Businger and Golub, 1965; Golub, 1965]. The matrices Q and R can be accurately and efficiently updated if a column of N_q is deleted or added [Bartels, et al., 1970]. Kalfon, et al. [1968*a*, 1968*b*], in their version of the gradient projection method, implemented the Householder transformation and achieved a very stable computing process.

Nonlinear Constraints

The Rosen gradient projection method has been found efficient for problems with linear constraints. There are, however, at least two ways in which this method can handle nonlinear constraints. One possibility is via the sequential transformation where the nonlinear constraints are absorbed by the redefined objective function and the linear constraints remain as side restrictions (Sec. 6-3). This transformation reduces the original problem with nonlinear constraints to the formulation with linear constraints.

Another technique is to linearize locally the active nonlinear constraints and consider a sequence of approximate problems with linearized constraints. There are unfortunately at least two reasons why this last technique is not as efficient here as when all constraints are linear. A major computational problem is introduced by the fact that, in general, we cannot use the recurrence formulas that relate $(N_q^T N_q)^{-1}$ and $(N_{q-1}^T N_{q-1})^{-1}$. When the new solution $\mathbf{x}^{k+1}$ has been obtained it is very likely that several columns of N_q have to be replaced by the new linear approximation to the constraints. Thus, the old inverse becomes almost useless and a completely new one has to be computed. This is true even when the new set of active constraints remains unchanged or differs only by one constraint from the last one.

Another difficulty is introduced by the problem of returning to the convex constraints after a move has been performed along the projected gradient on the intersection of the hypersurfaces tangential to the active set of constraints (Fig. 7-5). A return direction $\mathbf{r}$ is defined as

$$\mathbf{r} = N_q(N_q^T N_q)^{-1}\mathbf{t} \tag{7-106}$$

where $\mathbf{t}$ is a vector whose q components are preassigned numbers according to the rule that the more we want to move away from some hypersurface, the larger is the corresponding number. Kalfon, et al. [1968*b*] avoided this difficulty by forming a linear combination of $\mathbf{r}$ and $P\nabla M(\mathbf{x}^k)$,

$$\mathbf{v} = -P_q\nabla M(\mathbf{x}^k) + \beta\mathbf{r} \tag{7-107}$$

such that $\mathbf{v}$ is a feasible direction.

Modifications

The gradient projection method can be viewed as the steepest descent method with the capability of handling constraints. Therefore, it is reasonable

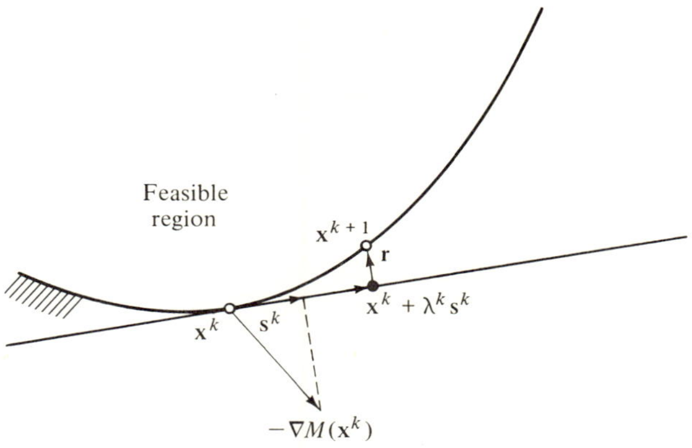

Figure 7-5 Gradient projection method

to expect that the method may be improved by using the conjugate gradient vectors instead of gradients in step 2 of the iteration. Other acceleration devices mentioned in Chap. 5 would also be appropriate for interior points.

The design of such a refinement has been attempted by Goldfarb [1969] and Goldfarb and Lapidus [1968], and their method has proved to be quite successful. The capability of the method is limited to a case with linear constraints and its derivation is based on the quadratic objective function

$$M(\mathbf{x}) = a + \mathbf{b}^T\mathbf{x} + \tfrac{1}{2}\mathbf{x}^T A\mathbf{x} \tag{7-108}$$

It follows from (7-108) that

$$\mathbf{x}^{k+1} - \mathbf{x}^k = A^{-1}[\nabla M(\mathbf{x}^{k+1}) - \nabla M(\mathbf{x}^k)] \tag{7-109}$$

and if $\mathbf{x}^{k+1}$ gives a minimum of $M(\mathbf{x})$ on the cross-section of the hyperplanes $g_{j_p}(\mathbf{x}) = 0,\ p = 1, 2, \ldots, q$, then

$$\nabla M(\mathbf{x}^{k+1}) = N_q\boldsymbol{\gamma} \tag{7-110}$$

where N_q is defined as before and $\boldsymbol{\gamma}$ is a vector to be determined. From (7-109), (7-110), and the condition that $N_q(\mathbf{x}^{k+1} - \mathbf{x}^k) = 0$, we get

$$\mathbf{x}^{k+1} = \mathbf{x}^k - \hat{P}_q A^{-1}\nabla M(\mathbf{x}^k) \tag{7-111}$$

where

$$\hat{P}_q = I - A^{-1}N_q(N_q^T A^{-1}N_q)^{-1}N_q^T. \tag{7-112}$$

Formula (7-111) is an extension of the Newton method where, by using matrix $\hat{P}$, we restrict the search for a minimum to the feasible region defined by the linear constraints of the problem. Due to the well-known disadvantages of the Newton method it is preferred to implement the conjugate directions method of Davidon. This idea leads to a version of the variable metric method capable of optimizing a nonlinear function subject to linear constraints. This method uses positive definite matrices H_k [to compute

$\mathbf{s}^k = -H_k \nabla M(\mathbf{x}^k)$] which approximate $-\hat{P}_q A^{-1}$ in the case of quadratic programming. The matrix H_k is modified as in the variable metric method (Davidon) if a minimum of $M(\mathbf{x}^k + \lambda \mathbf{s}^k)$ is found before a new constraint is reached. If, however, the search direction encounters a constraint before, then a new conjugate direction cannot be found by the Davidon modification of H_k and a new set of conjugate directions is started.

Murtagh and Sargent [1969] have made a further contribution to the conjugate gradient projection method. They suggest a class of methods in which the inverse of the Hessian is modified for steps of arbitrary lengths and directions. Their versions of the gradient projection method can also handle nonlinear constraints and objective functions (not necessarily quadratic).

Another extension of the Goldfarb and Lapidus method that handles nonlinear constraints has been proposed by Davies [Davies and Swann, 1969].

Discussion

The Rosen gradient projection method is efficient if the problem constraints are linear and becomes less attractive if nonlinear constraints are involved. However, successful applications of the method to the problems with nonlinear constraints have been reported [Brown and Ang, 1966; Davies and Swann, 1969]. The method can handle nonconvex problems.

Among its disadvantages are the rather complicated code and potential difficulties in inverting $(N_q^T N_q)$ due to the ill-conditioning of this matrix.

The most promising versions of the gradient projection method seem to be the conjugate gradient method by Goldfarb and Lapidus [1968] and the class of techniques suggested by Murtagh and Sargent [1969].

7-8 REDUCED GRADIENT METHOD

Wolfe [1963, 1967] proposed a method that extends the simplex procedure to problems with nonlinear (convex) objective functions. We consider the minimization of $M(\mathbf{x})$ subject to the linear constraints

$$A\mathbf{x} - \mathbf{b} = 0 \qquad \text{and} \qquad \mathbf{x} \geq 0 \tag{7-113}$$

where A is a $(m \times n)$ matrix and $m < n$. The variables $\mathbf{x}$ can be partitioned into basic (dependent) variables $\mathbf{x}_B = (x_{B1}, \ldots, x_{Bm})^T$ and nonbasic (independent) variables $\mathbf{x}_N = (x_{N1}, \ldots, x_{N,n-m})^T$. If the nonsingular square matrix B $(m \times m)$ consists of the columns of A representing the coefficients of the basic variables $\mathbf{x}_B$, then we have from (7-113)

$$\mathbf{x}_B = -B^{-1}C\mathbf{x}_N + B^{-1}\mathbf{b} \tag{7-114}$$

where C is the $m \times (n - m)$ matrix which consists of the columns of A related to the nonbasic variables $\mathbf{x}_N$.

In this case, the Kuhn-Tucker conditions for optimality [Sec. 1-7; Eqs. (1-31)] are

$$\begin{aligned}
&\mathbf{v} = \nabla_{\mathbf{x}} M(\mathbf{x}) + A^T\mathbf{u} \geq 0 \\
&\mathbf{v}^T\mathbf{x} = 0 \\
&\mathbf{x} \geq 0 \\
&A\mathbf{x} = \mathbf{b}
\end{aligned} \tag{7-115}$$

Since the basic variables are positive, $\mathbf{x}_B > 0$; therefore, the corresponding part of $\mathbf{v}$ is

$$\mathbf{v}_B = \nabla_{\mathbf{x}_B} M(\mathbf{x}) + B^T\mathbf{u} = 0 \tag{7-116}$$

from which we have

$$\mathbf{u} = -(B^T)^{-1}\nabla_{\mathbf{x}_B} M(\mathbf{x}) \tag{7-117}$$

Furthermore, we get

$$\begin{aligned}
dM(\mathbf{x}) &= [\nabla_{\mathbf{x}_B} M(\mathbf{x})]^T d\mathbf{x}_B + [\nabla_{\mathbf{x}_N} M(\mathbf{x})]^T d\mathbf{x}_N \\
&= [\nabla_{\mathbf{x}_N} M(\mathbf{x}) - C^T(B^T)^{-1}\nabla_{\mathbf{x}_B} M(\mathbf{x})]^T d\mathbf{x}_N \\
&= [\nabla_{\mathbf{x}_N} M(\mathbf{x}) + C^T\mathbf{u}]^T d\mathbf{x}_N = \mathbf{v}_N^T d\mathbf{x}_N
\end{aligned} \tag{7-118}$$

where $\mathbf{v}_N$ is the part of $\mathbf{v}$ which corresponds to the nonbasic variables $\mathbf{x}_N$. From (7-118) we get the reduced gradient

$$\frac{dM(\mathbf{x})}{d\mathbf{x}_N} = \mathbf{v}_N \tag{7-119}$$

which is used to define the descent direction $\mathbf{s}^k$ in the kth iteration

$$\mathbf{x}^{k+1} = \mathbf{x}^k + \lambda^k\mathbf{s}^k \tag{7-120}$$

This direction is defined as follows:

$$s_{Ni}^k = 0 \qquad \text{if } x_{Ni}^k = 0 \qquad \text{and} \qquad v_{Ni}^k > 0 \tag{7-121}$$

otherwise,

$$s_{Ni}^k = -v_{Ni} \tag{7-122}$$

and

$$\mathbf{s}_B^k = -B^{-1}C\mathbf{s}_N^k \tag{7-123}$$

From (7-120) and (7-123) it follows that if $\mathbf{x}^k$ belongs to the manifold defined by the linear constraints $A\mathbf{x} = \mathbf{b}$, then $\mathbf{x}^{k+1}$ also belongs to the manifold. We have

$$\begin{aligned}
A\mathbf{x}^{k+1} &= B\mathbf{x}_B^{k+1} + C\mathbf{x}_N^{k+1} \\
&= B\mathbf{x}_B^k + \lambda^k B\mathbf{s}_B^k + C\mathbf{x}_N^k + \lambda^k C\mathbf{s}_N^k \\
&= B\mathbf{x}_B^k - \lambda^k C\mathbf{s}_N^k + C\mathbf{x}_N^k + \lambda^k C\mathbf{s}_N^k \\
&= B\mathbf{x}_B^k + C\mathbf{x}_N^k = A\mathbf{x}^k = \mathbf{b}
\end{aligned} \tag{7-124}$$

Relationships (7-121) through (7-123) yield the descent directions $\mathbf{s}^k \neq 0$ unless $\mathbf{v}^T\mathbf{s}^k = 0$. If the latter equation holds, then the Kuhn-Tucker conditions are satisfied and $\mathbf{x}^k$ is optimal.

The steplength λ^k is found from the following two relations:

$$\lambda_1^k = \min\left\{-\frac{x_i^k}{s_i^k}\middle| s_i^k < 0 \qquad i = 1, \ldots, n\right\} \tag{7-125}$$

since $x_i^{k+1} = x_i^k + \lambda^k s_i^k \geq 0. \qquad i = 1, \ldots, n$
and λ_2^k solves the problem

$$\min_{\lambda} \{M(\mathbf{x}^k + \lambda \mathbf{s}^k), \qquad 0 < \lambda \leq \lambda_1^k\} \tag{7-126}$$

(Methods to solve this problem are described in Chap. 3.)

Before the next iteration is attempted the following three cases should be considered:

1 $\lambda_2^k < \lambda_1^k$. In this case all the new basic variables $\mathbf{x}_B^{k+1} > 0$, and we can perform the next step without any changes.
2 $\lambda_2^k = \lambda_1^k$, but $\mathbf{x}_B^{k+1} > 0$. The course of action is as in (**1**).
3 $\lambda_2^k = \lambda_1^k$, but there is a basic variable x_{Bp} such that $x_{Bp}^{k+1} = 0$. Now we have to perform a simplex pivot operation (appendix) interchanging the vanishing variable x_{Bp} with some nonvanishing nonbasic variable x_{Nq}.

With the new updated partitioning between the basic and nonbasic variables we proceed to the next iteration.

Iteration Steps

1 Initialize, set $k = 0$; $\mathbf{x}^0$ is given ($A\mathbf{x}^0 = \mathbf{b}$).
2 Compute the gradient vector $\nabla M(\mathbf{x}^k)$ and

$$\mathbf{v}_N^k = \nabla_{\mathbf{x}_N} M(\mathbf{x}^k) + C^T \mathbf{u}^k \tag{7-127}$$

where
$$\mathbf{u}^k = -(B^T)^{-1} \nabla_{\mathbf{x}_B} M(\mathbf{x}^k) \tag{7-128}$$

3 Define the direction $\mathbf{s}^k$ as:

$$s_{Ni}^k = 0 \qquad \text{if } x_{Ni}^k = 0 \quad \text{and} \quad v_{Ni}^k > 0$$

otherwise, $\quad s_{Ni}^k = -v_{Ni}^k$

and

$$\mathbf{s}_B^k = -B^{-1} C \mathbf{s}_N^k$$

4 If $\|\mathbf{s}_N^k\| < \epsilon$, where $\epsilon > 0$ is a predetermined small number, then we assume that $\mathbf{x}^k$ is optimal and the computation is terminated, otherwise proceed to step 5.
5 Find

$$\lambda_1^k = \min\left\{-\frac{x_i^k}{s_i^k}\middle| s_i^k < 0, \qquad i = 1, \ldots, n\right\}$$

$$\lambda_2^k \text{ minimizing } \{M(\mathbf{x}^k + \lambda \mathbf{s}^k), \qquad 0 < \lambda \leq \lambda_1^k\}$$

and set
$$\lambda^k = \lambda_2^k$$

$$\mathbf{x}^{k+1} = \mathbf{x}^k + \lambda^k \mathbf{s}^k$$

6 If all new basic variables $\mathbf{x}_B^{k+1} > 0$, then return to step 2 with $k := k + 1$. Otherwise, perform a simplex pivot operation interchanging the vanishing basic variable with some nonvanishing nonbasic variable, and return to step 2 with $k := k + 1$ and updated basic matrix B.

Discussion

We may remark that the above procedure becomes an ordinary simplex method if $M(\mathbf{x})$ is linear and if we define in step 3 an index l by

$$|v_l| = \max_{i \in I} |v_i| \tag{7-129}$$

where I is the set of all indices for which equation (7-122) holds, and we set all $s_{Ni}^k = 0$ except $s_{Nl}^k = -v_{Nl}$.

The reduced gradient method has been extensively tested by Faure and Huard [1965] [see also Colville, 1968], and performed quite satisfactorily. Faure and Huard also describe the necessary conditions for the method to converge.

A further development is due to Abadie and Carpentier [1969] who generalized the method to also handle nonlinear constraints. The constraints are repetitively linearized, which permits using the basic concept of the reduced gradient method of Wolfe. They have also applied the conjugate gradient directions instead of the gradient directions in order to accelerate the convergence. The Colville [1968] report indicates that the generalized reduced gradient method is the leading technique among several other methods tested and compared in the study, which included complex, Rosenbrock, MAP, method of feasible directions, Davidon with SUMT, and conjugate gradient projection techniques.

A method similar to the reduced gradient method has been developed and implemented by Dommel and Tinney [1968]. In their method the sets of basic and nonbasic variables are fixed and the simplex-like exchange technique is avoided. Lower and upper bounds on the nonbasic variables $\mathbf{x}_N$, $\mathbf{x}_N^{\min} \leq \mathbf{x}_N \leq \mathbf{x}_N^{\max}$ are enforced by the following projection technique: when nonbasic variable x_{Ni} has reached its limit, the next step proceeds along the projection of the negative gradient onto the constraining equation $x_{Ni} = x_{Ni}^{\min}$ or $x_{Ni} = x_{Ni}^{\max}$. The bounds on the basic variables $\mathbf{x}_B$ are handled by the penalty functions of the Courant type (see Sec. 2-1).

7-9 COMPUTING SYSTEMS

Computing systems applicable to constrained problems are listed in tables 7-1 and 7-2.

Table 7-1 METHODS USING DERIVATIVES

Title and Reference	*Type, Machine and Storage*	*Experience*	*Remarks*
POP–II 7090–H9 IBM 0021; technique similar to the method of approximate programming by Griffith and Stewart [1961]	FORTRAN IV, MAP, 7090, 7094 Operating system IBSYS; 11.7K words for program storage plus 11.4K for COMMON; plus user supplied subroutine	Contributor reported successful tests with a wide variety of nonlinear problems	1. Contributor: H. V. Smith 2. Maximum problem size: 50 independent variables, 50 constraints
NLP SHARE SDA 3199, Zoutendijk's method of feasible directions [Dorn, 1963]	FORTRAN II, 7090, 7094, $1439 + m^2 + mn + 7(m + n)$ words n = Number of variables m = Number of constraints	NLP has been tested quite extensively by the contributor; more than 20 problems have been successfully solved	1. Contributor: N. J. Anthony 2. Problem size: limited by core size
OPTM SHARE SDA 3541 Gradient projection method; also known as GGS (generalized gradient search)	FORTRAN IV 7094 Operating system IBSYS; approximately 14824 words	Tested by the contributors; success on Wood and Colville problems	1. Contributors: K. E. Cross and W. L. Kephart 2. Program handles problems with nonlinear objective function and nonlinear equality or inequality constraints 3. Problem size: limited by core size
E4 EDF PHIMAX VIM* Reduced gradient method	FORTRAN program for CDC 6600		1. Author: E. Meallier 2. For linear, nonlinear or quadratic objective function and linear constraints

*VIM, Users Organization for Control Data Corporation Computers, *Program Library Catalog*, 1971.

7-10 NUMERICAL RESULTS

A number of factors have to be considered to establish the practical value of a nonlinear programming algorithm (code). Some of these are:

1 Class of problems the code can handle—for example, general nonlinear

Table 7-2 COMPLEX METHOD

Title and Reference	*Type, Machine and Storage*	*Experience*	*Remarks*
H2–NPCS, BOXPLEX VIM* Minimization of an arbitrary function	FORTRAN IV CDC 6600, 3521 words plus users' subroutines FORTRAN 63 for CDC 1604	Program has been checked by the contributor	1. Contributor: R. R. Hilleary 2. Complex method described by Box [1965]

*VIM, Users Organization for Control Data Corporation Computers, *Program Library Catalog*, 1971.

formulation, convex problems, problems with linear constraints, unconstrained objective functions, and specially structured problems.

2 Ease of setting up and programming the problems for the code.

3 Necessity of supplying information in addition to the objective function and constraint values, such as first or second partial derivatives. In many practical problems, these derivatives are not available in analytical form.

4 Efficiency of the code measured in terms of the computer time required for the solution of the problem. Another possible measure of efficiency is the number of function and constraint evaluations. This criterion, however, is meaningful only for those problems and methods where the time required for the function and constraint evaluations strongly dominates other components of the optimization process.

5 Accuracy of the solution value of the objective function and independent variables. Frequently, in real-life applications, only fair approximations to the solution are required.

6 Other features of the code such as quality of coding, programming language, and the like can also be of importance.

Empirical results invariably reflect the composite performance of the algorithm, the program and possibly the analyst (who picks ad hoc parameters) and the machine. Although we would like to be able to separate out these various components, it is clearly not feasible.

One of the most extensive numerical evaluations of nonlinear programming codes was conducted by Colville [1968] of IBM. About 30 different computer codes were tested in this evaluation study. Colville has classified the methods which he experimented with as follows:

1 Search methods (simplex, complex, Rosenbrock method of rotating coordinates).

2 Small-step gradient methods (MAP).

3 Large-step gradient methods (generalized reduced gradient, feasible directions, conjugate gradient projection, gradient projection).
4 Second-order methods (SUMT in conjunction with the Newton method).
5 Miscellaneous.

The test problems suggested by Colville were small or medium size (3 to 16 variables and 4 to 20 constraints) but quite complex. Two of the nine Colville problems are used in this book for the purpose of providing numerical illustrations (Wood and Colville problems). Solution times have been chosen as the main basis for comparing results. No information is available concerning the accuracy of the results. Colville's ranking system uses the time ratio number

$$t_{MP} = \frac{t_{MPC}}{t_c} \tag{7-130}$$

where t_{MPC} is the computing time for method M, on problem P using computer C, and t_c is a standard time attributed to every computer in order to standarize and compare solution times produced by different computers. In summarizing the results of his study, Colville has reached the following conclusions:

1 Efficiency and performance of a nonlinear programming code can be substantially affected by the method of coding. Quite different performances were shown by some of the codes based on an identical mathematical method.
2 The third and fourth groups of methods performed better than the other three major classes of methods. Colville remarks that groups 3 and 4 utilized analytically calculated derivatives, whereas group 2 used derivatives computed numerically. This may be one of the factors contributing to the difference in the performance of the methods. We may also add that methods in groups 3 and 4 have better ultimate convergence properties.
3 Colville noticed that some of the methods showed a bad average performance, but were quite efficient with respect to one or more specific problems. Therefore, it is undesirable to draw any firm conclusions concerning the numerical efficiency of any optimization method from just a few numerical cases.

Among the individual participating algorithms the best performers were all the variants of the reduced gradient method and the gradient projection method. This observation has also been confirmed by Abadie and Guigou [1970].

To illustrate performance of three available SHARE codes (POP, NLP, and OPTM) we have used them to solve the Rosen-Suzuki and Colville problems (Sec. 1-10). Results are summarized in Tables 7-3 to 7-7. It is possible that better convergence can be obtained by more careful setting of some of the input parameters in these codes, but more extensive experimentation would be required.

Table 7-3 METHOD OF APPROXIMATE PROGRAMMING (POP); [Rosen-Suzuki], $\mathbf{x}^0 = (0,0,0,0)^T$, $M(\mathbf{x}^0) = 0$.

Number of Evaluations of Objective Function and Constraints	$M(\mathbf{x})$	x_1	x_2	x_3	x_4
10	−33.000	1.0000	1.0000	1.0000	1.0000
28	−45.844	−0.4349	0.5348	3.0000	−0.1192
56	−41.055	0.0195	1.0561	2.2258	0.0058
73	−42.080	0.0194	1.0561	2.1714	−0.2442
82	−42.922	−0.2306	0.8061	2.2841	−0.4942
100	−44.087	−0.0117	0.8061	2.0617	−0.9942
118	−44.010	0.0508	0.9936	1.9674	−1.0411

Table 7-4 METHOD OF APPROXIMATE PROGRAMMING (POP); [Colville], $\mathbf{x}^0 = (0,0,0,0,1)^T$, $M(\mathbf{x}^0) = 20$.

Number of Evaluations of Objective Function and Constraints	$M(\mathbf{x})$	x_1	x_2	x_3	x_4	x_5
12	−10.635	0.1857	0.1000	0.2000	0.2000	0.8000
24	−19.767	0.3001	0.1000	0.4000	0.8000	0.6000
56	−31.1581	0.3000	0.2556	0.4000	0.4000	0.4643
100	−32.006	0.3000	0.3600	0.4000	0.5000	0.1643
166	−32.3481	0.3000	0.3321	0.4000	0.4250	0.2271

Table 7-5 FEASIBLE DIRECTIONS METHOD OF ZOUTENDIJK [1960] (NLP); [Colville], $\mathbf{x}^0 = (0,0,0,0,1)^T$, $M(\mathbf{x}^0) = 20$.

Number of Objective Function Evaluations	$M(\mathbf{x})$	x_1	x_2	x_3	x_4	x_5
2	−19.183	0.1338	0.0224	0.3169	0.5423	0.3403
5	−23.915	0.3000	0.3327	0.4000	0.6728	0.0287
10	−31.604	0.2568	0.3221	0.3784	0.4893	0.2978
20	−31.780	0.2739	0.3474	0.3870	0.5147	0.2192
45	−32.286	0.3000	0.3234	0.4000	0.4521	0.2648
100*	−32.3485	0.3000	0.3328	0.4000	0.4268	0.2254

*19 evaluations of the objective function gradient and 19 evaluations of the constraint set.

Table 7-6 GRADIENT PROJECTION METHOD (OPTM); [Rosen-Suzuki], $\mathbf{x}^0 = (0,0,0,0)^T$, $M(\mathbf{x}^0) = 0$.

Number of Objective Function Evaluations	$M(\mathbf{x})$	x_1	x_2	x_3	x_4
6	−19.235	0.0119	0.0119	1.0000	−0.0167
8	−23.225	0.0565	0.0565	1.1875	−0.0793
10	−31.629	0.1568	0.1568	1.6094	−0.2203
12	−41.672	−0.3115	0.4111	2.2175	−0.7536
16	−43.576	−0.1727	1.2628	2.0124	−0.9191
20	−43.886	−0.0558	1.1530	1.9872	−0.9798
24	−43.988	0.0496	1.0552	1.9489	−1.0503
27*	−43.999(9)	0.0034	1.0004	1.9974	−1.0030

*56 evaluations of the constraint set plus 56 evaluations of the constraint gradients.

Table 7-7 GRADIENT PROJECTION METHOD (OPTM); [Colville], $\mathbf{x}^0 = (0,0,0,0,1)^T$, $M(\mathbf{x}^0) = 20$.

Number of Objective Function Evaluations	$M(\mathbf{x})$	x_1	x_2	x_3	x_4	x_5
7	11.214	0.0301	0	0.0485	0.0500	0.9527
10	−24.362	0.2534	0	0.3767	0.4063	0.6160
15	−24.825	0.2534	0	0.3867	0.4063	0.6160
20	−31.583	0.3000	0.2735	0.4000	0.3887	0.4025
26	−32.042	0.3000	0.3556	0.4000	0.4923	0.1719
31	−32.329	0.3000	0.3402	0.4000	0.4448	0.2081
45*	−32.3485	0.3000	0.3341	0.4000	0.4298	0.2226

*67 evaluations of the constraint system plus 67 evaluations of the constraint gradients.

References

ABADIE, J. and J. CARPENTIER: "Generalization of the Wolfe Reduced Gradient Method to the Case of Nonlinear Constraints," in R. Fletcher (ed.), *Optimization*, Academic Press Inc., New York, Ch. 4, 37, 1969.

——— and J. GUIGOU: "Numerical Experiments with the GRG Method," in J. Abadie (ed.), *Integer and Nonlinear Programming*, North-Holland Publishing Co., Amsterdam, Appendix III, 529, 1970.

BALINSKI, M. L.: "Integer Programming: Methods, Uses, Computation," *Management Science*, Vol. 12, 253, 1965.

BARTELS, R. H., GOLUB, G. H. and M. A. SAUNDERS: "Numerical Techniques in Mathematical Programming," in J. B. Rosen, et al. (eds.), *Nonlinear Programming*, Academic Press Inc., New York, 123, 1970.

BEALE, E. M. L.: "Numerical Methods," in J, Abadie (ed.), *Nonlinear Programming*, North-Holland Publishing Co., Amsterdam, Ch. VII, 135, 1967.

BOX, M. J.: "A New Method of Constrained Optimization and a Comparison with Other Methods," *Computer Journal*, Vol. 8, 42, 1965.

BROWN, D. M. and A. H. S. ANG: "Structural Optimization by Nonlinear Programming," *Proceedings of the ASCE, Structural Division*, Vol. 92, 319, 1966.

BUSINGER, P. and G. H. GOLUB: "Linear Least Squares Solutions by Householder Transformations," *Numer. Math.*, Vol. 7, 269, 1965.

CHENEY, E. W. and A. A. GOLDSTEIN: "Newton's Method for Convex Programming and Tchebycheff Approximation," *Numer. Math.*, Vol. 1, 253, 1959.

COLLATZ, L. and W. WETTERLING: *Optimierungsaufgaben*, Springer-Verlag, Berlin, 1966.

COLVILLE, A. R.: *A Comparative Study on Nonlinear Programming Codes*, Report No. 320–2949, IBM Scientific Center, 1968.

DAVIES, D. and W. H. SWANN: "Review of Constrained Optimization," in R. Fletcher, (ed.), *Optimization*, Academic Press Inc., New York, Ch. 12, 187, 1969.

DOMMEL, H. W. and W. F. TINNEY: "Optimal Power Flow Solutions," *IEEE Transactions on Power Apparatus and Systems*, Vol. PAS–87, 1866, 1968.

DORN, W. S., "Nonlinear Programming A Survey," *Management Science*, Vol. 9, 171, 1963.

FAURE, P. and P. HUARD: "Résolution des Programmes Mathématiques á Fonction Nonlineaire par la Méthode du Gradient Réduit," *Revue Française de Recherche Operationnelle*, Vol. 9, 167, 1965.

FLETCHER, R.: *An Efficient, Globally Convergent, Algorithm for Unconstrained and Linearily Constrained Optimization Problems*, 7th International Mathematical Programming Symposium, The Hague, 1970.

GOLDFARB, D.: "Extension of Davidon's Variable Metric Method to Maximization under Linear Inequality and Equality Constraints," *SIAM Journal Appl. Math.*, Vol. 17, 739, 1969.

——— and L. LAPIDUS: "Conjugate Gradient Method for Nonlinear Programming Problems with Linear Constraints," *Industrial and Engineering Chemistry Fundamentals*, Vol. 7, 142, 1968.

GOLUB, G. H.: "Numerical Methods for Solving Linear Least Squares Problems," *Numer. Math.*, Vol. 7, 206, 1965.

GRIFFITH, R. E. and R. A. STEWART: "Nonlinear Programming Technique for the Optimization of Continuous Processing Systems," *Management Science*, Vol. 7, 379, 1961.

GUIN, J. A.: "Modification of the Complex Method of Constrained Optimization," *Computer Journal*, Vol. 10, 416, 1968.

HARTLEY, H. O. and R. R. HOCKING: "Convex Programming by Tangential Approximation," *Management Science*, Vol. 9, 600, 1963.

KALFON, P., G. RIBIERE and J. C. SOGNO: *Methode Du Gradient Projete Utilisant La Triangularisation Unitare*, Centre National de la Recherche Scientifique, Institut Blaise Pascal, Paris, 1968*a*.

———: "A Method of Feasible Directions Using Projection Operators," *Proceedings of the IFIP Congress, Math.*, Booklet A, 123, 1968*b*.

KELLEY, J. E.: "The Cutting Plane Method for Solving Convex Programs," *SIAM Journal of Appl. Math.*, Vol. 8, 703, 1960.

KLEIBOHM, K: *Ein Verfahren zur Approximation Lösung von Konvexen Programmen*, Diss. Universität Zürich, 1966.

MURTAGH, B. A. and R. W. H. SARGENT: "A Constrained Minimization Method with Quadratic Convergence," in R. Fletcher (ed.), *Optimization*, Academic Press Inc., New York, Ch. 14, 215, 1969.

POPE, G. G.: "The Application of Linear Programming Techniques in the Design of Optimum Structures," *Conference Proceedings No. 36 of the AGARD Symposium on Structural Optimization*, Istanbul, Turkey, 1969.

REINSCHMIDT, K. F., C. A. CORNELL and J. F. BROTCHIE: "Iterative Design and Structural Optimization," *Proceedings of the ASCE, Structural Division*, Vol. 92, ST. 6, 281, 1966.

ROSEN, J. B.: "The Gradient Projection Method for Nonlinear Programming, Part I, Linear Constraints," *SIAM Journal*, Vol. 8, 181, 1960.

———: "The Gradient Projection Method for Nonlinear Programming, Part II, Nonlinear Constraints," *SIAM Journal*, Vol. 9, 514, 1961.

SCHWARTZ, L. E.: "Large Step Methods for Decomposable Nonlinear Programming Problems," in R. Fletcher (ed.), *Optimization*, Academic Press Inc., New York, Ch. 8, 99, 1969.

TOPKIS, D. M. and A. F. VEINOTT, Jr.: "On the Convergence of Some Feasible Direction Algorithms for Nonlinear Programming," *SIAM Journal, Cont. Theory and Applic.* Vol. 5, 268, 1967.

TRAUTH, C. A. and R. E. WOOSLEY: "Integer Programming: A Study in Computational Efficiency," *Management Science*, Vol. 15, 481, 1969.

VEINOTT, A. F., JR.: "The Supporting Hyperplane Method for Unimodal Programming, *Journal of ORSA*, Vol. 15, 147, 1967.

WILSON, R. B.: *A Simplicial Algorithm for Concave Programming*, Doctoral Thesis, Graduate School of Business Administration, Harvard University, Boston, Mass., 1963.

WOLFE, P.: "Accelerating the Cutting Plane Method for Nonlinear Programming," *SIAM Journal* Vol. 9, 481, 1961.

———: "Methods of Nonlinear Programming," in R. L. Graves and P. Wolfe (eds.), *Recent Advances in Mathematical Programming*, McGraw-Hill Book Company, New York, 1963.

———: *On the Convergence of Gradient Methods Under Constraints*, IBM Research Paper RZ–204, Zurich, Switzerland, 1966.

———: "Methods of Nonlinear Programming," in J. Abadie (ed.), *Nonlinear Programming*, North-Holland Publishing Co., Amsterdam, Ch. VI, 97, 1967.

Zangwill, W. I.: "A Decomposable Nonlinear Programming Approach," *Journal of ORSA*, Vol. 15, 1068, 1967.

———: "Convergence Conditions for Nonlinear Programming Algorithms," *Management Science*, Vol. 16, 1, 1969.

Zoutendijk, G.: *Methods of Feasible Directions*, Elsevier Publishing Co., Amsterdam, 1960.

———: "Nonlinear Programming: A Numerical Survey," *SIAM Journal, Cont. Theory and Applic.*, Vol. 4, 194, 1966.

APPENDIX
Auxiliary Techniques

A-1 LINEAR PROGRAMMING

The linear programming problem may arise directly from a linear mathematical model, as a reformulated separable or quadratic problem (Sec. 2-2), or as a subproblem generated inside the iterative cycle of a nonlinear programming method, such as the cutting plane, approximate programming, and feasible directions method techniques (Chap. 7).

The formal statement of a linear program, or programming problem, is

$$\min \{x_0 = \mathbf{c}^T\mathbf{x}\} \tag{A-1}$$

subject to

$$A\mathbf{x} = \mathbf{b} \qquad \text{and} \qquad \mathbf{x} \geq 0 \tag{A-2}$$

where $\mathbf{x} = (x_1, \ldots, x_n)^T$ is a vector of structural and slack variables; $\mathbf{c} = (c_1, \ldots, c_n)^T$ and $\mathbf{b} = (b_1, \ldots, b_m)^T$ are vectors of constants referred to as *cost vector* and *right-hand vector*, respectively; A is a given matrix of constants referred to as the *coefficient matrix*; and $n > m$.

Any linear inequality constraint can be easily expressed as an equation by adding a positive or negative *slack variable*. The system (A-2) of equations is undetermined since $m < n$. It has a nonzero solution, consisting of at most m positive values, that minimizes the linear objective function (A-1).

The formulation and the process of solving the linear programming problem (LP) were first provided by Dantzig in 1947 [Dantzig, 1963]. The general linear programming algorithm is referred to as the *simplex method.* Note that an evolutionary operations method for unconstrained nonlinear minimization problems is also called the simplex method (Sec. 4-5), but it is not related to the linear programming method considered here. To avoid confusion, the method for linear programming will be referred to as the LP simplex method.

The LP simplex method selects iteratively trial subsets of the variables $\mathbf{x}$ and checks whether they are optimal—that is, minimize the objective function x_0. If they are optimal, the computations are terminated; if they are not, the method drops one variable from the current subset and adds a new one in such a manner that the objective function value is improved. If we knew the optimal subset of m variables x_{Bj}, $j = 1, 2, \ldots, m$, the solution could be obtained by solving the linear equation system $B\mathbf{x}_B = \mathbf{b}$, where $\mathbf{x}_B = (x_{B1}, x_{B2}, \ldots, x_{Bm})^T$ and B is an $m \times m$ square matrix of coefficients of the variables $x_{B1}, \ldots, x_{Bm}$. Since, in general, this subset is not known, we start from a feasible but nonoptimal solution and perform usually $p > m$ iterations to find the optimal solution. This number of iterations varies in practice from $2m$ to $4m$.

Let us assume that the current trial set of nonnegative feasible variables is $\mathbf{x}_B = (x_{B1}, \ldots, x_{Bm})^T$ (*basic variables*). The remaining variables $\mathbf{x}_N = (x_{N1}, x_{N2}, \ldots, x_{N(n-m)})^T$ (*nonbasic variables*) have zero values. Since $\mathbf{x}_N = 0$, we get from (A-2)

$$B\mathbf{x}_B = \mathbf{b} \qquad \text{and} \qquad \mathbf{x}_B = B^{-1}\mathbf{b} \tag{A-3}$$

It is convenient to reformulate our problem in the following way:

$$\min x_0$$

$$\text{subject to} \qquad x_0 - \mathbf{c}^T\mathbf{x} = 0 \qquad A\mathbf{x} = \mathbf{b} \qquad \mathbf{x} \geq 0 \tag{A-4}$$

or, in matrix notation, $\min x_0$ subject to

$$\begin{bmatrix} 1 & -\mathbf{c}^T \\ 0 & A \end{bmatrix} \begin{bmatrix} x_0 \\ \mathbf{x} \end{bmatrix} = \begin{bmatrix} 0 \\ \mathbf{b} \end{bmatrix} \qquad \mathbf{x} \geq 0 \tag{A-5}$$

Introducing the matrix

$$B_0 = \begin{bmatrix} 1 & -\mathbf{c}_B^T \\ 0 & B \end{bmatrix} \tag{A-6}$$

where $\mathbf{c}_B$ is a vector of components of $\mathbf{c}$ that correspond to the current basic variables, we have from (A-5)

$$\begin{bmatrix} x_0 \\ \mathbf{x}_B \end{bmatrix} = B_0^{-1} \begin{bmatrix} 0 \\ \mathbf{b} \end{bmatrix} \tag{A-7}$$

and we may verify that

$$B_0^{-1} = \begin{bmatrix} 1 & \mathbf{c}_B^T B^{-1} \\ 0 & B^{-1} \end{bmatrix} \tag{A-8}$$

Premultiplying (A-5) by B_0^{-1} we get

$$\begin{bmatrix} 1 & \mathbf{c}_B^T B^{-1} A - \mathbf{c}^T \\ 0 & B^{-1} A \end{bmatrix} \begin{bmatrix} x_0 \\ \mathbf{x} \end{bmatrix} = \begin{bmatrix} \mathbf{c}_B^T B^{-1} \mathbf{b} \\ B^{-1} \mathbf{b} \end{bmatrix} \tag{A-9}$$

From (A-9) we can derive the iterative steps of the *revised simplex method* due to Dantzig and Orchard-Hays [1954], which is currently the most useful method for solution of linear programming problems:

1 Test whether the current basic solution $\mathbf{x}_B$ is optimal by checking coefficients of x_0, which can be computed from (A-9). If for all of the nonbasic variables x_i the reduced costs

$$\hat{c}_i = \mathbf{c}_B^T B^{-1} \operatorname{col}_i(A) - c_i \tag{A-10}$$

are nonpositive, that is

$$\mathbf{c}_B^T B^{-1} \operatorname{col}_i(A) - c_i \leq 0 \tag{A-11}$$

where $\operatorname{col}_i(A)$ is the ith column of A, which corresponds to the nonbasic variable x_i, then by introducing any of these variables into the basis with a positive value we would not improve the value of the objective function x_0. Therefore, the current basic solution $\mathbf{x}_B$ is optimal and the computations are terminated. Otherwise, the nonbasic variable, say x_e, with the largest positive reduced cost is selected to enter the basis.

2 Calculate the coefficients of x_e in the current set of equations (A-9)

$$\boldsymbol{\alpha} = (\alpha_1, \ldots, \alpha_m)^T = B^{-1}\operatorname{col}_e(A) \tag{A-12}$$

and the right-hand side of these equations

$$\boldsymbol{\beta} = (\beta_1, \ldots, \beta_m)^T = B^{-1}\mathbf{b} \tag{A-13}$$

3 Perform the ratio test between the elements of $\boldsymbol{\alpha}$ and $\boldsymbol{\beta}$ to determine

$$\frac{\beta_l}{\alpha_l} = \min_{\alpha_k > 0, k} \left\{ \frac{\beta_k}{\alpha_k} \right\} \tag{A-14}$$

where $k = 1, \ldots, m$. The basic variable x_{Bl} leaves the set of basic variables since it is the first basic variable that decreases to zero as x_e is increased to the value β_l/α_l. This step determines the pivotal row l.

4 Update the inverse matrix B^{-1} and the set of basic variables. The former operation can be done by the Gauss-Jordan method of elimination used in solving linear equations. In matrix form we have

$$B_{\text{updated}}^{-1} = E B_{\text{current}}^{-1} \tag{A-15}$$

where E is the elementary matrix with the following structure

$$E = \{\mathbf{e}_1, \ldots, \mathbf{e}_{l-1}, \boldsymbol{\eta}, \mathbf{e}_{l+1}, \ldots, \mathbf{e}_m\} \tag{A-16}$$

where $\mathbf{e}_i$ is the ith unit vector and $\boldsymbol{\eta}$ is given by

$$\boldsymbol{\eta} = \{-\alpha_1/\alpha_l, \ldots, -\alpha_{l-1}/\alpha_l, \alpha_l^{-1}, -\alpha_{l+1}/\alpha_l, \ldots, -\alpha_m/\alpha_l\}^T \quad \text{(A-17)}$$

5 Repeat from step 1.

A feasible solution is needed to initiate the iterative process described in steps 1 to 5. A simple way of producing such a feasible solution is to make all right-hand-side vectors $\mathbf{b}$ positive. Then replace all inequality constraints with equations by adding positive slack variables X_i so that $\sum_j a_{ij}x_j \leq b_i$ becomes $\sum_j a_{ij}x_j + X_i = b_i$, and negative slack variables $-X_i$ so that $\sum_j a_{ij}x_j \geq b_i$ becomes $\sum_j a_{ij}x_j - X_i = b_i$. Supplement the latter equation with an artificial variable A_i so that it becomes $\sum_j a_{ij}x_j - X_i + A_i = b_i$. Artificial variables should also be added to the equality constraints which do not require slacks. The positive slacks and artificial variables can be used as an initial basic set of variables and the simplex method can be applied to minimize an auxiliary objective function defined as the sum of infeasibilities, $x_o = \sum_i A_i$. A feasible solution is achieved when the sum of infeasibilities is reduced to zero. An efficient version of this procedure, Wolfe's composite simplex method [Wolfe, 1965], can handle negative values of the linear programming problem variables. This method is particularly useful when there is a starting basis that is close to the optimal basis but produces an infeasible solution. This situation occurs frequently when nonlinear programming problems are solved by a sequence of linearized subproblems that differ slightly from each other.

A further improvement related to the technique of finding an initial feasible solution, known as crashing, is presented by Beale [1969*b*]. The crashing process is an attempt to introduce into the basis as many structural variables as possible. The computation is initialized with a unit basis created by introducing slacks to all inequality constraints and artificials to all equality constraints. Then every structural variable x_i is tested in turn and is made basic, without creating new infeasible variables, provided this removes an artifical variable from the basis, has a positive reduced cost, or reduces the number of infeasible variables.

An early method of producing a feasible solution consisted of assigning large costs to artificial variables and handling them as structural variables. The use of this method must be strongly discouraged, since the presence of large cost coefficients corresponding to artificial variables introduces serious rounding errors.

The first step in the LP simplex method—that of finding a variable leaving the set of nonbasic variables—is very time-consuming. A technique known as multiple column selection can be applied to accelerate this first step in the revised simplex method. In this technique, we select $k > 1$ nonbasic columns with the largest positive reduced costs, update them in step 2, and perform the ratio test in step 3. The variable x_i which causes the greatest reduction in the objective function is removed from the nonbasic set. Then step 4 is performed, and the columns that are selected are updated. The computation returns to step 3, where we seek a new variable that can be removed beneficially from the nonbasic set. This process is continued as long as it is possible to achieve a reduction in the objective function. It is also possible to select the $k > 1$ nonbasic columns with the largest positive reduced costs and form a restricted problem involving the current basis and these k columns. In this suboptimization process, we allow interchange of variables between the basic and nonbasic sets. Multiple column selection and suboptimization are very useful for solving problems which are so large that the complete problem data cannot reside in the computer core, while the restricted problem data can.

The general strategy for solving LP problems is to find a feasible solution and work toward optimality. Sometimes it may be better to start with an infeasible solution that satisfies the optimality criterion and to work toward feasibility. This concept motivates the *dual simplex method* of Lemke [1954], which is particularly suitable when minor changes are made in the formulation of a previously solved problem—for example, when a new constraint is added to the previous set of constraints. If the former optimal solution is no longer feasible, then the dual simplex method can be applied quite efficiently.

The linear programming problems frequently have some or all variables constrained by *lower or upper bounds* of the form $x_i \geq b_i$, $x_i \leq u_i$. These simple constraints can be handled more efficiently than ordinary linear constraints. The lower bounded constraints may be removed from the formulation by a small adjustment in the problem. If $x_i \geq b_i$ where $b_i > 0$, we substitute $x_i = \bar{x}_i + b_i$ for x_i in our problem. The new variable $\bar{x}_i$ has to be nonnegative, and this requirement is automatically satisfied by the simplex method. The upper bounds $x_i \leq u_i$, $u_i > 0$ are taken care of by modifying our rule of selecting the basic variable which leaves the basis. By rule (A-14), the leaving basic variable is the one that reaches zero value first as the entering variable is increased. If the leaving variable reached zero before the entering variable reached the upper bound, then the simplex method proceeds in its usual way. If, however, the entering variable x_i reached its upper bound u_i before any basic variable decreased to zero, then in the next iterations we keep $x_i = u_i$ and introduce an extra slack variable with zero value, $\bar{x}_i = 0$. We replace x_i in the set of equations by $u_i - \bar{x}_i$. In this manner, the entering variable x_i is removed from the basic set and the variable $\bar{x}_i$ becomes a new nonbasic variable. In computational practice, this reduces to a simple "housekeeping" process that keeps track of nonbasic variables at their upper bound.

As can be seen, the LP simplex iteration involves operations of two kinds:

(1) selection of a nonbasic variable to enter the basis and of a basic variable to exit the basis, and (2) updating of the information corresponding to the replacement of the basic variable. Two additional special problems can be solved with the LP simplex algorithm with a modified basis entry/exit rule. These are the reformulated separable and quadratic problems discussed in Sec. 2-2. In the discussion of the reformulation of these problems, we also show the appropriate modifications of the basis entry/exit rule.

The revised simplex method as described in this appendix has several important properties. The matrix B^{-1} may be obtained as a product of elementary matrices. It is known that, in general, the explicit inverse of B may be dense (has many nonzero elements) even if B itself is sparse, and that the density of the product form is comparable to the density of B. This difference in densities is essential to solution of large sparse linear programming problems in which the over-all density of nonzero elements in matrix A may be less than 1%. We may also observe that in the revised simplex method, each iteration adds another vector $\boldsymbol{\eta}$, and the representation of B^{-1} becomes cumbersome and is influenced by round-off errors. To avoid this difficulty, B is reinverted periodically. We compute m (or fewer) vectors $\boldsymbol{\eta}$ from the initial unit matrix by replacing unit vectors with the columns of A corresponding to the current (known) basic set of variables. In the reinversion process we have a considerable degree of freedom because the product form of the inverse is not unique; therefore, it is possible to achieve compactness in the B^{-1} product form representation by designing an optimal pivoting process.

One practical pivoting strategy is based on the following observation. If matrix B can be reduced to a triangular matrix by permuting its rows and columns, then pivoting on the diagonal elements in order yields a product form representation that has no more nonzero elements than the original matrix B. It is usually impossible to triangularize B completely by permuting its rows and columns, but the permutation can frequently yield a matrix B_1 in the form

$B_1 =$ 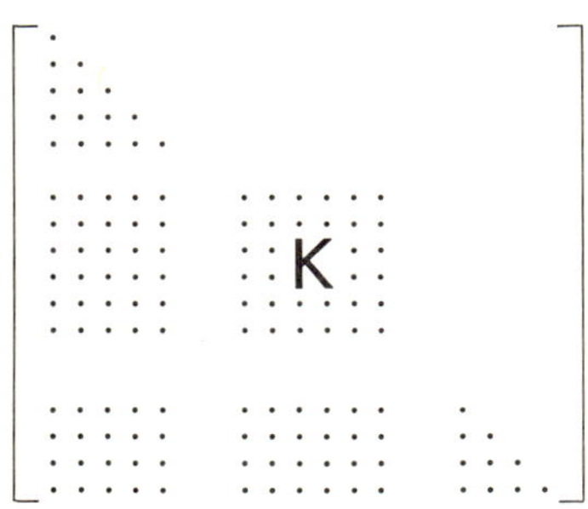

where the area outside the triangular, square and rectangular submatrices contains zeros only. If the product form inverse is produced by pivoting on the diagonal elements in order, then additional nonzeros can be created only in the kernel matrix K and below it. Several procedures have been suggested to limit the proliferation of nonzeros in these two submatrices, and some remarks on the practical implementation of such schemes can be found in De Buchet [1971]. A different approach to the inversion and updating of B has been suggested by Bartels and Golub [1969*a*, 1969*b*] and Bartels, et al. [1970]. In updating matrix B, the computing inaccuracies due to round-off errors are propagated from iteration to iteration. In addition, if pivot element α_l is small relative to other elements α_i in vector $\boldsymbol{\eta}$, then the iterative step involving the premultiplication EB^{-1} can be quite inaccurate. They recommend the LU decomposition of B (where L is a lower and U is an upper triangular matrix) together with row interchanges to select optimal pivots. They also show how to update these matrices from iteration to iteration. This method has good behavior with respect to round-off error, but it is difficult to apply to large sparse problems. The LU decomposition of B produces a more compact inverse than the Gauss-Jordan product form. In order to use LU decomposition, matrix B must be restructured by row and column permutations so that it becomes

$B_2 =$

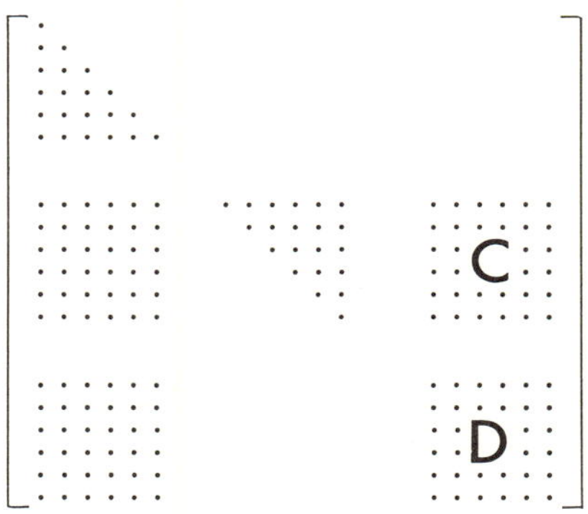

Now the LU decomposition of B affects only matrices C and D, but, in general, the triangular factors of these matrices will produce fewer new nonzeros than Jordan elimination applied to B_1. After decomposing B_2 we can either add product form elementary transformations in the B^{-1} updating process (step 4) or update the LU factors. The first strategy has been practically implemented in a commercial system by Beale [1971]. The second idea has given rise to several algorithms reviewed by Tomlin [1970]. Forrest and Tomlin [1971] also tried one of the methods for updating triangular factors on several numerical examples and found that it produced very compact inverses. See also Orchard-Hays [1968*b*] and Dantzig, et al. [1968] for further information related to the treatment of problems with sparse matrices.

We have focused our attention on the most essential features of the simplex iteration and ignored certain computational complications which can arise when using this technique. For more detailed presentations of the simplex method, see Dantzig [1963], Beale [1968, 1969*a*], and Orchard-Hays [1968*a*].

LP codes for small- to medium-size problems are listed in Table A-1.

Table A-1 LP CODES FOR SMALL- TO MEDIUM-SIZE PROBLEMS

Title and Reference	*Type, Machine and Storage*	*Remarks*
7040–HO, 3147M–40* (the M–40 program for 7040/44)	IBM 7040/44 IBSYS	All in core; flexible dimensions; revised simplex with product form of inverse
7040–H1, 3384 LUSB*	IBM 7040/44	
360A–CO–14X MPS/360†	OS/360 assembler language program. IBM/360 with $\geq$ 65,536 bytes and other minimal options	Maximum problem size depends on memory; program storage is $\sim$30,000 bytes
360A–CO–18X LPS/360†	FORTRAN IV and some assembler language routines. IBM/360 with $\geq$32K ($\leq$ 200 rows) $\geq$ 64K ($\leq$ 1500 rows) and DOS and some other minimal requirements	
360D–15.2006 System 360 Linear Programming System†	FORTRAN. IBM/360 with $\geq$ 32K, BPS, DOS, TOS or OS FORTRAN	Uses "contracted tableau" form of simplex; in core $m \times n \leq 40 \times 40$; m = number of constraints, $n = m$ + number of variables
360D–15.2.007 MFOR 360 Linear Programming Code†	FORTRAN IV and one assembler language routine. IBM/360–65 OS Version II	All in core; revised simplex with product form of inverse
890146 SIGMA 517 MFOR Linear Programming Code‡	FORTRAN H (version B compiler). 32K Basic control monitor (RAD–75–BCM)	All in core; revised simplex with product form of inverse $\leq$ 511 rows, $\leq$2000 columns, $\leq$ 6000 matrix entries
H1 UTAB REGINA VIM§ General purpose linear programming code	3L00 COMPASS program for CDC 3100 with MSOS operating system	1. Author: T. Sontag 2. $\leq$ 250 rows with 16-24-bit words; $\leq$ 1200 rows with sufficient core. No limit on number of variables

Table A-1 (Cont.)

Title and Reference	*Type, Machine and Storage*	*Remarks*
H1 CODA CDM2 VIM§ Linear programming system 2	FORTRAN 60 program for CDC 1604, CO-OP operating system	For straight and separable LP, uses product form revised simplex, ≤ 200 rows, ≤ 599 matrix entries, ≤ 14000 transformation entries.
H1 CODA CDM3 VIM§ Linear programming system 3	FORTRAN 62 program for CDC 1604, CO-OP operating system	For straight and separable LP, uses product form revised simplex, ≤ 400 rows, ≤ 999 variables, ≤ 4000 nonzero matrix entries, ≤ 10000 transformation entries.

*IBM Systems Reference Library, Publication C–20–1604–9, Catalog of Programs for IBM 704–709–7040–7044–7090 and 7094, IBM Corporation, Hawthorne, N. Y.

†IBM Systems Reference Library, Publication GC20–1619–7, Catalog of Programs for IBM System/360 Models 25 and above, IBM Corporation, Hawthorne, N. Y.

‡Xerox Data Systems (XDS), Software Distribution, El Segundo, Calif.

§VIM, Users Organization for Control Data Corporation Computers, *Program Library Catalog*, 1971.

A-2 NUMERICAL DIFFERENTIATION

Optimization techniques may be classified into two groups: (1) direct search methods requiring only the evaluation of the function $M(\mathbf{x})$; and (2) descent methods, which usually require the evaluation of both the function $M(\mathbf{x})$ and the first derivatives $\partial M(\mathbf{x})/\partial x_i$, $i = 1, 2, \ldots n$. Descent methods may also require the calculation of second derivatives $\partial^2 M(\mathbf{x})/\partial x_i \partial x_j$, $i = 1, \ldots, n; j = 1, \ldots, n$. Thus, differentiation is an integral part of many optimization techniques. In discussing some of the methods for differentiation of a computer, we will be concerned with the calculation and evaluation of the first- and higher-order derivatives of a function at specific points. This operation can be performed either analytically or numerically.

Numerical differentiation provides an estimate of the derivative at a given point. If the function is differentiable and if its derivatives are evaluated along the coordinate directions, the first-order gradient vector can be calculated using the following first-term divided-differences formulas:

Newton's forward formula

$$\nabla M_j(\mathbf{x}) \simeq \frac{M(\mathbf{x} + \delta_j \mathbf{e}_j) - M(\mathbf{x})}{\delta_j} \tag{A-19}$$

Sterling's central difference formula

$$\nabla M_j(\mathbf{x}) \cong \frac{M(\mathbf{x} + \delta_j \mathbf{e}_j) - M(\mathbf{x} - \delta_j \mathbf{e}_j)}{2\delta_j} \tag{A-20}$$

Newton's backward formula

$$\nabla M_j(\mathbf{x}) \cong \frac{M(\mathbf{x}) - M(\mathbf{x} - \delta_j \mathbf{e}_j)}{\delta_j} \tag{A-21}$$

where $\nabla M_j(\mathbf{x})$ is $\partial M(\mathbf{x})/\partial x_j$, $j = 1, 2, \ldots, n$; and $\mathbf{e}_j$ is the vector with all components zero except the jth component equal to unity; and δ_j is the step-length. Formulas (A-19) and (A-21) are similar; both have the disadvantage of being onesided and therefore not as accurate as the central difference formula (A-20). On the other hand, the central difference formula usually requires an additional function evaluation for each component of the gradient which means additional computer time.

If an optimization method requires the calculation of the second derivatives, we can approximate these in a manner similar to the above. For second derivatives of the form $\nabla^2 M_j = \partial^2 M(\mathbf{x})/\partial x_j^2$ we have:
Newton's forward formula

$$\begin{aligned}\nabla^2 M_j(\mathbf{x}) &\cong \frac{\nabla M_j(\mathbf{x} + \delta_j \mathbf{e}_j) - \nabla M_j(\mathbf{x})}{\delta_j} \\ &= \frac{M(\mathbf{x} + 2\delta_j \mathbf{e}_j) - 2M(\mathbf{x} + \delta_j \mathbf{e}_j) + M(\mathbf{x})}{\delta_j^2}\end{aligned} \tag{A-22}$$

Sterling's central difference formula

$$\begin{aligned}\nabla^2 M_j(\mathbf{x}) &\cong \frac{\nabla M_j(\mathbf{x} + \frac{1}{2}\delta_j \mathbf{e}_j) - \nabla M(\mathbf{x} - \frac{1}{2}\delta_j \mathbf{e}_j)}{\delta_j} \\ &= \frac{M(\mathbf{x} + \delta_j \mathbf{e}_j) - 2M(\mathbf{x}) + M(\mathbf{x} - \delta_j \mathbf{e}_j)}{\delta_j^2}\end{aligned} \tag{A-23}$$

Newton's backward formula

$$\begin{aligned}\nabla^2 M_j(\mathbf{x}) &\cong \frac{\nabla M_j(\mathbf{x}) - \nabla M_j(\mathbf{x} - \delta_j \mathbf{e}_j)}{\delta_j} \\ &= \frac{M(\mathbf{x}) - 2M(\mathbf{x} - \delta_j \mathbf{e}_j) + M(\mathbf{x} - 2\delta_j \mathbf{e}_j)}{\delta_j^2}\end{aligned} \tag{A-24}$$

For mixed partials we have a more complicated equation. Newton's forward formula:

$$\frac{\partial}{\partial x_i}\left(\frac{\partial M}{\partial x_j}\right) = \frac{M(\mathbf{x} + \delta_i \mathbf{e}_i + \delta_j \mathbf{e}_j) - M(\mathbf{x} + \delta_i \mathbf{e}_i) - M(\mathbf{x} + \delta_j \mathbf{e}_j) + M(\mathbf{x})}{\delta_i \delta_j} \tag{A-25}$$

and other formulas could be developed in a similar fashion.

An approach to differentiation that promises greater precision and also offers a better estimate for the magnitude of the error is *differentiation of higher-order interpolating polynomials* $P(\mathbf{x})$. Thus, we are replacing the given function $M(\mathbf{x})$ by some interpolating function $P(\mathbf{x})$ which can be differentiated analytically. This approach is particularly attractive when the function $M(\mathbf{x})$ is given only at discrete (tabular) points. One of the well-known polynomials used to derive formulas for numerical differentiation is the Lagrangian interpolation polynomial

$$P(\lambda) = \sum_{j=1}^{m} l_j(\lambda)M(\lambda^j) \tag{A-26}$$

where

$$l_j(\lambda) = \frac{p_m(\lambda)}{(\lambda - \lambda^j)p'_m(\lambda^j)}$$

$$p_m(\lambda) = \prod_{j=1}^{m} (\lambda - \lambda^j)$$

$$p'_m(\lambda^j) = \left.\frac{dp_m(\lambda)}{d\lambda}\right|_{\lambda=\lambda^j}$$

and $\mathbf{x} + \lambda^j \mathbf{e}_j$, $j = 1, \ldots, m$ with $m \geq 2$ are the tabular points and $M(\lambda) = M(\mathbf{x} + \lambda \mathbf{e}_j)$. Expression (A-26) is exact if $M(\lambda)$ is a polynomial of degree $(m - 1)$ or less, and it fits precisely any function M at the tabular points. For nontabular points we have

$$M(\lambda) = P(\lambda) + E(\lambda) \tag{A-27}$$

The truncation error is

$$E(\lambda) = \frac{p_m(\lambda)}{m!} M^{(m)}(\Lambda) \tag{A-28}$$

where Λ, an unknown function of λ, lies in the interval spanned by $\lambda^1, \lambda^2, \ldots, \lambda^m$ and λ.

Numerical differentiation formulas can be obtained by differentiating (A-27)

$$M'(\lambda) = \sum_{j=1}^{m} l'_j(\lambda)\, M(\lambda^j) + \frac{dE(\lambda)}{d\lambda} \tag{A-29}$$

When the tabular points are equally spaced—that is, $\lambda^j = \lambda^0 + hj$, $j = 1, 2, \ldots, m$—it can be shown that at a tabular point

$$\frac{dE}{d\lambda} = h^{m-1}\, c(\lambda) \tag{A-30}$$

where $c(\lambda)$ does not depend on h. We see that the truncation error in (A-30) is proportional to a power of h. On the other hand, the round-off error that occurs in computing the first term of $M'(\lambda)$ is inversely proportional to h. The ideal steplength yields approximately equal bounds on the round-off and truncation error. Examples of how to calculate the optimal steplengths h in some simple cases can be found in Hildebrand [1956], Kopal [1955], and Ralston [1965].

In general, to find the optimal steplength for a given differentiation formula we need to know bounds on the errors in the evaluation of the function $M(\mathbf{x})$ and on the second- or higher-order derivatives of this function in the range of interest. To illustrate the optimal step-finding problem [Newbery,

1970], we introduce the computed secant slope S_C intercepting $M(\lambda)$ at the points λ and $\lambda + \delta\lambda$, $S_C = \delta M/\delta\lambda$; and the true secant slope S_T,

$$S_T = (\delta M \pm 2\theta\epsilon)/\delta\lambda \qquad 0 \leq \theta \leq 1,$$

where $\epsilon > 0$ is a bound on the error in the evaluation of the function M. It is assumed that the error in computing δM for two given values of M is negligible in comparison to ϵ. From the mean-value theorem (valid for continuous function) we have

$$|S - S_T| \leq B\left|\frac{\delta\lambda}{2}\right| \tag{A-31}$$

where S is the true slope at the midpoint, and B bounds the second derivative of $M(\lambda)$. Finally, we estimate the difference between S and S_C:

$$|S - S_C| \leq |S - S_T| + |S_T - S_C| \leq B\left|\frac{\delta\lambda}{2}\right| + \left|\frac{2\epsilon}{\delta\lambda}\right| \tag{A-32}$$

The optimal stepsize can then be found from the condition

$$\min_{\delta\lambda} \left\{B\left|\frac{\delta\lambda}{2}\right| + \left|\frac{2\epsilon}{\delta\lambda}\right|\right\} \tag{A-33}$$

which yields a unique value $\delta\lambda$ for given B and ϵ

$$\delta\lambda_{\text{opt}} = 2\epsilon^{1/2}B^{-1/2} \tag{A-34}$$

Stewart [1967] suggested a modification of Davidon's method that uses difference approximations of derivatives. Stewart's more elaborate scheme uses information available in the metric matrix to compute the optimal steplengths, which balance off the effect of truncation and round-off error.

Computational experience has shown that another more inituitive approach to determining the steplength in numerical differentiation is frequently feasible. This is applicable in descent methods when a nonoptimal steplength strategy is used—for example, in conjunction with the steepest descent or second-order gradient methods. As indicated in Sec. 3-7, the nonoptimal steplength may be adjusted according to the progress of the iteration. Thus, it is reasonable to assume that these steplengths do not vary drastically from step to step. Derivatives (gradient, Hessian) are used in descent methods to obtain "locally steepest" directions. Steps of finite length are usually taken in the descent direction and it makes sense (intuitively) to explore the function behavior at a finite (not infinitesimal) distance from the current point. This distance should not be too different from the current steplength. This suggests using the (nonoptimal) descent steplength from the previous iteration in the numerical differentiation of the current iteration. A scheme that has worked well is the use of $\lambda^k/\sqrt{n}$. Further numerical experimentation with this approach is recommended, so that one can better assess the impact of the deviation from the theoretically justified descent formula.

A-3 MINIMIZATION OF SUMS OF SQUARES

Here we briefly review the special methods for minimization of functions having the form of a sum of squares. Functions of this type arise in a number of applications, such as that of solving a system of nonlinear equations, curve-fitting or regression analysis.

The linear least-squares problem is

$$\min_{\mathbf{x}} \{M(\mathbf{x}) = (A\mathbf{x} - \mathbf{b})^T(A\mathbf{x} - \mathbf{b}) = \mathbf{x}^T A^T A\mathbf{x} - 2\mathbf{b}^T A\mathbf{x} + \mathbf{b}^T\mathbf{b}\} \tag{A-35}$$

where A is an $m \times n$ matrix, $\mathbf{b}$ is a vector with m components, and $m > n$. If the rank of A is n, the solution to (A-35) satisfies the normal equations $A^T A\mathbf{x} = A^T\mathbf{b}$, which are derived from the condition $\partial M(\mathbf{x})/\partial\mathbf{x} = 0$. These equations are frequently difficult to solve numerically since they are inherently ill-conditioned, and accurate solutions require algorithms that use special techniques. The linear least-squares problem has been recently extensively studied; the theoretical background is discussed and several numerical algorithms are described by Golub [1965], Golub and Saunders [1970], and Peters and Wilkinson [1970]. For numerical evaluation of linear least-squares computer programs see Wampler [1969]. Golub and Saunders [1970] also discuss linear least-squares problems with linear constraints

$$\min_{\mathbf{x}} (A\mathbf{x} - \mathbf{b})^T(A\mathbf{x} - \mathbf{b}) \tag{A-36}$$

subject to $G\mathbf{x} = \mathbf{h}$ or $G\mathbf{x} \geq \mathbf{h}$. The least-squares problems with the quadratic constraint $\mathbf{x}^T\mathbf{x} = \alpha^2$, where α is a constant, have been considered by Forsythe and Golub [1965].

In a more general case, the objective function is

$$M(\mathbf{x}) = [\mathbf{f}(\mathbf{x})]^T\mathbf{f}(\mathbf{x}) \tag{A-37}$$

where $f_i(\mathbf{x})$, $i = 1, 2, \ldots, m$ are nonlinear functions of $\mathbf{x}$. To minimize function (A-37) we can disregard its special structure and use any direct search or descent method described in Chaps. 4 and 5. This approach is not recommended, however, since better results often can be obtained by taking into account the fact that $M(\mathbf{x})$ is a sum of squares.

We assume that $m > n$ and reduce the problem of minimizing $M(\mathbf{x}) = [\mathbf{f}(\mathbf{x})]^T\mathbf{f}(\mathbf{x})$ to a linear problem by expanding $\mathbf{f}(\mathbf{x})$ in a Taylor series about the current iteration point $\mathbf{x}^k$ and ignoring nonlinear terms,

$$\mathbf{f}(\mathbf{x}^k + \Delta\mathbf{x}^k) \cong \mathbf{f}(\mathbf{x}^k) + J(\mathbf{x}^k)\Delta\mathbf{x}^k \tag{A-38}$$

where $J(\mathbf{x}^k)$ is the Jacobian matrix with elements $\partial f_i(\mathbf{x}^k)/\partial x_j$.

The normal equations corresponding to this linearized problem are

$$[J(\mathbf{x}^k)]^T[J(\mathbf{x}^k)]\Delta\mathbf{x}^k = -[J(\mathbf{x}^k)]^T\mathbf{f}(\mathbf{x}^k) \tag{A-39}$$

and they yield the correction vector $\Delta\mathbf{x}^k$. Relationship (A-39) establishes the basis for the Gauss method (also called the generalized least-squares method):

1 Compute $J(\mathbf{x}^k)$.
2 Form $[J(\mathbf{x}^k)]^T J(\mathbf{x}^k) = T(\mathbf{x}^k)$.
3 Compute $\Delta\mathbf{x}^k$ from $T(\mathbf{x}^k)\Delta\mathbf{x}^k = -[J(\mathbf{x}^k)]^T \mathbf{f}(\mathbf{x}^k)$. (A-40)
4 Set $\mathbf{x}^{k+1} = \mathbf{x}^k + \Delta\mathbf{x}^k$, $k := k + 1$, and repeat from step 1.

The method is descent—that is, $[\Delta\mathbf{x}^k]^T \nabla M(\mathbf{x}^k) < 0$ since

$$\nabla M(\mathbf{x}^k) = 2[J(\mathbf{x}^k)]^T \mathbf{f}(\mathbf{x}^k) \tag{A-41}$$

so that
$$[\Delta\mathbf{x}^k]^T \nabla M(\mathbf{x}^k) = -2[\mathbf{f}(\mathbf{x}^k)]^T J(\mathbf{x}^k)[T(\mathbf{x}^k)]^{-1}[J(\mathbf{x}^k)]^T \mathbf{f}(\mathbf{x}^k) \tag{A-42}$$

and
$$[\Delta\mathbf{x}^k]^T \nabla M(\mathbf{x}^k) < 0 \tag{A-43}$$

as $[T(\mathbf{x}^k)]^{-1}$ is positive definite. Algorithm (A-40) is potentially divergent [Kowalik and Osborne, 1968; Kowalik, 1967], however, and has to be modified. This modification requires a search along the correction vector $\Delta\mathbf{x}^k$ to find $\lambda^{\min}$ satisfying $\min_{\lambda} M(\mathbf{x}^k + \lambda\Delta\mathbf{x}^k)$. The new approximation to the solution is $\mathbf{x}^{k+1} = \mathbf{x}^k + \lambda^{\min}\Delta\mathbf{x}^k$ instead of that in (A-40).

This modification of the Gauss method is undoubtedly better than the unmodified algorithm (A-40), but it still has a serious weakness. Numerical experiments performed with the Gauss method indicate that the direction vector $\Delta\mathbf{x}^k$ given in (A-40), step 3, frequently makes large angles (nearly 90°) with the steepest descent direction $-\nabla M = -2[J(\mathbf{x}^k)]^T \mathbf{f}(\mathbf{x}^k)$. This results in a highly oscillatory path of iterations (short steplengths $\lambda^{\min}$) and very slow convergence.

Levenberg [1944] and Marquardt [1963] attempted to overcome this difficulty by introducing a new parameter $\mu > 0$ and solving

$$\{[J(\mathbf{x}^k)]^T J(\mathbf{x}^k) + \mu I\}\,\Delta\mathbf{x} = -2[J(\mathbf{x}^k)]^T \mathbf{f}(\mathbf{x}^k) \tag{A-44}$$

instead of the direction vector of (A-40). Clearly, if $\mu \longrightarrow \infty$, then $\Delta\mathbf{x}^k$ approaches the steepest descent direction $-\nabla M(\mathbf{x}^k)$. It can also be demonstrated that as μ increases from zero, $\|\Delta\mathbf{x}^k\|$ decreases monotonically, as does the angle between $\Delta\mathbf{x}^k$ and $-\nabla M(\mathbf{x}^k)$. The parameter μ in the normal equations causes the correction vector $\Delta\mathbf{x}^k$ to interpolate between the steepest descent direction and that one given by the Gauss method. The method's major drawback is the necessity of choosing the parameter μ. There is little information on how to choose and adjust μ in each iteration in order to speed up convergence.

Jones [1970] proposed a method that searches roughly the same area as the Marquardt method does, but the search points are generated by vector additions $\mathbf{x}^{k+1} = \mathbf{x}^k + \mathbf{s}^k$ rather than by solving (A-44). The vectors $\mathbf{s}^k$ interpolate between the steepest descent and Gauss directions, and the method does not require a minimization along $\mathbf{s}^k$.

For $m = n$ the problem of minimizing $M(\mathbf{x}) = [\mathbf{f}(\mathbf{x})]^T \mathbf{f}(\mathbf{x})$ becomes that of solving a system of nonlinear equations $\mathbf{f}(\mathbf{x}) = 0$, and the Gauss method reduces to the Newton method, which proceeds as follows (the optimum step variant):

1 Compute $\Delta\mathbf{x}^k$ from $J(\mathbf{x}^k)\Delta\mathbf{x}^k = -\mathbf{f}(\mathbf{x}^k)$.

2 Find $\lambda^{\min}$ satisfying $\min_{\lambda} M(\mathbf{x}^k + \lambda\Delta\mathbf{x}^k)$. (A-45)

3 Set $\mathbf{x}^{k-1} = \mathbf{x}^k + \lambda^{\min}\Delta\mathbf{x}^k$, $k := k + 1$, and repeat from step 1.

The Newton method has second-order convergence and, hence, converges rapidly if an initial guess $\mathbf{x}^0$ is close to the solution. Unfortunately, it can fail when the Jacobian matrix is singular or diverge if it is started from a poor estimate of the solution, and the full-length steps $\lambda = 1$ are taken. The cost of performing each iteration is dominated by (1) evaluation of the Jacobian matrix, (2) solution of the linear system of equations (A-45), and (3) search along the correction vector $\Delta\mathbf{x}^k$.

Several attempts have been made to economize the method. The most obvious improvement is to keep the Jacobian $J(\mathbf{x})$ fixed for several iterations. This is justified by the fact that in the vicinity of the solution the matrix $J(\mathbf{x})$ changes very little. This strategy, however, slows down the convergence, which at best is of first order if $J(\mathbf{x})$ is kept constant. Another alternative is to use difference approximations to $J(\mathbf{x})$ and update the inverse of $J(\mathbf{x})$ from step to step; techniques have been provided by Barnes [1965], Broyden [1965], and Powell [1965]. (The Powell variant applies to the general case with $m \geq n$.) A comprehensive description of these schemes can be found in Kowalik and Osborne [1968].

A unified approach to solving the nonlinear least-squares problem is given by Fletcher [1968]. This approach is based on the generalized inverse of the Jacobian matrix. The matrices $[J(\mathbf{x}^k)]^{-1}$ used in the Newton method and $\{[J(\mathbf{x}^k)]^T J(\mathbf{x}^k)\}^{-1}[J(\mathbf{x}^k)]^T$ used in the Gauss method are generalized inverses of $J(\mathbf{x}^k)$ when this matrix is of rank n. Fletcher also discusses the problem of setting up difference approximations to the generalized inverses, methods of updating these inverses and ways of handling cases when the rank of $J(\mathbf{x}^k)$ is less than n.

Powell [1968] described a FORTRAN subroutine of a hybrid method for solving systems of nonlinear equations. It has been tested on practical problems and performed reasonably well. The method has the following basic features:

1 The correction vector $\Delta\mathbf{x}^k$ interpolates between the Newton direction $\mathbf{v} = -[J(\mathbf{x}^k)]^{-1}\mathbf{f}(\mathbf{x}^k)$ and the steepest descent direction $\mathbf{u} = -[J(\mathbf{x}^k)]^T\mathbf{f}(\mathbf{x}^k)$

$$\Delta\mathbf{x}^k = (1 - \theta)\mathbf{u} + \theta\mathbf{v} \qquad 0 \leq \theta \leq 1 \tag{A-46}$$

2 The Jacobian matrix and its inverse are approximated numerically and updated from step to step by the computational scheme, which takes only $0(n^2)$ operations, and thus is faster than the direct solution of equations (A-45), which requires $0(n^3)$ operations.

3 The method does not search for the optimal step along the correction vector $\Delta \mathbf{x}^k$, and every iteration requires only one evaluation of $\mathbf{f}(\mathbf{x})$.

A constrained least-squares problem

$$\min \{M(\mathbf{x}) = \sum_{i=1}^{m} f_i^2(\mathbf{x}) \mid A\mathbf{x} \geq \mathbf{b}\}$$

has been considered by Shanno [1970]. He has proposed to use the gradient projection method to compute the search directions

$$\mathbf{s}^k = -P\psi(\mathbf{x}^k) \tag{A-47}$$

where P is the projection matrix (Sec. 7-7) and $\psi(\mathbf{x}^k)$ is the normalized vector

$$\delta(\mathbf{x}^k, \mu) = \{[J(\mathbf{x}^k)]^T J(\mathbf{x}^k) + \mu I\}^{-1} \nabla M(\mathbf{x}^k) \tag{A-48}$$

The scalar parameter $\mu > 0$ improves computational stability of the Gauss method and guarantees that the directions $\mathbf{s}^k$ are descent; that is, $(\mathbf{s}^k)^T \nabla M(\mathbf{x}^k) < 0$.

A method for minimizing a sum of squares of nonlinear functions without using derivatives was given by Peckham [1970]. The function $M(\mathbf{x}) = \sum_{i=1}^{m} f_i^2(\mathbf{x})$ is replaced by a quadratic approximation whose minimum can be easily computed. The method reduces to a sequence of solutions of the normal equations of linear least-squares problems.

Codes for nonlinear least square problems are listed in Table A-2.

Table A-2 CODES FOR NONLINEAR LEAST-SQUARES PROBLEMS

Title and Reference	*Type, Machine and Storage*	*Remarks*
E 4 BKY SSQMIN VIM[1]	FORTRAN IV for CDC 6600, $1156 + n + (m + 3n/2)\,n + 1$ words	1. Author: E. R. Beals 2. Powell's method Powell [1965]
E 2 WISC GAUSHAUS VIM[1]	CODAP 1 FORTRAN 63 COMPASS 36 program for CDC 1604	3. Author P. Wolfe

References

BARNES, J. G. P.: "An Algorithm for Solving Nonlinear Equations Based on the Secant Method," *Computer Journal*, Vol. 8, 66, 1965.

BARTELS, R. H. and G. H. GOLUB: "The Simplex Method of Linear Programming Using LU Decomposition," *Communications of the ACM*, Vol. 12, 266, 1969*a*.

———: "Simplex-Method Procedure Employing LU Decomposition," *Communications of the ACM*, Vol. 12, 275, 1969*b*.

BARTELS, R. H., G. H. GOLUB, and M. A. SAUNDERS: *Numerical Techniques In Mathematical Programming*, Stanford University Report No. CS–70–162, 1970.

BEALE, E. M. L.: *Mathematical Programming in Practice*, John Wiley & Sons, Inc., New York, 1968.

———: "Nonlinear Optimization by Simplex-Like Methods," in R. Fletcher (ed.), *Optimization*, Academic Press Inc., New York, Ch. 3, 21, 1969*a*.

———: "Mathematical Programming: Algorithms," *Progress in Operations Research*, John Wiley & Sons, Inc., New York, Vol. 3, 135, 1969*b*.

———: "Sparseness in Linear Programming," J. K. Reid (ed.), *Large Sparse Sets of Linear Equations*, Academic Press Inc., New York, 1, 1971.

BROYDEN, C. G.: "A Class of Methods for Solving Nonlinear Equations," *Math. of Computation*, Vol. 19, 577, 1965.

DANTZIG, G. B.: *Linear Programming and Extensions*, Princeton University Press, Princeton, N. J., 1963.

——— and W. ORCHARD-HAYS: "The Product Form of the Inverse in the Simplex Method," *Math. of Computation*, Vol. 8, 64, 1954.

———, R. P. HARVEY, R. D. MCKNIGHT and S. S. SMITH: "Sparse Matrix Techniques in Two Mathematical Programming Codes," *Proceedings of the Symposium on Sparse Matrices and their Applications*, IBM Watson Research Center, 1968.

DE BUCHET, J.: "How to Take into Account the Low Density of Matrices to Design a Mathematical Programming Package, Relevant Effects on Optimization and Inversion Algorithms," J. K. Reid, (ed.), *Large Sparse Sets of Linear Equations*, Academic Press Inc., New York, 211, 1971.

FLETCHER, R.: "Generalized Inverse Methods for the Best Least Squares Solution of Systems of Nonlinear Equations," *Computer Journal*, Vol. 10, 392, 1968.

FORREST, J. J. H. and J. A. TOMLIN: *Updating Triangular Factors of the Basis to Maintain Sparsity in the Product Form Simplex Method*, Scientific Control Systems Ltd., London, 1971.

FORSYTHE, G. R. and G. H. GOLUB: "On the Stationary Values of a Second-Degree Polynomial on the Unit Sphere," *SIAM Journal*, Vol. 13, 1050, 1965.

GOLUB, G. H.: "Numerical Methods for Solving Linear Least-Squares Problems," *Numer. Math.*, Vol. 7, 206, 1965.

——— and M. A. SAUNDERS: "Linear Least Squares and Quadratic Programming," in J. Abadie (ed.), *Integer and Nonlinear Programming*, North-Holland Publishing Co., Amsterdam, Ch. 10, 229, 1970.

HILDEBRAND, F. B.: *Introduction to Numerical Analysis*, McGraw-Hill Book Company., New York, 1956.

JONES, A.: "Spiral—A New Algorithm for Non-linear Parameter Estimation Using Least Squares," *Computer Journal*, Vol. 13, 301, 1970.

KOPAL, Z.: *Numerical Analysis*, Chapman and Hall, Ltd., London, 1955.

KOWALIK, J.: "A Note on Nonlinear Regression Algorithm," *The Australian Computer Journal*, Vol. 1, 51, 1967.

——— and M. R. OSBORNE: *Methods for Unconstrained Optimization Problems*, American Elsevier Publishing Company, New York, 72, 1968.

LEMKE, C. E.: "The Dual Method of Solving the Linear Programming Problem," *Naval Research Logistics Quarterly*, Vol. 1, 36, 1954.

LEVENBERG, K. A.: "A Method for the Solution of Certain Nonlinear Problems in Least Squares," *Quarterly of Appl. Math.*, Vol. 2, 164, 1944.

MARQUARDT, D. W.: "An Algorithm for Least Squares Estimation of Nonlinear Parameters," *SIAM Journal*, Vol. 11, 431, 1963.

NEWBERY, A. C. R.: Private Discussion, 1970.

ORCHARD-HAYS, W.: *Advanced Linear Programming Computing Techniques*, McGraw-Hill Book Company, New York, 1968*a*.

———: "MP Systems Technology for Large Sparse Matrices," *Proceedings of the Symposium on Sparse Matrices and Their Applications*, IBM Watson Research Center, 1968*b*.

PECKHAM, G.: "A New Method for Minimizing a Sum of Squares Without Calculating Gradients," *Computer Journal*, Vol. 13, 418, 1970.

PETERS, G. and J. H. WILKINSON: "The Least Squares Problem and Pseudo-Inverses," *Computer Journal*, Vol. 13, 309, 1970.

POWELL, M. J. D.: "A Method for Minimizing a Sum of Squares of Nonlinear Functions without Calculating Derivatives," *Computer Journal*, Vol. 7, 303, 1965.

———: *A Fortran Subroutine for Solving Systems of Nonlinear Algebraic Equations*, Report No. AERE—R5947, Atomic Energy Research Establishment, Harwell, England, 1968.

RALSTON, A.: *A First Course in Numerical Analysis*, McGraw-Hill Book Company, New York, 1965.

SHANNO, D. F.: "An Accelerated Gradient Projection Method for Linearly Constrained Nonlinear Estimation," *SIAM Journal*, Vol. 18, 322, 1970.

STEWART, G. W.: "A Modification of Davidon's Minimization Method to Accept Difference Approximation of Derivatives," *Journal of the ACM*, Vol. 14, 72, 1967.

TOMLIN, J. A.: *Maintaining a Sparse Inverse in the Simplex Method*, Techn. Report No. 70–16, Stanford, Calif. Stanford University, 1970.

WAMPLER, R. H.: "An Evaluation of Linear Least Squares Computer Programs," *Journal of Research of the National Bureau of Standards –B. Mathematical Sciences*, Vol, 73B, 59, 1969.

WOLFE, P.: "The Composite Simplex Method," *SIAM Review*, Vol. 7, 42, 1965.

NOTATION

A	Matrix
B	Matrix
D	Determinant or diagonal matrix
E	Heaviside step function
F_k	kth Fibonacci number
I	Identity matrix
J	Jacobian matrix
L	Lagrangian function
M	Objective function
P	Transformation function or projection matrix
R	Feasible region
S	Set or matrix of direction vectors $\mathbf{s}^k$
T	Superscript indicating transposition of matrix or vector
$\bar{X}$	Matrix of descent steps $\Delta\mathbf{x}^k$
Y	Matrix of gradient differences $\mathbf{y}^k$
e	Basis of natural logarithms
f	Function in general
g	Constraint function
h	Constraint function or elements of Hessian matrix
i	Index used for variables
j	Index used for constraints
k	Superscript indicating iteration number
l	Linear constraint function
m	Number of constraints
n	Number of variables

r	Perturbation parameter
$\mathbf{s}$	Vector $(s_1, \ldots, s_n)^T$ indicating direction
$\mathbf{u}$	Vector $(u_1, \ldots, u_n)^T$ of Lagrange multipliers
$\mathbf{v}$	Vector $(v_1, \ldots, v_m)^T$ of Lagrange multipliers
$\mathbf{x}$	Vector $(x_1, \ldots, x_n)^T$ of variables
γ	Eigenvalue of a matrix
δ	Small increment
ϵ	Small positive number
λ	Variable in univariate minimization, steplength
ξ	Eigenvector of a matrix
τ	Predetermined accuracy tolerance
ϕ	Barrier function
ψ	Penalty function

Special Notation

A^{-1}	Superscript indicates the inverse of the matrix A.
∇	Denotes gradient [e.g., $\nabla M(x_1, x_2, \ldots, x_n) = (\partial M/\partial x_1, \partial M/\partial x_2, \ldots, \partial M/\partial x_n)^T$].
Hess	Denotes Hessian matrix of the objective function, i.e., Hess($\mathbf{x}$) has elements $\partial^2 M(\mathbf{x})/\partial x_i \partial x_j$.
Δ	Denotes increment.
Bar |	Used before an expression, denotes "such that"—for example, $\mathbf{x} \mid p(\mathbf{x})$ means "$\mathbf{x}$ such that property $p(\mathbf{x})$ is satisfied."
Bar |	Used after an expression, means "evaluated at," [e.g., $\nabla M(\mathbf{x})\|_{\mathbf{x}^0}$ means $\nabla M(\mathbf{x}^0)$].
Bars || ||	Denotes the Euclidean distance between the points $\mathbf{x}$ and $\mathbf{y}$, $\|\mathbf{x} - \mathbf{y}\| = \left[\sum_{j=1}^{n} (x_j - y_j)^2\right]^{1/2}$.
$k := k + 1$	Means "substitute $k + 1$ for k."
$\in$	Indicates "belongs to" (is an element of).

GLOSSARY

Absolute minimum See *minimum.*

Active constraint See *constraint.*

Algorithm An ordered sequence of mathematical operations (e.g., numerical, algebraic) leading to a required problem solution.

Barrier function See *transformation* for solution by interior point techniques.

Binding constraint See *constraint.*

Bound, upper and lower A real number u is an upper bound of a set S of real numbers if it satisfies $x \leq u$ for all $x \in S$; a real number b is a lower bound of a set S of real numbers if it satisfies $x \geq b$ for all $x \in S$.

A vector $\mathbf{u}$ whose n components are real numbers is an upper bound of a set S of vectors whose n components are real numbers if it satisfies $\mathbf{x} \leq \mathbf{u}$ for all $\mathbf{x} \in S$; a vector $\mathbf{b}$ whose n components are real numbers is a lower bound of a set S of vectors whose n components are real numbers if it satisfies $\mathbf{x} \geq \mathbf{b}$ for all $\mathbf{x} \in S$ (i. e., $x_i \leq u_i$ and $x_i \geq b_i$ for $i = 1, \ldots, n$).

Bounded set (*set bounded from above/below*) A set is bounded if it has an upper and lower bound; it is bounded from above (below) if it has an upper (lower) bound. If the set is bounded from above (below) it has one upper (lower) bound that is the smallest (greatest), referred to as the *smallest upper* (*greatest lower*) *bound.*

Boundary The set of all boundary points of the feasible region.

Boundary point A feasible point that lies on at least one hypersurface corresponding to a constraint.

Characteristic root, value See *Eigenvalue.*

Characteristic vector See *Eigenvector.*

Concave function A function whose negative is a convex function.

Condition number of a matrix See *matrix*.

Conjugate directions A set of n nonzero direction vectors $\mathbf{s}^0, \ldots, \mathbf{s}^{n-1}$ are said to be conjugate to each other with respect to a given positive definite matrix A if

$$(\mathbf{s}^k)^T A \mathbf{s}^l \begin{cases} = 0 & \text{for } 0 \leq k \neq l \leq n-1 \\ > 0 & \text{for } 0 \leq k = l \leq n-1 \end{cases}$$

Connected set A set with the property that each pair of its points can be joined by a curve all of whose points are in the set.

Constrained nonlinear program See *mathematical programming problem*.

Constraint, constraint function A relationship, inequality $g_j(\mathbf{x}) \geq 0$ or equality $h_j(\mathbf{x}) = 0$, which restricts the solution of a (constrained) mathematical programming problem. At any given feasible point $\mathbf{x}^k$, the inequality constraints are said to be *active* if $g_j(\mathbf{x}^k) = 0$ and *inactive* if $g_j(\mathbf{x}^k) > 0$. A constraint is said to be *binding* if it is active at the solution.

Convergence, order of The power $p \geq 1$ (if it exists) in the expression

$$\lim_{k \to \infty} \frac{\| \mathbf{x}^{k+1} - \hat{\mathbf{x}} \|}{\| \mathbf{x}^k - \hat{\mathbf{x}} \|^p} = C$$

where $(\hat{\mathbf{x}})$ is the solution to the problem, C is a (asymptotic error) constant, and k denotes the iteration number. Convergence of order 1, i.e., $p = 1$, with $C < 1$ is *linear convergence*; convergence of order greater than 1, i.e., $p > 1$, is *superlinear* convergence. *Geometric convergence* is said to occur if $C < 1$ exists such that

$$\frac{\| \mathbf{x}^{k+1} - \hat{\mathbf{x}} \|}{\| \mathbf{x}^k - \hat{\mathbf{x}} \|} = C$$

An algorithm is said to have *quadratic* (*cubic*) *convergence* if it can find the solution of an unconstrained quadratic (cubic) function in a finite number of iterations. This should not be confused with the convergence of order 2 or 3.

Convex function A function $f(\mathbf{x})$, defined on a convex set S if $f[\theta \mathbf{x} + (1 - \theta)\mathbf{y}] \leq \theta f(\mathbf{x}) + (1 - \theta) f(\mathbf{y})$ for all $\mathbf{x}, \mathbf{y} \in S$ and $0 \leq \theta \leq 1$. The function $f(\mathbf{x})$ is a *strictly convex* function if strict inequality holds in the above definition.

Convex set A set S of points such that $[\theta \mathbf{x} + (1 - \theta)\mathbf{y}] \in S$ whenever $\mathbf{x}, \mathbf{y} \in S$ and $0 \leq \theta \leq 1$.

Convex programming problem See *mathematical programming problem: convex optimization problem*.

Cubic convergence See *convergence, order of*.

Dense matrix See *matrix*.

Descent direction A direction vector $\mathbf{s}^k$ with the property, with respect to the objective function $M(\mathbf{x})$, that at $\mathbf{x}^k$ there is a scalar $\bar{\lambda} > 0$ such that for all scalars λ satisfying $0 \leq \lambda \leq \bar{\lambda}$ we have $M(\mathbf{x}^{k+1}) = M(\mathbf{x}^k + \lambda \mathbf{s}^k) < M(\mathbf{x}^k)$.

Descent steplength The scalar λ measuring the distance along the descent direction $\mathbf{s}^k$ between two successive iteration points $\mathbf{x}^k$ and $\mathbf{x}^{k+1}$; i.e., $\mathbf{x}^{k+1} = \mathbf{x}^k + \lambda \mathbf{s}^k$.

Determinant of an $n \times n$ matrix A is the number

$$|A| = \sum (-1)^k \prod_{i=1}^{n} a_{ij_i}$$

where in each term the subscripts $j_1, \ldots, j_n$ are a permutation of the numbers $1, \ldots, n$, where the sum is taken over all $(n!)$ possible permutations of these numbers, and where for each term the exponent k is the number of inversions of the subscripts $j_1, \ldots, j_n$.

Diagonal elements See *matrix*.

Diagonal matrix See *matrix*.

Difference formula A formula for estimating a derivative of a function at a point.

Differentiation, numerical A process that provides an estimate of a derivative of a function at a point by means of a difference formula.

Direction, direction vector An n-dimensional unit vector $\mathbf{s}^k = (s_1^k, \ldots, s_n^k)^T$ defining a direction (from a point $\mathbf{x}^k$) in the n-dimensional space.

Directional derivative The derivative of a function $f(\mathbf{x}^k)$, at a point $\mathbf{x}^k$, in the direction $\mathbf{s}^k$; i.e.,

$$\left.\frac{df(\mathbf{x}^k + \lambda \mathbf{s}^k)}{d\lambda}\right|_{\lambda=0} = (\mathbf{s}^k)^T \nabla f(\mathbf{x}^k)$$

where $\nabla f(\mathbf{x}^k)$ is the gradient of $f(\mathbf{x})$ at $\mathbf{x}^k$.

Distance Between two points $\mathbf{x}^1$ and $\mathbf{x}^2$ in n-dimensional space,

$$d(\mathbf{x}^1, \mathbf{x}^2) = [(\mathbf{x}^1 - \mathbf{x}^2)^T A (\mathbf{x}^1 - \mathbf{x}^2)]^{1/2}$$

where A is a positive definite $n \times n$ matrix. The matrix A entering the definition of distance is referred to as the *metric matrix*.

Distribution, probability The function whose domain is the set of real numbers between $-\infty$ and $+\infty$, and whose value is the probability that a random variable (e.g., number) assumes one of a specified set of values in its domain.

Distribution, probability, rectangular or uniform A constant distribution function stating that the random variable (e.g., random numbers) assumes values within a given interval (e.g., the unit interval [0, 1]) with equal likelihood.

Dual problem $\max_{\mathbf{u} \geq 0} \{\min_{\mathbf{x}} L(\mathbf{x}, \mathbf{u})\}$ where L is the Lagrangian function. See also *primal problem*.

Eigenvalue problem Characteristic equation $|\lambda I - A| = 0$.

Eigenvalues (of an $n \times n$ matrix A) The solutions (roots) of the eigenvalue problem (characteristic equation) $|\lambda I - A| = 0$.

Eigenvector A vector $\boldsymbol{\xi} \neq 0$ corresponding to an eigenvalue γ of an $n \times n$ matrix A, satisfying $A\boldsymbol{\xi} = \gamma\boldsymbol{\xi}$.

Error, round-off An error caused by rounding numbers that have infinite representation (e.g., in a computer a fixed number of digits is used, and an arithmetic operation will result in a round-off error).

Error, truncation The result of truncating an infinite process to get a finite process (e.g., by using the first n terms of an infinite Taylor series expansion of a function as an approximation).

Euclidean norm See *norm of a vector.*

Exterior point See *infeasible point.*

Feasible direction A direction vector $\mathbf{s}^k = (s_1^k, \ldots, s_n^k)^T$ with the property that, at a feasible point $\mathbf{x}^k$, a scalar $\bar{\lambda} > 0$ exists such that for any scalar λ satisfying $0 \leq \lambda \leq \bar{\lambda}$ the point $(\mathbf{x}^k + \lambda\mathbf{s}^k)$ also is feasible.

Feasible point Any point in the feasible region.

Feasible region The set of all points satisfying the constraints.

Feasible solution See *feasible point.*

Fibonacci numbers Numbers $F_p, p = 0, 1, \ldots$, given by the recurrence relation $F_p = F_{p-1} + F_{p-2}$ for $p \geq 2$ and $F_0 = F_1 = 1$, the solution of which is

$$F_p = \frac{1}{\sqrt{5}}\left[\left(\frac{1+\sqrt{5}}{2}\right)^{p+1} - \left(\frac{1-\sqrt{5}}{2}\right)^{p+1}\right] \qquad p = 0, 1, \ldots$$

Function A mathematical expression $f(\mathbf{x})$ describing (stating) a correspondence between points $\mathbf{x}$ of a set S and points $f(\mathbf{x})$. The function is said to be defined over a set S (the *domain* of the function). The points $f(\mathbf{x})$ are referred to as function values or elements in the function *range*. The domains of definition of functions treated in this book are assumed to be multivariable (multidimensional) and real in general. Function values are assumed to be real numbers. If x is a single variable, the function is said to be *one-dimensional.*

Function, linearity of A function $f(\mathbf{x})$, said to be *linear* if for any real constant θ, $f(\theta\mathbf{x}) = \theta f(\mathbf{x})$ and for any two points $\mathbf{x}, \mathbf{y} \in S$, the domain of function definition, $f(\mathbf{x} + \mathbf{y}) = f(\mathbf{x}) + f(\mathbf{y})$. Otherwise, the function is said to be *nonlinear.*

Geometric convergence See *convergence, order of.*

Geometric programming problem See *mathematical programming problem.*

Global minimum See *minimum.*

Gradient vector The vector denoted by $\nabla f(\mathbf{x})$ whose components are the first partial derivatives of a differentiable function $f(\mathbf{x})$; that is,

$$\nabla f(\mathbf{x}) = \left(\frac{\partial f}{\partial x_1}, \ldots, \frac{\partial f}{\partial x_n}\right)^T$$

Gram-Schmidt orthogonalization procedure The procedure of constructing a set of n orthonormal direction vectors $\mathbf{v}^0, \ldots, \mathbf{v}^{n-1}$ from a given set of n linearly independent direction vectors $\mathbf{q}^0, \ldots, \mathbf{q}^{n-1}$, as follows: $\mathbf{v}^0 = \mathbf{q}^0$, $\mathbf{s}^0 = \mathbf{v}^0/\|\mathbf{v}^0\|$; and for $j = 1, \ldots, n-1$, $\mathbf{v}^j = \mathbf{q}^j - \sum_{i=0}^{j-1} [(\mathbf{q}^j)^T\mathbf{s}^i]\mathbf{s}^i$, and $\mathbf{s}^i = \mathbf{v}^i/\|\mathbf{v}^i\|$.

Greatest lower bound See *Bounded set.*

Heaviside step function A function $E(G)$ defined as

$$E(G)\begin{cases} = 0 & \text{if } G \geq 0 \\ = 1 & \text{if } G < 0 \end{cases}$$

Hessian matrix The matrix whose elements are the second partial derivatives of a twice differentiable function $f(\mathbf{x})$ denoted by

$$\text{Hess}(\mathbf{x}) = \begin{bmatrix} h_{11} & \cdots & h_{1n} \\ \cdots & & \cdots \\ h_{n1} & \cdots & h_{nn} \end{bmatrix}$$

where

$$h_{ij} = \frac{\partial^2 f(\mathbf{x})}{\partial x_i \, \partial x_j}$$

Hypersurface A geometric representation of $f(\mathbf{x}) = C$ where $f(\mathbf{x})$ is a function of n variables and C is a constant. The hypersurface is said to be *singular* at a point if $f(\mathbf{x})$ is infinite there.

Identity matrix See *matrix.*

Ill-conditioning, Ill-conditioned matrix See *matrix.*

Inactive constraint See *constraint.*

Infeasible point A point that is not in the feasible region or on its boundary; frequently also referred to as *exterior point.*

Infeasible solution See *infeasible point.*

Inverse of a matrix See *matrix.*

Integer programming problem See *mathematical programming problem.*

Interior point A feasible point that is not a boundary point.

Iteration, iteration step The typical step of an iterative method whereby the $(k + 1)$st approximation to the solution is obtained given the kth.

Iteration number The integer k (see *iteration*), usually $k = 0, 1, \ldots$.

Iterative method (technique) A type of algorithm wherein the problem solution is obtained numerically by successive approximations. A new (and usually better) approximation to the solution can be obtained by such a technique, if one or more such approximations are given.

Jacobian matrix The matrix whose elements are the first partial derivatives of m differentiable functions $f_1(\mathbf{x}), \ldots, f_m(\mathbf{x})$ denoted by

$$J = \begin{bmatrix} \frac{\partial f_1}{\partial x_1} & \frac{\partial f_1}{\partial x_2} & \cdots & \frac{\partial f_1}{\partial x_n} \\ \frac{\partial f_2}{\partial x_1} & \cdots & \cdots & \frac{\partial f_2}{\partial x_n} \\ \cdots & \cdots & \cdots & \cdots \\ \frac{\partial f_m}{\partial x_1} & \cdots & \cdots & \frac{\partial f_m}{\partial x_n} \end{bmatrix}$$

Lagrangian, Lagrangian function The function

$$L(\mathbf{x}, \mathbf{u}) = M(\mathbf{x}) - \sum_{j=1}^{m} u_j g_j(\mathbf{x})$$

corresponding to the constrained minimization problem

$$\min_{\mathbf{x}} \{M(\mathbf{x}) \mid [g_j(\mathbf{x}) \geq 0, \qquad j = 1, \ldots, m]\}$$

where $\mathbf{u} = (u_1, \ldots, u_m)^T$ is a vector of *Lagrangian multipliers.*

Lagrangian multipliers See *Lagrangian, Lagrangian function.*

Limit point of a sequence The point $\mathbf{x}^*$ to which a sequence of points $\mathbf{x}^k$, $k = 0, 1, 2, \ldots$ converges for $k \longrightarrow \infty$; i.e., given an arbitrary ϵ, there is an integer K such that for $k > K$ we have $\| \mathbf{x}^k - \mathbf{x}^* \| < \epsilon$.

Linear convergence See *convergence, order of.*

Linear dependence A set of n vectors $\mathbf{u}^1, \ldots, \mathbf{u}^n$ are said to be linearly dependent if n scalars α_i, not all zero, exist such that

$$\sum_{i=1}^{n} \alpha_i \mathbf{u}^i = 0$$

Linear function See *function.*

Linear independence The vectors $\mathbf{u}^i$ are said to be linearly independent if they are not linearly dependent; see *linear dependence.*

Linear manifold See *manifold, linear.*

Linear programming problem (program) See *mathematical programming problem.*

Local minimum See *minimum*

Lower bound See *bound, upper and lower.*

Manifold, linear The point

$$\mathbf{x} = \sum_{i=1}^{r} a_i \mathbf{v}^i$$

where $-\infty < a_i < +\infty$ and $\mathbf{v}^i$ are linearly independent vectors, generates an r-dimensional *subspace.* The translation of a subspace is said to be a linear manifold; i.e.,

$$\mathbf{y} = \mathbf{v}^0 + \sum_{i=1}^{r} a_i \mathbf{v}^i$$

generates an r-dimensional manifold.

Mathematical model A set of mathematical expressions (functions) describing the behavior of a system.

Mathematical programming problem, mathematical program, optimization problem The problem of finding the values of n variables $\mathbf{x} = (x_1, \ldots, x_n)^T$ satisfying $\min_{\mathbf{x}} \{M(\mathbf{x}) \mid [g_j(\mathbf{x}) \geq 0, \quad j = 1, \ldots, m_1;$ and $g_j(\mathbf{x}) = 0, \quad j = m_1 + 1, \ldots, m_2;$ and $l_j(\mathbf{x}) \geq 0, \quad j = m_2 + 1, \ldots, m_3;$ and $l_j(\mathbf{x}) = 0, \quad j = m_3 + 1, \ldots, m]\}$.

This problem is in general *nonlinear* (*a nonlinear program*), $M(\mathbf{x})$ is in general a nonlinear function of $\mathbf{x}$ and is referred to as the *objective function*, g_j are nonlinear functions of $\mathbf{x}$, and l_j are linear functions of $\mathbf{x}$; g_j and l_j are referred to as *constraints* or *constraint functions.*

The above-defined problem is *constrained*, but in the special case in which there are no constraints (i.e., $m = 0$) the problem is referred to as *unconstrained.*

Convex optimization problem. The above-defined problem if the objective function is convex over a convex feasible region.

Integer programming problem The above-defined problem if all the variables are restricted to assume only integer values.

Linear programming problem The above-defined problem if the objective function and all constraint functions are linear.

Mixed integer programming problem The above-defined problem if some of the variables are restricted to assume only integer values.

Quadratic programming problem The above-defined problem if the objective function is quadratic $M(\mathbf{x}) = a + \mathbf{b}^T\mathbf{x} + \frac{1}{2}\mathbf{x}^T A\mathbf{x}$, where a is a constant, $\mathbf{b}$ an n-dimensional vector and A an $n \times n$ matrix; and the constraints are linear.

Separable programming problem The above-defined problem if the nonlinear functions $M(\mathbf{x})$ and $g_j(\mathbf{x})$ can be rewritten as sums of functions of one variable.

Matrix An ordered array of elements, which may be numbers or functions. Denoting these elements by a_{ij}, the first subscript, i, indicates the *row number* and the second subscript, j, indicates the *column number.*

If $i = 1, \ldots, m$ and $j = 1, \ldots, n$, the matrix is said to be $m \times n$ where in general $m \neq n$. The matrix is said to be *square* if $m = n$. Elements with $i = j$, i.e., $a_{11}, a_{22}, \ldots, a_{ii} \ldots$, are called the *diagonal elements.* Two matrices A, with elements a_{ij}, and B, with elements b_{ij}, are *equal* if and only if $a_{ij} = b_{ij}$ for all i and j. A is a *zero matrix* if $a_{ij} = 0$, for all i

and j. A is a *diagonal matrix* if all elements are zero except the diagonal elements. The special case of a square diagonal matrix with elements equal to 1 is an *identity matrix*, usually denoted by I. A *lower* (*upper*) *triangular matrix* is a square matrix with elements $a_{ij} = 0$ for $i < j$ $(i > j)$. A matrix A is said to be *symmetric* if $A^T = A$ where the superscript T denotes "*transpose*" and means that the elements of A^T are a_{ji} if the elements of A are a_{ij}. A matrix is said to be *sparse* if it has many zero elements and *dense* if it has few zero elements.

The *inverse of a matrix A*, denoted by A^{-1}, is defined by $AA^{-1} = I$. If the determinant of A is not zero, the matrix A is said to be *nonsingular* and the inverse A^{-1} exists. Otherwise, the matrix A is said to be *singular*. If $A^T = A^{-1}$, A is said to be an *orthogonal matrix*. The *rank* of the matrix A is the largest integer $r \geq 0$ such that A has r linearly independent columns.

For every nonsingular matrix A we define the *condition number* C_N as $C_N = \|A\|\,\|A^{-1}\|$, where $\|A\| = [\gamma_{\max}(AA^T)]^{1/2}$ is called the *spectral norm* of A, and $\gamma_{\max}(AA^T)$ is the largest eigenvalue of AA^T. If A is symmetric and positive definite, then $C_N = \gamma_{\max}(A)/\gamma_{\min}(A)$. If the condition number C_N is large, then A is said to be *ill-conditioned* with respect to the optimization problem with the quadratic objective function $\frac{1}{2}\mathbf{x}^T A\mathbf{x} + \mathbf{b}^T\mathbf{x}$.

A *positive definite* (*semidefinite*) matrix is a square matrix $A \neq 0$ if the quadratic form $\mathbf{x}^T A\mathbf{x}$ is positive definite (semidefinite).

Maximum The mathematical programming problem with $\min\,[M(\mathbf{x})]$ is equivalent to the problem with $\max\,[-M(\mathbf{x})]$; See *minimum*.

Metric matrix See *distance*.

Minimum The value of the objective function $M(\mathbf{x})$ at the solution point $\mathbf{x}^*$ of the mathematical programming problem. It is said to be a *global* (*absolute*) *minimum*, denoted by $M(\mathbf{x}_G)$, if for every feasible point $\mathbf{x}$, $M(\mathbf{x}_G) \leq M(\mathbf{x})$. A minimum is said to be a *local* (*relative*) *minimum* denoted by $M(\mathbf{x}_L)$ if there is a feasible *neighborhood* of $\mathbf{x}_L$ (defined by $\|\mathbf{x}_L - \mathbf{x}\| < \delta$, $\mathbf{x}$ feasible and $\delta > 0$) such that for all $\mathbf{x}$ in that neighborhood $M(\mathbf{x}_L) \leq M(\mathbf{x})$.

Mixed integer problem See *mathematical programming problem*.

Monotone increasing function A function $f(x)$ in an interval $[b, u]$ if, given any two points $x_1, x_2 \in [b, u]$ with $x_1 < x_2$, we have $f(x_1) \leq f(x_2)$. The function is *strictly monotone increasing* if $x_1 < x_2$ implies $f(x_1) < f(x_2)$.

Neighborhood See *minimum*.

Nonlinear function See *function*.

Nonlinear programming problem (*program*) See *mathematical programming problem*.

Norm of a vector The Euclidean norm of a vector $\mathbf{x} = (\mathbf{x}_1, \ldots, x_n)^T$ denoted by $\|\mathbf{x}\|$, and defined as $\|\mathbf{x}\| = (\mathbf{x}^T\mathbf{x})^{1/2}$. Other norms of a vector can be defined.

Normalized vector A vector $\mathbf{v}$ such that $\|\mathbf{v}\| = 1$.

Objective function See *mathematical programming problem.*

Optimality conditions The necessary (and/or sufficient) conditions for a point $\mathbf{x}^*$ to be a solution of a mathematical programming problem.

Optimization problem See *mathematical programming problem.*

Optimum A maximum or a minimum.

Orthogonal vectors, set of A set of n vectors $\mathbf{x}^1, \ldots, \mathbf{x}^n$ with the property that

$$(\mathbf{x}^i)^T\mathbf{x}^j \begin{cases} = 0 & \text{for } 1 \le i \ne j \le n \\ \ne 0 & \text{for } 1 \le i = j \le n \end{cases}$$

Orthonormal vectors, set of A set of n vectors $\mathbf{x}^1, \ldots, \mathbf{x}^n$ that are orthogonal and satisfy $\|\mathbf{x}^i\| = 1,\ i = 1, \ldots, n.$

Penalty coefficient See *penalty transformation.*

Penalty function See *penalty transformation.*

Penalty transformation for solution by exterior point technique An objective function, denoted by $P_E(\mathbf{x}, r)$, of an unconstrained optimization problem derived from a constrained problem by combining the objective function and constraints. This combination is such that the function is penalized; i.e., its value is increased when constraints are violated in a minimization problem. Usually $P_E(\mathbf{x}, r) = M(\mathbf{x}) + r^{-1} \sum_j \psi[g_j(\mathbf{x})]$, where ψ is referred to as the *penalty function*, $g_j(\mathbf{x})$ are the constraint functions, and r is the *penalty parameter.*

Positive definite, semidefinite quadratic form See *quadratic form.*

Predetermined tolerance A small value, predetermined before the computation, which is the acceptable error that will be tolerated at the termination of the computation.

Primal problem $\min\limits_{\mathbf{x}} \{M(\mathbf{x}) | [g_j(\mathbf{x}) \ge 0, \quad j = 1, \ldots, m]\}$; see also *dual problem.*

Probability distribution See *distribution.*

Pseudoconcave function A function whose negative is a pseudoconvex function.

Pseudoconvex function A differentiable function $f(\mathbf{x})$ defined on a convex set S if for all points $\mathbf{x}$ and $\mathbf{y}$ in S, $[\nabla f(\mathbf{x})]^T(\mathbf{y} - \mathbf{x}) \ge 0$ implies that $f(\mathbf{y}) \ge f(\mathbf{x})$.

Pseudorandom number See *random number.*

Quadratic convergence See *convergence, order of.*

Quadratic form A function $f(\mathbf{x})$ of the form $f(\mathbf{x}) = \mathbf{x}^T A\mathbf{x}$, where $\mathbf{x}$ is an n vector and A an $n \times n$ symmetric matrix.

Positive definite quadratic form. A quadratic form such that $f(\mathbf{x}) > 0$ for every $\mathbf{x} \ne 0$.

Positive semidefinite quadratic form. A quadratic form such that $f(\mathbf{x}) \ge 0$ for every $\mathbf{x} \ne 0$.

Quadratic programming problem See *mathematical programming problem.*

Quasiconcave function A function whose negative is a quasiconvex function.

Quasiconvex function A function $f(\mathbf{x})$ defined on some set S if $f[\theta\mathbf{x} + (1 - \theta)\mathbf{y}] \leq \max[f(\mathbf{x}), f(\mathbf{y})]$ for all $\mathbf{x}$, $\mathbf{y} \in S$ and $0 \leq \theta \leq 1$.

Random number A number equal to one of a specified set of numbers with a known probability. This probability varies, in general from number to number in the specified set. The function defining this probability is referred to as the probability distribution; see *distribution.*

Pseudorandom Number. A number in a sequence produced by a computer program for generating random numbers. Such sequences pass standard statistical tests for randomness although, given a number in the sequence and the computer program, the next number in the sequence is always predictable. Truly random numbers should not be predictable or quantitatively definable.

Rectangular (probability) distribution See *distribution.*

Recursion formula A relation between successive terms in a sequence, enabling the calculation of the $(k + 1)$st term if the kth is known.

Regular simplex See *simplex.*

Relative minimum See *minimum.*

Round-off error See *error, round-off.*

Saddlepoint A point $(\mathbf{x}^*, \mathbf{u}^*)$ that satisfies $L(\mathbf{x}^*, \mathbf{u}) \leq L(\mathbf{x}^*, \mathbf{u}^*) \leq L(\mathbf{x}, \mathbf{u}^*)$ where $L(\mathbf{x}, \mathbf{u})$ is the Lagrangian function.

Separable programming problem See *mathematical programming problem.*

Simplex A set of $(n + 1)$ points in n-dimensional space.

Regular simplex. A set of $(n + 1)$ equidistant points in n-dimensional space.

Singular hypersurface See *hypersurface.*

Singular matrix See *matrix.*

Slack variable See *variable.*

Smallest upper bound See *bounded set.*

Sparse matrix See *matrix.*

Spectral norm of a matrix See *matrix.*

Square matrix See *matrix.*

Starting point (solution) The initial value of the vector $\mathbf{x}$ of independent variables used in the optimization iteration, denoted by $\mathbf{x}^0$.

Steplength See *descent steplength.*

Strictly convex function See *convex function.*

Subspace See *manifold, linear.*

Superlinear convergence See *convergence, order of.*

Symmetric matrix See *matrix.*

Tangent hyperplane Hyperplane $(\mathbf{x} - \mathbf{x}^k)^T\nabla f(\mathbf{x}^k) = 0$ (to a function $f(\mathbf{x})$ at a point $\mathbf{x}^k$).

Taylor's series expansion Expansion of a differentiable function $f(\mathbf{x})$ about a point $\mathbf{x}^k$ is

$$f(\mathbf{x}) = f(\mathbf{x}^k) + \frac{1}{1!}(\mathbf{x} - \mathbf{x}^k)^T \nabla f(\mathbf{x}^k) + \frac{1}{2!}(\mathbf{x} - \mathbf{x}^k)^T \operatorname{Hess}(\mathbf{x}^k)(\mathbf{x} - \mathbf{x}^k) + \cdots$$

[$f(\mathbf{x})$ must be differentiable to a sufficiently high order, depending on the number of terms retained.]

Tolerance, predetermined See *predetermined tolerance.*

Transformation for solution by interior point techniques An objective function, denoted by $P_I(\mathbf{x}, r)$, of an unconstrained optimization problem derived from a constrained problem by combining the objective function and constraints. This combination is such that the function is penalized; i.e., its value is increased when constraints are approached from within the feasible region. Usually $P_I(\mathbf{x}, r) = M(\mathbf{x}) + r \sum_j \phi[g_j(\mathbf{x})]$ where ϕ is referred to as the *barrier function*, $g_j(\mathbf{x})$ are the constraint functions, and $r > 0$ are parameters.

Transpose See *matrix.*

Triangular (lower, upper) matrix See *matrix.*

Truncation error See *error, truncation.*

Unconstrained mathematical programming problem, unconstrained optimization problem See *mathematical programming problem.*

Uniform (probability) distribution See *distribution.*

Unimodal function A one-dimensional function is said to be unimodal over an interval $[b, u]$ if it has only one minimum or maximum in $[b, u]$.

Unit vector A vector $\mathbf{s}$ of unit length; i.e., $||\mathbf{s}|| = 1$.

Upper bound See *bound, upper and lower.*

Usable direction A feasible direction $\mathbf{s}^k$ if there exists a scalar $\bar{\lambda} > 0$ such that for any scalar λ satisfying $0 < \lambda \leq \bar{\lambda}$ we have $M(\mathbf{x}^k + \lambda \mathbf{s}^k) \leq M(\mathbf{x}^k)$; or equivalently, if $M(\mathbf{x})$ is differentiable, $(\mathbf{s}^k)^T \nabla M(\mathbf{x}^k) \leq 0$.

Variable, slack A variable $\bar{x}_j \geq 0$ introduced into an inequality $g_j(\mathbf{x}) \geq 0$ or $g_j(\mathbf{x}) \leq 0$ to convert it into an equality $g_j(\mathbf{x}) - \bar{x}_j = 0$ or $g_j(\mathbf{x}) + \bar{x}_j = 0$.

ANNOTATED BIBLIOGRAPHY

Optimization, Background Material

GOLDSTEIN, A. A.: *Constructive Real Analysis*, Harper & Row, Publishers, New York, 1967.

HADLEY, G.: *Nonlinear and Dynamic Programming*, Addison-Wesley Publishing Co., Inc., Reading, Mass., 1964. [Selected Chapters.]

Practical Approaches to Optimization

BEALE, E. M. L.: *Mathematical Programming in Practice*, John Wiley & Sons, Inc., New York, 1968. [Linear programming methods and extensions; practical difficulties with formulating problems for digital computers.]

BOX, M. J., D. DAVIES, and W. H. SWANN: *Nonlinear Optimization Techniques*, ICI Ltd., Monograph N. 5, Oliver and Boyd, Ltd., Edinburgh, 1969. [Primarily unconstrained optimization problems.]

FOX, R. L.: Optimization Methods for Engineering Design, Addison-Wesley Publishing Co., Inc., Reading, Mass., 1971.

Unconstrained Optimization and Nonlinear Programming; Optimization of Multistage Problems and Optimal Control Theory

KOWALIK, J. and M. R. OSBORNE: *Methods for Unconstrained Optimization Problems*, American Elsevier Publishing Company, New York, 1968.

LUENBERGER, D. G.: *Optimization by Vector Space Methods*, John Wiley & Sons, Inc., New York, 1969. [Unifies the optimization theory by geometrical principles of linear vector space theory.]

WILDE, D. J. and C. S. BEIGHTLER: *Foundation of Optimization*, Prentice-Hall, Inc., Englewood Cliffs, N.J., 1967.

ZANGWILL, W. I.: *Nonlinear Programming: A Unified Approach*, Prentice-Hall, Inc., Englewood Cliffs, N.J., 1969.

LASDON, L. S., *Optimization Theory for Large Systems*, The Macmillan Company, New York, 1970.

Fundamental Theory of Nonlinear Programming for Applied Mathematicians

MANGASARIAN, O. L.: *Nonlinear Programming*, McGraw-Hill Book Company, New York, 1969.

Sequential Transformation Techniques

FIACCO, A. V. and G. MCCORMICK: *Nonlinear Programming: Sequential Unconstrained Minimization Techniques*, John Wiley & Sons, Inc., New York, 1968.

Unconstrained and Constrained Optimization: Recent Theory and Methodology

FLETCHER, R. (ed.): *Optimization*, Academic Press Inc., London and New York, 1969.

ABADIE, J. (ed.): *Integer and Nonlinear Programming*, North-Holland Publishing Company, Amsterdam, 1970.

Reviews of Optimization Methods

GEOFFRION, A. M.: "Elements of Large Scale Mathematical Programming, Part I: Concepts," *Management Science*, Vol. 16, 652, 1970*a*.

———: "Elements of Large Scale Mathematical Programming, Part II: Synthesis of Algorithms and Bibliography," *Management Science*, Vol. 16, 676, 1970*b*.

KOWALIK, J.: "Feasible Direction Methods," *Proceedings of Symposium on Structural Optimization*, NATO Advisory Group for Aerospace Research and Development (AGARD), Istanbul, 2–1, 1969.

PEARSON, J. D.: "Variable Metric Methods of Minimization," *Computer Journal*, Vol. 12, 171, 1969.

POWELL, M.J. D.: "Finding Minima of Functions of Several Variables," in J. Walsh, *Numerical Analysis—An Introduction*, Academic Press Inc., New York, Ch. 8, 143, 1966.

——— : "A Survey of Numerical Methods for Unconstrained Optimization," *SIAM Review*, Vol. 12, 79, 1970.

SORENSEN, H. W.: "Comparison of Some Conjugate Direction Procedures for Function Minimization," *Journal of The Franklin Institute*, Vol. 288, 421, 1969.

WOLFE, P.: "Methods of Nonlinear Programming," in J. Abadie (ed.), *Nonlinear Programming*, North-Holland Publishing Company, Amsterdam, Ch. VI, 97, 1967.

ZOUTENDIJK, B.: "Nonlinear Programming: A Numerical Survey," *SIAM Journal, Cont. Theory and Applic.* Vol. 4, 194, 1966.

Bibliography

KUNZI, H. P. and W. OETTLI: *Nichtlineare Optimierung: Neuere Verfahren Bibliographie*, Springer-Verlag, Berlin, Heidelberg, New York, 1969.

LEON, A.: "A Classified Bibliography on Optimization," in A. Lavi and T. P. Vogl (eds.), *Recent Advances in Optimization Techniques*, John Wiley & Sons, Inc., New York, 599, 1966.

Methods for Problems in Linear Algebra

FORSYTHE, G. E. and C. B. MOLER: *Computer Solution of Linear Algebraic Systems*, Prentice-Hall, Inc., Englewood Cliffs, N. J., 1967.

WESTLAKE, J. R.: *A Handbook of Numerical Matrix Inversion and Solution of Linear Equations*, John Wiley & Sons, Inc., New York, 1968.

Sources of Generally Available (no fee for usage) Computer Programs

ACM Collected Algorithms, a service of the Association for Computing Machinery, New York.

BIT, Nordisk Tidskrift for Informasjons Behandling, Copenhagen, Denmark.

Centre d'études Nucléaires de Saclay, Department de Calcul Electronique, France.

COSMIC, Computer Center, University of Georgia, Athens, Ga, (prepared under sponsorship of NASA).

European Organization for Nuclear Research (CERN), Genève, Switzerland.

KUNZI, H. P., H. G. TZSCHACH and C. A. ZEHNDER: *Numerical Methods of Mathematical Optimization*, Academic Press Inc., New York, 1968.

Numerische Mathematik, *Handbook for Automatic Computation*, Springer-Verlag, New York.

RALSTON and WILF: *Mathematical Methods for Digital Computers*, John Wiley & Sons, Inc., New York, 1966.

SHAPIRO, M. and M. GOLDSTEIN: *A Collection of Mathematical Computer Routines*, AEC Computing and Applied Mathematics Center, Courant Institute of Math. Sciences, New York University, New York, 1965.

The Computer Journal, British Computer Society, London.

Computer programs are also available through the following users' organizations:

CUBE, Burroughs Users' Group, Computing Center, University of Denver, Denver, Colo.

IBM Systems Reference Library, Publication GC20–1619–7, Catalog of Programs for IBM System/360 Models 25 and above, IBM Corporation, Hawthorne, N. Y.

IBM Systems Reference Library (SHARE), Publication C–20–1604–9, Catalog of Programs for IBM 704–709–7040–7044–7090 and 7094, IBM Corporation, Hawthorne, N. Y.

National Computer Centre, Index of Computer Programs, N. C., Manchester, England.

SSP, System/360 Scientific Subroutine Package (SSP) (360A–CM–03X) Version III, Programmer's Manual H20 — 0205 — 3, IBM Corporation, White Plains, N.Y.

USE Program Library Interchange, Univac Division, Sperry Rand Corporation, Philadelphia.

VIM Inc., Users Organization for Control Data Corporation 6000 and 7000 series computers, VIM Distribution Agency, Control Data Corporation, Palo Alto, Calif.

Xerox Data Systems (XDS), Software Distribution, El Segundo, Calif.

Computer programs that may be obtained for a fee are listed in publications such as:

ICP Quarterly, International Computer Programs, Inc., 2511 East 46th St., Indianapolis.

Software Packages: An Encyclopedic Guide, System Interaction Corporation, 8th W. 40th St., New York.

The following organizations specialize in developing optimization routines:

Atomic Energy Research Establishment, Mathematics Branch, Harwell, Didcot, Berks, England.

Boeing Computer Services Inc., P.O. Box 24346, Seattle, Wash. 98124.

National Physical Laboratory, Division of Numerical Analysis and Computing, Teddington, Middlesex, England.

Numerical Optimization Centre, Hatfield College of Technology, Hatfield, Herts, England.

INDEX